CRM Short Courses

The volumes in the **CRM Short Courses** series have a primarily instructional aim, focusing on presenting topics of current interest to readers ranging from graduate students to experienced researchers in the mathematical sciences. Each text is aimed at bringing the reader to the forefront of research in a particular area or field, and can consist of one or several courses with a unified theme. The inclusion of exercises, while welcome, is not strictly required. Publications are largely but not exclusively, based on schools, instructional workshops and lecture series hosted by, or affiliated with, the *Centre de Researches Mathématiques* (CRM). Special emphasis is given to the quality of exposition and pedagogical value of each text.

More information about this series at http://www.springer.com/series/15360

Daniel W. Stroock

Elements of Stochastic Calculus and Analysis

CENTRE DE RECHERCHES MATHÉMATIQUES

Springer

Daniel W. Stroock
Department of Mathematics
Massachusetts Institute of Technology
Cambridge, MA
USA

ISSN 2522-5200 ISSN 2522-5219 (electronic)
CRM Short Courses
ISBN 978-3-030-08354-0 ISBN 978-3-319-77038-3 (eBook)
https://doi.org/10.1007/978-3-319-77038-3

Mathematics Subject Classification (2010): 60XX

Printed on acid-free paper

This Springer imprint is published by the registered company Springer International Publishing AG part of Springer Nature
The registered company address is: Gewerbestrasse 11, 6330 Cham, Switzerland

Preface

Like many other branches of mathematics, stochastic calculus and analysis were created originally to address specific problems and later took on a life of their own. Although the later versions usually contain improvements in both clarity and breadth, all too often the thinking from which those improvements derive gets lost. As a consequence, what began as an deep insight into a mathematical problem looks like a theory that sprung fully formed from the head of some Zeus. For those who are content to master and apply the resulting theory, the origins of the theory are of little importance. On the other hand, for those who are hoping to solve new problems, the initial insight can be of greater value than the finished product. With this in mind, in this book I have tried to present the ideas here in a way that does not ignore their origins. This decision has its drawbacks. In particular, it means that the material is presented in a less than efficient manner and at times has resulted in repetition. For example, it is reasonable to think that the theory of stochastic integration given in Chapter 4 makes the content of Chapter 3 unnecessary and that, once one has Itô's formula, Chapter 1 seems like a waste of time. Thus, those who are looking for efficiency should look elsewhere.

What is *stochastic analysis and the associated calculus*? The name implies that it is a collection of randomly chosen topics in analysis. Even though to some extent that is what it is, that is not all that it is. Indeed, during the twentieth century it became increasingly clear that there are many aspects of probability theory that are inextricably bound to analysis and that the two subjects have a lot to contribute to one another. Thus, there now exists a reasonably well-defined amalgam of probabilistic and analytic ideas and techniques that, at least among the cognoscenti, are easily recognized as stochastic analysis. Nonetheless, the term continues to defy a precise definition, and an understanding of it is best acquired by way of examples. For this reason, this book should be viewed as an attempt to provide its readers with enough examples of stochastic analysis that they will be able to recognize it whenever they encounter it elsewhere.

Because they are the origin of most of what follows, I have devoted the first chapter to an explanation of Kolmogorov's equations and Itô's interpretation of them. I begin by describing the class of Lévy operators, the operators that appear in Kolmogorov's equations and that Itô thought of as the tangent to a path with values in the space of probability measures on $\mathbb{R}^N$. I then show how one goes about solving Kolmogorov's forward equation, which, from Itô's perspective, means recovering the path from its tangent field. Initially, my treatment applies to all Lévy operators, but I restrict my attention to local ones as soon as I turn to the problem of solving Kolmogorov's equation with variable coefficients. The main result there is Theorem 1.2.1, which is the analog of the general existence theorem for first-order ordinary differential equations with continuous coefficients.

Like the corresponding theorem in ordinary differential equations, Theorem 1.2.1 is an existence result that uses an Euler approximation scheme and relies on compactness to prove that there is a convergent subsequence but not that there is only one limit. To provide a corresponding uniqueness result, in the second chapter I lay the groundwork for and develop Itô's approach to proving convergence of the approximations by lifting everything to a pathspace setting. The first step is to construct Brownian motion and prove a few of its elementary properties. Once this is done, I show how to interpret the proof of Theorem 1.2.1 in terms of random variables constructed from Brownian motion and show that, under suitable conditions, they lead to a proof of convergence.

Although the ideas in Chapter 2 are basically Itô's, introduction of his theory of stochastic integration is postponed until Chapter 3. The reason for my decision to postpone its introduction is that I want my readers to understand that his theory was motivated by the sort of considerations in Chapter 2. Before presenting Itô's integral, I discuss the Paley–Wiener integral and apply it to prove the Cameron–Martin–Segal formula and an integration by parts formula for functions on Wiener space. This is followed by Itô's theory of stochastic integration, the derivation of a few of its elementary properties, and its application to the construction carried out in Chapter 2. I then derive Itô's formula and demonstrate its power by using it to prove one-fourth of Burkholder's inequality as well as Tanaka's formula for Brownian local time. The next topic is properties of solutions to Itô's equation as a function of the starting point, and the chapter concludes with a proof of Wiener's decomposition of the space of square-integrable functions on Wiener space into subspaces of homogeneous chaos.

Chapter 4 deals with extensions of and variations on Itô's theory of stochastic integration. Specifically, I present Kunita and Watanabe's theory of stochastic integration with respect to continuous martingales and then introduce Stratonovich's theory in order to explain the Wong–Zakai approximation of solutions to stochastic integral equations by integral curves.

The concluding chapter is a somewhat eclectic selection of topics. I begin by addressing the question of uniqueness. Namely, I investigate to what extent the operator L that appears in Kolmogorov's equations uniquely determines the distribution of a Markov process. To pose this problem in a precise manner, I first formulate it in terms of a "martingale problem" and then give a couple a criteria for deciding when the martingale problem is well posed in the sense that it has one and only one solution. The second topic is the use of exponential semi-martingales as integrating factors to derive the Feynman–Kac and the Cameron–Martin–Girsanov formulae. This is followed by a construction of Brownian motion on an embedded submanifold by projecting the Brownian on the ambient space, and the fourth topic is the Kalman–Bucy filter for real-valued Gaussian processes. In the final section of this chapter, I introduce the reader to Malliavin's calculus. In no sense do I give a definitive account of that challenging topic. Instead, using examples, I try to convey a few of the ideas involved in the application of Malliavin's calculus to the derivation of regularity results for solutions to Kolmogorov's equations.

Those who have read the preceding may well be asking for what audience this book was written. The answer is that it started as a set of notes that I wrote for the students in a graduate-level course at MIT entitled "Stochastic Calculus." My own introduction to stochastic calculus was in a seminar at which H.P. McKean was presenting the material in the final draft of his book *Stochastic Integrals*, copies of which are now available for free on the Internet. Aside from Itô's memoir [7] and Doob's brief treatment in [3], McKean's book was the only source of this material in the English language at that time. Since then a raft of books on the subject have appeared, especially after F. Black, M. Scholes, and R. Merton convinced Wall Street and the economics Nobel Prize committee that stochastic calculus held the key to financial success. Among the most popular of these books is the one by B. Økensendal [14], which is now in its sixth printing. Øksendal's book is a carefully written, excellent introduction to stochastic calculus and its many applications, and I initially thought that I would base my course on it. However, I soon realized that it lacked the background material that I wanted for my students, and that is why I decided to write my own version.[1] As a result, I am confident that this book will not have six printings, and, if it has any, its audience will be much smaller than Øksendal's. Nonetheless, I hope that it will appeal to readers who, even if they have encountered some of the material here elsewhere, want to acquire a deeper understanding of its origins.

Nederland, CO, USA
November 2017

Daniel W. Stroock

---

[1] Unavailable at the time was the recent book [11] by J.-F. Le Gall. Le Gall is one of the masters of the subject, and his book reflects his mastery.

Notation

Sets, functions, and derivatives	
$\mathbb{R}$, $\mathbb{C}$, & $\mathbb{Q}$	The real, the complex, and the rational numbers.
$\mathbb{Z}$ & $\mathbb{Z}^+$	The sets of integers and strictly positive integers.
$C(E;F)$	The space of F-valued, continuous functions on E.
$C^n(\mathbb{R}^N;F)$	The space of n times continuously differentiable F-valued functions on $\mathbb{R}^N$.
$C^n_{\mathrm{b}}(\mathbb{R}^N;F)$	The bounded elements of $C^n(\mathbb{R}^N;F)$ with bounded derivatives.
$C^n_{\mathrm{c}}(\mathbb{R}^N;F)$	The compactly supported elements of $C^n(\mathbb{R}^N;F)$.
$\dot{f}$	The time derivative of a function f.
$C^{1,2}([0,\infty)\times\mathbb{R}^N;F)$	Continuously differentiable on $[0,\infty)\times\mathbb{R}^N$, once in time and twice in space.
$\mathscr{S}(\mathbb{R}^N;F)$	The Schwartz class of smooth F-valued functions all of whose derivatives decrease faster than the reciprocal of any polynomial.
$\mathcal{P}(\mathbb{R}^N)$	The pathspace $C\big([0,\infty);\mathbb{R}^N\big)$ with the topology of uniform convergence on compact intervals.
$\mathcal{B}_t$	The σ-algebra $\sigma\big(\{\psi(\tau):\tau\in[0,t]\}\big)$ over $\mathcal{P}(\mathbb{R}^N)$.
$\mathbb{W}(\mathbb{R}^N)$	The Banach space of $w\in\mathcal{P}(\mathbb{R}^N)$ such that $w(0)=\mathbf{0}$ and $\lim_{t\to\infty}t^{-1}\lvert w(t)\rvert=0$ with norm $\displaystyle\lVert w\rVert_{\mathbb{W}(\mathbb{R}^N)}=\sup_{t\geq 0}(1+t)^{-1}\lvert w(t)\rvert.$
$f\restriction S$	The restriction of a function f to the set S.
$\nabla^n\varphi(\mathbf{x})$	The multilinear functional on $\mathbb{R}^{N^n}$ given by

$$\big[\nabla^n\varphi(\mathbf{x})\big](\boldsymbol{\xi}_1,\dots,\boldsymbol{\xi}_n)=\frac{\partial^n}{\partial_{t_1}\cdots\partial t_n}\varphi\left(\mathbf{x}+\sum_{m=1}^n t_m\boldsymbol{\xi}_m\right)\Bigg|_{t_1=\cdots=t_n=0}.$$

$\mathcal{L}_V$	The directional derivative determined by the vector V.

Related to measures

$\mathbf{1}_\Gamma$	The indicator function of Γ.
$\mathbb{E}^{\mathbb{P}}[X,\Gamma]$	The expected value $\int_\Gamma X\,d\mathbb{P}$ of X under $\mathbb{P}$ on Γ.
$\mu \ll \nu$	The measure μ is absolutely continuous with respect to the measure ν.
$\mu \perp \nu$	The measure μ is singular to the measure ν.
$\mathbf{M}_1(E)$	The space of Borel probability measures on a metric space with the weak topology: $\mu_n \longrightarrow \mu \iff \langle\varphi,\mu_n\rangle \longrightarrow \langle\varphi,\mu\rangle$ for all $C_b(E;\mathbb{R})$.
$\langle\varphi,\mu\rangle$	The integral of a function φ with respect to a measure μ.
$\hat{\mu}$	The Fourier transform or characteristic function of the measure μ.
$F_*\mu$	The pushforward of the measure μ under the map F: $F_*\mu(\Gamma) = \mu(F^{-1}\Gamma)$.
$\mathcal{B}_E$	The sigma-algebra of Borel subsets of E.
$\sigma(S)$	The sigma-algebra generated by the elements of S.
$\mathcal{B}_t$	The sigma-algebra $\sigma(\{\psi(\tau) : \tau \in [0,t]\})$ over $\mathcal{P}(\mathbb{R}^N)$.
$\mathcal{W}$	Wiener measure on $\mathbb{W}(\mathbb{R}^N)$.
$\mathcal{W}_t$	The $\mathcal{W}$-completion of the sigma-algebra $\sigma(\{w(\tau) : \tau \in [0,t]\})$ over $\mathbb{W}(\mathbb{R}^N)$.
$\gamma_{\mathbf{m},C}$	The Gaussian measure with mean $\mathbf{m}$ and covariance C.
$PM^2(\mathbb{R}^N)$	The space of square-integrable, progressively measurable functions.
$PM^2_{\mathrm{loc}}(\mathbb{R}^N)$	The space of locally square-integrable, progressively measurable functions.
$M^2(\mathbb{P};H)$	The space of continuous, $\mathbb{P}$-square-integrable martingales with valued in the Hilbert space H.

Norms and miscellany

$A := B$	The quantity A is defined by the quantity B.
$\|\cdot\|_{\mathrm{u}}$	The uniform or supremum norm.
$B_E(a,r)$	The open ball of radius r around a in a metric space E. When the subscript E is missing, it is assumed that $E = \mathbb{R}^N$.
$\Gamma\complement$	The complement of the set Γ.
$\mathbb{S}^{N-1}$	The unit sphere in $\mathbb{R}^N$.
$\|\cdot\|_S$	The uniform norm on the set S.
$\|\cdot\|_{\mathrm{Lip}}$	The Lipschitz norm.
$\mathrm{var}_I(\psi)$	The variation of ψ on the interval I.
$(h_1,h_2)_H$	The inner product of h_1 and h_2 in the Hilbert space H.
$\mathrm{Hom}(E;F)$	The set of linear maps from E into F.
$\|A\|_{\mathrm{H.S.}}$	The Hilbert–Schmidt norm $\sqrt{\sum_{i,j} \|a_{i,j}\|^2}$ of $A \in \mathrm{Hom}(\mathbb{R}^M;\mathbb{R}^N)$ identified as $\mathbb{R}^M \otimes \mathbb{R}^N$.
$\|A\|_{\mathrm{op}}$	The operator norm $\sup_{\|x\|_E=1} \|Ax\|_F$ of $A \in \mathrm{Hom}(E;F)$.
$\lfloor t \rfloor_n$	The largest number $m2^{-n} \le t$.

Contents

Chapter 1
Kolmogorov's Equations

A stochastic process is a parameterized family of random variables. For the most part, in this book the stochastic processes with which we will deal are parameterized by $t \in [0, \infty)$, to be thought of as "time," and take values in some Euclidean space $\mathbb{R}^N$. When attempting to analyze such a family $\{X(t) : t \geq 0\}$ of random variables on a probability space $(\Omega, \mathcal{F}, \mathbb{P})$, the first step is to understand the distribution μ_t of $X(t)$ for each $t \geq 0$. Thus one is forced to consider paths $t \rightsquigarrow \mu_t$ in the space $\mathbf{M}_1(\mathbb{R}^N)$ of probability measures on $\mathbb{R}^N$, and, as usual, this leads one to consider the "tangent field" along the path. With this in mind, the goal of this chapter is to first make precise exactly what the tangent to a $\mathbf{M}_1(\mathbb{R}^N)$-valued path is and to then develop a procedure for recovering the path from its tangent field.

1.1 Linear functionals that satisfy the minimum principle

Suppose that $t \in (0, \infty) \longmapsto \mu_t \in \mathbf{M}_1(\mathbb{R}^N)$ is a map with the properties that[1]

$$A\varphi := \lim_{t \searrow 0} \frac{\langle \varphi, \mu_t \rangle - \varphi(\mathbf{0})}{t}$$

exists for all[2] $\varphi \in \mathbb{D} := \mathbb{R} \oplus \mathscr{S}(\mathbb{R}^N; \mathbb{R}) = \{c + \psi : c \in \mathbb{R} \,\&\, \psi \in \mathscr{S}(\mathbb{R}^N; \mathbb{R})\}$ and

$$\varlimsup_{R \to \infty} \varlimsup_{t \searrow 0} \frac{\mu_t(B(\mathbf{0}, R)\complement)}{t} = 0. \tag{1.1.1}$$

[1] $\langle \varphi, \mu \rangle$ denotes the integral of φ with respect to μ.

[2] $\mathscr{S}(\mathbb{R}^N; F)$, where F equals $\mathbb{R}$ or $\mathbb{C}$, is the Schwartz test function class of infinitely differentiable functions which, together with all their derivatives, are rapidly decreasing in the sense that they tend to 0 at infinity faster that $|\mathbf{x}|^{-n}$ for all $n \geq 1$.

D. W. Stroock, *Elements of Stochastic Calculus and Analysis*,
CRM Short Courses, https://doi.org/10.1007/978-3-319-77038-3_1

Then

(i) A satisfies the **minimum principle**:

$$\varphi(\mathbf{0}) \leq \varphi \implies A\varphi \geq 0.$$

(ii) A is **quasi-local** in the sense that

$$\lim_{R\to\infty} A\varphi_R = 0 \quad \text{where } \varphi_R(\mathbf{x}) = \varphi\left(\frac{\mathbf{x}}{R}\right)$$

if φ is constant in a neighborhood of $\mathbf{0}$.

The first of these is obvious, since

$$\langle \varphi, \mu_t \rangle - \varphi(\mathbf{0}) = \langle \varphi - \varphi(\mathbf{0}), \mu_t \rangle \geq 0$$

if $\varphi(\mathbf{0}) \leq \varphi$. To check the second, choose $\delta > 0$ so that $\varphi(\mathbf{y}) = \varphi(\mathbf{0})$ for $|\mathbf{y}| < \delta$. Then

$$\begin{aligned} \big|\langle \varphi_R, \mu_t \rangle - \varphi(\mathbf{0})\big| &\leq \int_{B(\mathbf{0},R\delta)\complement} \left|\varphi\left(\frac{\mathbf{y}}{R}\right) - \varphi(\mathbf{0})\right| \mu_t(d\mathbf{y}) \\ &\leq 2\|\varphi\|_{\mathrm{u}} \mu_t\big(B(\mathbf{0}, R\delta)\complement\big), \end{aligned}$$

which, by (1.1.1), means that $\lim_{R\to\infty} A\varphi_R = 0$.

The goal here is to show that any linear functional that possesses properties (i) and (ii) has a very special structure. To begin with, notice that, by applying the minimum principle to both $\mathbf{1}$ and $-\mathbf{1}$, one knows that $A\mathbf{1} = 0$. Before going further, we have to introduce the following partition of unity for $\mathbb{R}^N \setminus \{\mathbf{0}\}$. Choose $\psi \in C^\infty(\mathbb{R}^N; [0,1])$ so that ψ has compact support in $B(\mathbf{0}, 2) \setminus \overline{B(\mathbf{0}, \frac{1}{4})}$ and $\psi(\mathbf{y}) = 1$ when $\frac{1}{2} \leq |\mathbf{y}| \leq 1$, and set $\psi_m(\mathbf{y}) = \psi(2^m\mathbf{y})$ for $m \in \mathbb{Z}$. Then, if $\mathbf{y} \in \mathbb{R}^N$ and $2^{-m-1} \leq |\mathbf{y}| \leq 2^{-m}$, $\psi_m(\mathbf{y}) = 1$ and $\psi_n(\mathbf{y}) = 0$ unless $m - 2 \leq n \leq m + 2$. Hence, if $\Psi(\mathbf{y}) = \sum_{m\in\mathbb{Z}} \psi_m(\mathbf{y})$ for $\mathbf{y} \in \mathbb{R}^N \setminus \{\mathbf{0}\}$, then Ψ is a smooth function with values in $[1, 5]$; and therefore, for each $m \in \mathbb{Z}$, the function χ_m given by $\chi_m(\mathbf{0}) = 0$ and $\chi_m(\mathbf{y}) = \frac{\psi_m(\mathbf{y})}{\Psi(\mathbf{y})}$ for $\mathbf{y} \in \mathbb{R}^N \setminus \{\mathbf{0}\}$ is a smooth, $[0,1]$-valued function that vanishes off of $B(\mathbf{0}, 2^{-m+1}) \setminus \overline{B(\mathbf{0}, 2^{-m-2})}$. In addition, for each $\mathbf{y} \in \mathbb{R}^N \setminus \{0\}$, $\sum_{m\in\mathbb{Z}} \chi_m(\mathbf{y}) = 1$.

If $n \in \mathbb{Z}^+$ and $\varphi \in C^n(\mathbb{R}^N; \mathbb{C})$, define $\nabla^n\varphi(\mathbf{x})$ to be the multilinear map on $(\mathbb{R}^N)^n$ into $\mathbb{C}$ given by

$$[\nabla^n\varphi(\mathbf{x})](\boldsymbol{\xi}_1, \ldots, \boldsymbol{\xi}_n) = \frac{\partial^n}{\partial t_1 \cdots \partial t_n} \varphi\left(\mathbf{x} + \sum_{m=1}^{n} t_m \boldsymbol{\xi}_m\right)\bigg|_{t_1=\cdots=t_n=0}.$$

Obviously, $\nabla\varphi$ and $\nabla^2\varphi$ can be identified as the gradient of φ and the Hessian of φ.

Finally, a Borel measure M on $\mathbb{R}^N$ is said to be a **Lévy measure** if

$$M(\{\mathbf{0}\}) = 0 \quad \text{and} \quad \int_{B(\mathbf{0},1)} |\mathbf{y}|^2 \, M(\mathbf{y}) + M\big(B(\mathbf{0},1)\complement\big) < \infty.$$

Theorem 1.1.1. *If $A\colon \mathbb{D} \longrightarrow \mathbb{R}$ is a quasi-local linear functional on $\mathbb{D}$ that satisfies the minimum principle, then there is a unique Lévy measure M such that $A\varphi = \int_{\mathbb{R}^N} \varphi(\mathbf{y})\, M(d\mathbf{y})$ whenever φ is an element of $\mathscr{S}(\mathbb{R}^N;\mathbb{R})$ for which $\varphi(\mathbf{0}) = 0$, $\nabla\varphi(\mathbf{0}) = \mathbf{0}$, and $\nabla^2\varphi(\mathbf{0}) = 0$. Next, given $\eta \in C_c^\infty(\mathbb{R}^N;[0,1])$ satisfying $\eta = 1$ in a neighborhood of $\mathbf{0}$, set $\eta_{\boldsymbol{\xi}}(\mathbf{y}) = \eta(\mathbf{y})(\boldsymbol{\xi},\mathbf{y})_{\mathbb{R}^N}$ for $\boldsymbol{\xi} \in \mathbb{R}^N$, and define $\mathbf{m}^\eta \in \mathbb{R}^N$ and $C \in \mathrm{Hom}(\mathbb{R}^N;\mathbb{R}^N)$ by*

$$\begin{aligned} (\boldsymbol{\xi}, \mathbf{m}^\eta)_{\mathbb{R}^N} &= A\eta_{\boldsymbol{\xi}} \\ \text{and} \quad (\boldsymbol{\xi}, C\boldsymbol{\xi}')_{\mathbb{R}^N} &= A(\eta_{\boldsymbol{\xi}}\eta_{\boldsymbol{\xi}'}) - \int_{\mathbb{R}^N} (\eta_{\boldsymbol{\xi}}\eta_{\boldsymbol{\xi}'})(\mathbf{y})\, M(d\mathbf{y}). \end{aligned} \tag{1.1.2}$$

Then C is symmetric, non-negative definite, and independent of the choice of η. Finally, for any $\varphi \in \mathbb{D}$,

$$\begin{aligned} A\varphi = {} & \tfrac{1}{2}\operatorname{Trace}\big(C\nabla^2\varphi(\mathbf{0})\big) + \big(\mathbf{m}^\eta, \nabla\varphi(\mathbf{0})\big)_{\mathbb{R}^N} \\ & + \int_{\mathbb{R}^N} \Big(\varphi(\mathbf{y}) - \varphi(\mathbf{0}) - \eta(\mathbf{y})\big(\mathbf{y}, \nabla\varphi(\mathbf{0})\big)_{\mathbb{R}^N}\Big)\, M(d\mathbf{y}). \end{aligned} \tag{1.1.3}$$

Conversely, if the action of A is given by (1.1.3), *then A satisfies the minimum principle and is quasi-local.*

Proof. The concluding converse assertion is easy. Indeed, if A is given by (1.1.3), then it is obvious that $A1 = 0$. Next suppose that $\varphi \in \mathscr{S}(\mathbb{R}^N;\mathbb{R})$ and that $\varphi(\mathbf{0}) \le \varphi$. Then $\nabla\varphi(\mathbf{0}) = \mathbf{0}$ and $\nabla^2\varphi(\mathbf{0})$ is non-negative definite. Hence $\big(\mathbf{m}^\eta, \nabla\varphi(\mathbf{0})\big)_{\mathbb{R}^N} = 0$, $\operatorname{Trace}\big(C\nabla^2\varphi(\mathbf{0})\big) \ge 0$, and

$$\varphi(\mathbf{y}) - \varphi(\mathbf{0}) - \eta(\mathbf{y})\big(\mathbf{y}, \nabla\varphi(\mathbf{0})\big)_{\mathbb{R}^N} \ge 0.$$

To see that A is quasi-local, suppose that $\varphi \in \mathscr{S}(\mathbb{R}^N;\mathbb{R})$ is constant in a neighborhood of $\mathbf{0}$, and choose $\delta > 0$ so that $\varphi = \varphi(\mathbf{0})$ on $B(\mathbf{0},\delta)$. Obviously, $\nabla\varphi(\mathbf{0}) = \mathbf{0}$ and $\nabla^2\varphi(\mathbf{0}) = 0$. In addition,

$$\begin{aligned} & \left| \int_{\mathbb{R}^N} \Big(\varphi_R(\mathbf{y}) - \varphi_R(\mathbf{0}) - \eta(\mathbf{y})\big(\mathbf{y}, \nabla\varphi_R(\mathbf{0})\big)_{\mathbb{R}^N}\Big) M(d\mathbf{y}) \right| \\ & \quad \le \int_{B(\mathbf{0},R\delta)\complement} |\varphi_R(\mathbf{y}) - \varphi(\mathbf{0})|\, M(d\mathbf{y}) \le 2\|\varphi\|_{\mathrm{u}} M\big(B(\mathbf{0},R\delta)\complement\big) \longrightarrow 0 \end{aligned}$$

as $R \to \infty$.

Now assume that A satisfies (i) and (ii). Referring to the partition of unity described above, for $\varphi \in C^\infty\big(B(\mathbf{0},2^{-m+1}) \setminus B(\mathbf{0},2^{-m-2})\big)$, define $\Lambda_m(\varphi) = A(\chi_m\varphi)$, where

$$\chi_m\varphi(\mathbf{y}) = \begin{cases} \chi_m(\mathbf{y})\varphi(\mathbf{y}) & \text{if } 2^{-m-2} \le |\mathbf{y}| \le 2^{-m+1} \\ 0 & \text{otherwise.} \end{cases}$$

Clearly Λ_m is linear. In addition, if $\varphi \geq 0$, then $\chi_m\varphi \geq 0 = \chi_m\varphi(\mathbf{0})$, and so, by (i), $\Lambda_m\varphi \geq 0$. Similarly, for any $\varphi \in C^\infty\big(\overline{B(\mathbf{0},2^{-m+1})} \setminus B(\mathbf{0},2^{-m-2});\mathbb{R}\big)$, $\|\varphi\|_{\mathrm{u}}\chi_m \pm \chi_m\varphi \geq 0 = \big(\|\varphi\|_{\mathrm{u}}\chi_m \pm \chi_m\varphi\big)(\mathbf{0})$, and therefore $|\Lambda_m\varphi| \leq K_m\|\varphi\|_{\mathrm{u}}$, where $K_m = A\chi_m$. Hence, Λ_m admits a unique extension as a continuous linear functional on $C\big(\overline{B(\mathbf{0},2^{-m+1})} \setminus B(\mathbf{0},2^{-m-2});\mathbb{R}\big)$ that is non-negativity preserving and has norm K_m. Therefore, by the Riesz representation theorem, we now know that there is a unique non-negative Borel measure M_m on $\mathbb{R}^N$ such that M_m is supported on $\overline{B(\mathbf{0},2^{-m+1})} \setminus B(\mathbf{0},2^{-m-2})$, $K_m = M_m(\mathbb{R}^N)$, and $A(\chi_m\varphi) = \int_{\mathbb{R}^N}\varphi(\mathbf{y})\,M_m(d\mathbf{y})$ for all $\varphi \in \mathscr{S}(\mathbb{R}^N;\mathbb{R})$.

Now define the non-negative Borel measure M on $\mathbb{R}^N$ by $M = \sum_{m\in\mathbb{Z}} M_m$. Clearly, $M(\{\mathbf{0}\}) = 0$. In addition, if $\varphi \in C_{\mathrm{c}}^\infty\big(\mathbb{R}^N \setminus \{\mathbf{0}\};\mathbb{R}\big)$, then there is an $n \in \mathbb{Z}^+$ such that $\chi_m\varphi = 0$ unless $|m| \leq n$. Thus,

$$\begin{aligned} A\varphi &= \sum_{m=-n}^{n} A(\chi_m\varphi) = \sum_{m=-n}^{n}\int_{\mathbb{R}^N}\varphi(\mathbf{y})\,M_m(d\mathbf{y}) \\ &= \int_{\mathbb{R}^N}\left(\sum_{m=-n}^{n}\chi_m(\mathbf{y})\varphi(\mathbf{y})\right)M(d\mathbf{y}) = \int_{\mathbb{R}^N}\varphi(\mathbf{y})\,M(d\mathbf{y}), \end{aligned}$$

and therefore

$$A\varphi = \int_{\mathbb{R}^N}\varphi(\mathbf{y})\,M(d\mathbf{y}) \tag{1.1.4}$$

for $\varphi \in C_{\mathrm{c}}^\infty\big(\mathbb{R}^N \setminus \{\mathbf{0}\};\mathbb{R}\big)$.

Before taking the next step, observe that, as an application of (i), if $\varphi_1, \varphi_2 \in \mathbb{D}$, then

$$(*) \qquad \varphi_1 \leq \varphi_2 \quad \text{and} \quad \varphi_1(\mathbf{0}) = \varphi_2(\mathbf{0}) \implies A\varphi_1 \leq A\varphi_2.$$

Indeed, by linearity, this reduces to the observation that, by (i), if $\varphi \in \mathbb{D}$ is non-negative and $\varphi(\mathbf{0}) = 0$, then $A\varphi \geq 0$.

With these preparations, we can show that, for any $\varphi \in \mathbb{D}$,

$$(**) \qquad \varphi \geq 0 = \varphi(\mathbf{0}) \implies \int_{\mathbb{R}^N}\varphi(\mathbf{y})\,M(d\mathbf{y}) \leq A\varphi.$$

To check this, apply $(*)$ to $\varphi_n = \sum_{m=-n}^{n}\chi_m\varphi$ and φ, and use (1.1.4) together with the monotone convergence theorem to conclude that

$$\int_{\mathbb{R}^N}\varphi(\mathbf{y})\,M(d\mathbf{y}) = \lim_{n\to\infty}\int_{\mathbb{R}^N}\varphi_n(\mathbf{y})\,M(d\mathbf{y}) = \lim_{n\to\infty}A\varphi_n \leq A\varphi.$$

Now let η be as in the statement of the theorem, and set $\eta_R(\mathbf{y}) = \eta(R^{-1}\mathbf{y})$ for $R > 0$. By $(**)$ with $\varphi(\mathbf{y}) = |\mathbf{y}|^2\eta(\mathbf{y})$ we know that

$$\int_{\mathbb{R}^N}|\mathbf{y}|^2\eta(\mathbf{y})\,M(d\mathbf{y}) \leq A\varphi < \infty.$$

At the same time, by (1.1.4) and $(*)$,

$$\int_{\mathbb{R}^N} (1-\eta(\mathbf{y}))\eta_R(\mathbf{y})\, M(d\mathbf{y}) \le A(\mathbf{1}-\eta)$$

for all $R > 0$, and therefore, by Fatou's lemma,

$$\int_{\mathbb{R}^N} (1-\eta(\mathbf{y}))\, M(d\mathbf{y}) \le A(\mathbf{1}-\eta) < \infty.$$

Hence, we have proved that M is a Lévy measure.

We are now in a position to show that (1.1.4) continues to hold for any $\varphi \in \mathscr{S}(\mathbb{R}^N;\mathbb{R})$ that vanishes along with its first and second order derivatives at $\mathbf{0}$. To this end, first suppose that φ vanishes in a neighborhood of $\mathbf{0}$. Then, for each $R > 0$, (1.1.4) applies to $\eta_R\varphi$, and so

$$\int_{\mathbb{R}^N} \eta_R(\mathbf{y})\varphi(\mathbf{y})\, M(d\mathbf{y}) = A(\eta_R\varphi) = A\varphi - A\big((\mathbf{1}-\eta_R)\varphi\big).$$

By $(*)$ applied to $\pm(\mathbf{1}-\eta_R)\varphi$ and $(\mathbf{1}-\eta_R)\|\varphi\|_{\mathrm{u}}$,

$$\big|A\big((\mathbf{1}-\eta_R)\varphi\big)\big| \le \|\varphi\|_{\mathrm{u}} A(\mathbf{1}-\eta_R) \longrightarrow 0 \quad \text{as } R \to \infty,$$

where (ii) was used to get the limit assertion. Thus,

$$A\varphi = \lim_{R\to\infty} \int_{\mathbb{R}^N} \eta_R(\mathbf{y})\varphi(\mathbf{y})\, M(d\mathbf{y}) = \int_{\mathbb{R}^N} \varphi(\mathbf{y})\, M(d\mathbf{y}),$$

because, since M is finite on the support of φ and therefore φ is M-integrable, Lebesgue's dominated convergence theorem applies. We still have to replace the assumption that φ vanishes in a neighborhood of $\mathbf{0}$ by the assumption that it vanishes to second order there. For this purpose, first note that, because M is a Lévy measure, φ is certainly M-integrable, and therefore

$$\int_{\mathbb{R}^N} \varphi(\mathbf{y})\, M(d\mathbf{y}) = \lim_{R\searrow 0} A\big((\mathbf{1}-\eta_R)\varphi\big) = A\varphi - \lim_{R\searrow 0} A(\eta_R\varphi).$$

By our assumptions about φ at $\mathbf{0}$, we can find a $C < \infty$ such that $|\eta_R\varphi(\mathbf{y})| \le CR|\mathbf{y}|^2\eta(\mathbf{y})$ for all $R \in (0,1]$. Hence, by $(*)$, there is a $C' < \infty$ such that $|A(\eta_R\varphi)| \le C'R$ for small $R > 0$, and therefore $A(\eta_R\varphi) \longrightarrow 0$ as $R \searrow 0$.

To complete the proof from here, let $\varphi \in \mathscr{S}(\mathbb{R}^N;\mathbb{R})$ be given, and set

$$\widetilde{\varphi}(\mathbf{x}) = \varphi(\mathbf{x}) - \varphi(\mathbf{0}) - \eta(\mathbf{x})\big(\mathbf{x}, \nabla\varphi(\mathbf{0})\big)_{\mathbb{R}^N} - \tfrac{1}{2}\eta(\mathbf{x})^2\big(\mathbf{x}, \nabla^2\varphi(\mathbf{0})\mathbf{x}\big)_{\mathbb{R}^N}.$$

Then, by the preceding, (1.1.4) holds for $\widetilde{\varphi}$. Further, if $(\mathbf{e}_1, \dots, \mathbf{e}_N)$ is the standard orthonormal basis in $\mathbb{R}^N$, then

$$\eta(\mathbf{y})^2\big(\mathbf{y},\nabla^2\varphi(\mathbf{0})\mathbf{y}\big)_{\mathbb{R}^N} = \sum_{i,j=1}^N \eta_{\mathbf{e}_i}(\mathbf{y})\eta_{\mathbf{e}_j}(\mathbf{y})\big(\mathbf{e}_i,\nabla^2\varphi(\mathbf{0})\mathbf{e}_j\big)_{\mathbb{R}^N}.$$

Hence

$$\begin{aligned}
A\varphi - \big(\mathbf{m}^\eta,\nabla\varphi(\mathbf{0})\big)_{\mathbb{R}^N} &- \frac{1}{2}\sum_{i,j=1}^N A\big(\eta_{\mathbf{e}_i}\eta_{\mathbf{e}_j}\big)\big(\mathbf{e}_i,\nabla^2\varphi(\mathbf{0})\mathbf{e}_j\big)_{\mathbb{R}^N}\\
= A\tilde{\varphi} = \int \Big(\varphi(\mathbf{y}) - \varphi(\mathbf{0}) &- \eta(\mathbf{y})\big(y,\nabla\varphi(\mathbf{0})\big)_{\mathbb{R}^N}\Big) M(d\mathbf{y})\\
&- \frac{1}{2}\sum_{i,j=1}^N \int \eta_{\mathbf{e}_i}(\mathbf{y})\eta_{\mathbf{e}_j}(\mathbf{y})\, M(d\mathbf{y})\big(\mathbf{e}_i,\nabla^2\varphi(\mathbf{0})\mathbf{e}_j\big)_{\mathbb{R}^N},
\end{aligned}$$

and so, after one rearranges terms, we have shown that (1.1.3) holds with $\mathbf{m}^\eta$ and C given by (1.1.2). Thus, the properties of C are all that remain to be proved. That C is symmetric requires no comment. In addition, from (**) applied to $\eta_{\boldsymbol{\xi}}^2$, it is clearly non-negative definite. Finally, to see that it is independent of the η chosen, let $\tilde{\eta}$ be a second choice, note that $\tilde{\eta}_{\boldsymbol{\xi}} = \eta_{\boldsymbol{\xi}}$ in a neighborhood of $\mathbf{0}$, and apply (1.1.4) to $\eta_{\boldsymbol{\xi}}\eta_{\boldsymbol{\xi}'} - \tilde{\eta}_{\boldsymbol{\xi}}\tilde{\eta}_{\boldsymbol{\xi}'}$. □

A triple $(\mathbf{m}, C, M)$, where $\mathbf{m} \in \mathbb{R}^N$, $C \in \mathrm{Hom}(\mathbb{R}^N;\mathbb{R}^N)$ is a symmetric and non-negative definite matrix, and M is a Lévy measure, is called a **Lévy system**.

A careful examination of the proof of Theorem 1.1.1 reveals a lot. Specifically, it shows why the operation performed by the linear functional A cannot be of order greater than 2. The point is, that, because of the minimum principle, A acts as a bounded, non-negative linear functional on the difference between φ and its second order Taylor polynomial, and, because of quasi-locality, this action can be represented by integration against a non-negative measure. The reason why the second order Taylor polynomial suffices is that second order polynomials are, apart from constants, the lowest order polynomials that can have a definite sign.

Another important observation is that there is enormous freedom in the choice of the function η in (1.1.3). Indeed, suppose that $\eta\colon \mathbb{R}^N \longrightarrow [0,1]$ is a Borel measurable function for which

$$\sup_{\mathbf{y}\in B(\mathbf{0},1)} |\mathbf{y}|^{-1}\big(1-\eta(\mathbf{y})\big) + \sup_{\mathbf{y}\notin B(\mathbf{0},1)} |\mathbf{y}|\eta(\mathbf{y}) < \infty,$$

and choose $\eta_0 \in C_{\mathrm{c}}^\infty\big(\mathbb{R}^N;[0,1]\big)$ so that $\eta_0 = 1$ on $\overline{B(\mathbf{0},1)}$ and $\eta_0 = 0$ off $B(\mathbf{0},2)$. Then there is a $C < \infty$ such that

$$|\eta_0(\mathbf{y}) - \eta(\mathbf{y})||\mathbf{y}| \le C\big(|\mathbf{y}|^2 \wedge 1\big),$$

and so

$$\int |\eta_0(\mathbf{y}) - \eta(\mathbf{y})||\mathbf{y}|\, M(d\mathbf{y}) < \infty.$$

Therefore, if $\varphi \in \mathscr{S}(\mathbb{R}^N;\mathbb{R})$, then $\mathbf{y} \rightsquigarrow \varphi(\mathbf{y}) - \varphi(\mathbf{0}) - \eta(\mathbf{y})\big(\mathbf{y}, \eta\varphi(\mathbf{0})\big)_{\mathbb{R}^N}$ is M-integrable and

$$\begin{aligned} A\varphi - \tfrac{1}{2}\operatorname{Trace}\big(C\nabla^2\varphi(\mathbf{0})\big) - \big(m^{\eta_0}, \nabla\varphi(\mathbf{0})\big)_{\mathbb{R}^N} \\ - \int \Big(\varphi(\mathbf{y}) - \varphi(\mathbf{0}) - \eta(\mathbf{y})\big(\mathbf{y}, \nabla\varphi(\mathbf{0})\big)_{\mathbb{R}^N}\Big) M(d\mathbf{y}) \\ = \int \big(\eta_0(\mathbf{y}) - \eta(\mathbf{y})\big)\big(\mathbf{y}, \nabla\varphi(\mathbf{0})\big)_{\mathbb{R}^N}\, M(d\mathbf{y}), \end{aligned}$$

and so (1.1.3) holds when

$$m^{\eta} = m^{\eta_0} + \int \big(\eta_0(\mathbf{y}) - \eta(\mathbf{y})\big)\mathbf{y}\, M(d\mathbf{y}).$$

In particular, there is no reason not to take $\eta = \mathbf{1}_{B(\mathbf{0},1)}$, and we will usually do so.

Finally, notice that A is **local**, in the sense that $A\varphi = 0$ whenever φ is constant in a neighborhood of $\mathbf{0}$, if and only if the corresponding $M = 0$.

1.1.1 *Canonical paths*

We have shown that if $\{\mu_t : t \geq 0\} \subseteq \mathbf{M}_1(\mathbb{R}^N)$ satisfies (1.1.1) and the limit

$$A\varphi = \lim_{t \searrow 0} t^{-1}\big(\langle\varphi, \mu_t\rangle - \varphi(\mathbf{0})\big)$$

exists for every $\varphi \in \mathscr{S}(\mathbb{R}^N;\mathbb{R})$, then there is a Lévy system $(\mathbf{m}, C, M)$ for which

$$\begin{aligned} A\varphi = (\mathbf{m}, \nabla\varphi) + \tfrac{1}{2}\operatorname{Trace}(C\nabla^2\varphi) \\ + \int_{\mathbb{R}^N} \Big(\varphi(\mathbf{y}) - \varphi(\mathbf{0}) - \mathbf{1}_{B(\mathbf{0},1)}(\mathbf{y})\big(\mathbf{y}, \nabla\varphi(0)\big)_{\mathbb{R}^N}\Big) M(d\mathbf{y}). \end{aligned}$$

The goal here is to show that, for each Lévy system, there is a canonical choice of $\{\lambda_t : t \geq 0\}$ such that

$$\begin{aligned} \lim_{t \searrow 0} \frac{\langle\varphi, \lambda_t(d\mathbf{y})\rangle - \varphi(\mathbf{0})}{t} \\ = \big(\mathbf{m}, \nabla\varphi(\mathbf{0})\big)_{\mathbb{R}^N} + \tfrac{1}{2}\operatorname{Trace}\big(C\nabla^2\varphi(\mathbf{0})\big) \\ + \int_{\mathbb{R}^N} \Big(\varphi(\mathbf{y}) - \varphi(\mathbf{0}) - \mathbf{1}_{B(\mathbf{0},1)}(\mathbf{y})\big(\mathbf{y}, \nabla\varphi(0)\big)_{\mathbb{R}^N}\Big) M(d\mathbf{y}). \end{aligned}$$

Lemma 1.1.2. *Set*

$$\ell(\boldsymbol{\xi}) = i(\mathbf{m},\boldsymbol{\xi})_{\mathbb{R}^N} - \tfrac{1}{2}\big(\boldsymbol{\xi}, C\boldsymbol{\xi}\big)_{\mathbb{R}^N} + \int_{\mathbb{R}^N}\Big(e^{i(\boldsymbol{\xi},\mathbf{y})_{\mathbb{R}^N}} - 1 - i\mathbf{1}_{B(\mathbf{0},1)}(\mathbf{y})\big(\boldsymbol{\xi},\mathbf{y}\big)_{\mathbb{R}^N}\Big)M(d\mathbf{y}).$$

Then $\mathfrak{Re}\big(\ell(\boldsymbol{\xi})\big) \le 0$ *and*

$$\lim_{|\boldsymbol{\xi}|\to\infty} |\boldsymbol{\xi}|^{-2}\big|\ell(\boldsymbol{\xi}) + \tfrac{1}{2}\big(\boldsymbol{\xi}, C\boldsymbol{\xi}\big)_{\mathbb{R}^N}\big| = 0.$$

Thus $|e^{t\ell(\boldsymbol{\xi})}| \le 1$ *for* $t \ge 0$, *and there exists a* $K < \infty$ *such that* $|\ell(\boldsymbol{\xi})| \le K(1+|\boldsymbol{\xi}|^2)$. *Finally, for each* $\varphi \in \mathscr{S}(\mathbb{R}^N;\mathbb{C})$,

$$\int_{\mathbb{R}^N} \ell(-\boldsymbol{\xi})\hat{\varphi}(\boldsymbol{\xi})\,d\boldsymbol{\xi} = \lim_{t\searrow 0} t^{-1}\int_{\mathbb{R}^N} \hat{\varphi}(\boldsymbol{\xi})\big(e^{t\ell(-\boldsymbol{\xi})} - 1\big)\,d\boldsymbol{\xi} = (2\pi)^N A\varphi.$$

Proof. Begin by noting that

$$\mathfrak{Re}\big(\ell(\boldsymbol{\xi})\big) = -\tfrac{1}{2}\big(\boldsymbol{\xi}, C\boldsymbol{\xi}\big)_{\mathbb{R}^N} + \int_{\mathbb{R}^N}\big(\cos(\boldsymbol{\xi},\mathbf{y})_{\mathbb{R}^N} - 1\big)M(d\mathbf{y}) \le 0.$$

Next, given $r \in (0,1]$, observe that

$$\begin{aligned}&\left|\int_{\mathbb{R}^N}\Big(e^{i(\boldsymbol{\xi},\mathbf{y})_{\mathbb{R}^N}} - 1 - i\mathbf{1}_{B(\mathbf{0},1)}(\mathbf{y})\big(\boldsymbol{\xi},\mathbf{y}\big)_{\mathbb{R}^N}\Big)M(d\mathbf{y})\right|\\ &\qquad \le \frac{|\boldsymbol{\xi}|^2}{2}\int_{B(\mathbf{0},r)}|\mathbf{y}|^2\,M(d\mathbf{y}) + (2+|\boldsymbol{\xi}|)M\big(B(\mathbf{0},r)\complement\big),\end{aligned}$$

and therefore

$$\begin{aligned}&\lim_{|\boldsymbol{\xi}|\to\infty}|\boldsymbol{\xi}|^{-2}\left|\int_{\mathbb{R}^N}\Big(e^{i(\boldsymbol{\xi},\mathbf{y})_{\mathbb{R}^N}} - 1 - i\mathbf{1}_{B(\mathbf{0},1)}(\mathbf{y})\big(\boldsymbol{\xi},\mathbf{y}\big)_{\mathbb{R}^N}\Big)M(d\mathbf{y})\right|\\ &\qquad \le \frac{1}{2}\int_{B(\mathbf{0},r)}|\mathbf{y}|^2\,M(d\mathbf{y}).\end{aligned}$$

Since $\int_{B(\mathbf{0},r)}|\mathbf{y}|^2\,M(d\mathbf{y}) \longrightarrow 0$ as $r \searrow 0$, this completes the proof of the initial assertions.

To prove the final assertion, note that, by Taylor's theorem,

$$\big|e^{t\ell(\boldsymbol{\xi})} - 1 - t\ell(\boldsymbol{\xi})\big| \le \frac{t^2|\ell(\boldsymbol{\xi})|^2}{2} \le \frac{K^2t^2(1+|\boldsymbol{\xi}|^2)^2}{2}.$$

Hence, since $\hat{\varphi}$ is rapidly decreasing[3], Lebesgue's dominated convergence theorem shows that

$$\lim_{t\searrow 0} t^{-1}\int_{\mathbb{R}^N}\hat{\varphi}(\boldsymbol{\xi})\big(e^{t\ell(-\boldsymbol{\xi})} - 1\big)\,d\boldsymbol{\xi} = \int_{\mathbb{R}^N}\hat{\varphi}(\boldsymbol{\xi})\ell(-\boldsymbol{\xi})\,d\boldsymbol{\xi}.$$

[3] That is, it tends to 0 at ∞ faster than $|\boldsymbol{\xi}|^{-n}$ for every $n \ge 0$.

At the same time, by the Fourier inversion formula,

$$\begin{aligned}(2\pi)^{-N}&\int_{\mathbb{R}^N}\left(-i(\mathbf{m},\boldsymbol{\xi})_{\mathbb{R}^N}-\tfrac{1}{2}(\boldsymbol{\xi},C\boldsymbol{\xi})_{\mathbb{R}^N}\right)\hat{\varphi}(\boldsymbol{\xi})\,d\boldsymbol{\xi}\\&=\big(\mathbf{m},\nabla\varphi(\mathbf{0})\big)_{\mathbb{R}^N}+\tfrac{1}{2}\operatorname{Trace}\big(C\nabla^2\varphi(\mathbf{0})\big).\end{aligned}$$

Finally, because

$$\iint_{\mathbb{R}^N\times\mathbb{R}^N}|\hat{\varphi}(\boldsymbol{\xi})|\left|e^{-i(\boldsymbol{\xi},\mathbf{y})_{\mathbb{R}^N}}-1+i\mathbf{1}_{B(\mathbf{0},1)}(\mathbf{y})(\boldsymbol{\xi},\mathbf{y})_{\mathbb{R}^N}\right|M(d\mathbf{y})d\boldsymbol{\xi}<\infty,$$

Fubini's theorem applies and says that

$$\begin{aligned}&\int_{\mathbb{R}^N}\hat{\varphi}(\xi)\left(\int_{\mathbb{R}^N}\left(e^{-i(\boldsymbol{\xi},\mathbf{y})_{\mathbb{R}^N}}-1+i\mathbf{1}_{B(\mathbf{0},1)}(\mathbf{y})(\boldsymbol{\xi},\mathbf{y})_{\mathbb{R}^N}\right)M(d\mathbf{y})\right)d\boldsymbol{\xi}\\&=\int_{\mathbb{R}^N}\left(\int_{\mathbb{R}^N}\left(e^{-i(\boldsymbol{\xi},\mathbf{y})_{\mathbb{R}^N}}-1+i\mathbf{1}_{B(\mathbf{0},1)}(\mathbf{y})(\boldsymbol{\xi},\mathbf{y})_{\mathbb{R}^N}\right)\hat{\varphi}(\boldsymbol{\xi})\,d\boldsymbol{\xi}\right)M(d\mathbf{y})\\&=(2\pi)^N\int_{\mathbb{R}^N}\left(\varphi(\mathbf{y})-\varphi(\mathbf{0})-\mathbf{1}_{B(\mathbf{0},1)}(\mathbf{y})\big(\mathbf{y},\nabla\varphi(\mathbf{0})\big)_{\mathbb{R}^N}\right)M(d\mathbf{y}),\end{aligned}$$

where we again used the Fourier inversion formula in the passage to the last line. □

In view of Lemma 1.1.2, we will be done once we show that, for each $t>0$, there exists a $\lambda_t\in\mathbf{M}_1(\mathbb{R}^N)$ such that $\widehat{\lambda_t}(\boldsymbol{\xi})=e^{t\ell(\boldsymbol{\xi})}$. Indeed, by Parseval's identity,

$$\langle\varphi,\lambda_t\rangle-\varphi(\mathbf{0})=(2\pi)^{-N}\int_{\mathbb{R}^N}\hat{\varphi}(\boldsymbol{\xi})\big(\widehat{\lambda_t}(-\boldsymbol{\xi})-1\big)\,d\boldsymbol{\xi}$$

for $\varphi\in\mathscr{S}(\mathbb{R}^N;\mathbb{C})$.

Since $t\ell(\boldsymbol{\xi})$ can be represented as the $\ell(\boldsymbol{\xi})$ for $(t\mathbf{m},tC,tM)$, it suffices to take $t=1$. Furthermore, because $\widehat{\mu*\nu}(\boldsymbol{\xi})=\hat{\mu}(\boldsymbol{\xi})\hat{\nu}(\boldsymbol{\xi})$, we can treat the terms in ℓ separately. The term corresponding to C is the **Gaussian component**, and the corresponding measure $\gamma_{\mathbf{0},C}$ is the distribution of $\mathbf{y}\rightsquigarrow C^{\frac{1}{2}}\mathbf{y}$ under the **standard Gauss measure**

$$\gamma_{\mathbf{0},\mathbf{I}}(d\mathbf{y})=(2\pi)^{-\frac{N}{2}}e^{-\frac{|\mathbf{y}|^2}{2}}\,d\mathbf{y}.$$

To deal with the term corresponding to M, initially assume that M is finite, and consider the Poisson measure

$$\Pi_M=e^{-M(\mathbb{R}^N)}\sum_{k=0}^{\infty}\frac{M^{*k}}{k!},$$

where $M^{*0}:=\delta_{\mathbf{0}}$ ($\delta_{\mathbf{a}}$ denotes the unit point mass at $\mathbf{a}$) and, for $k\geq 1$, $M^{*k}=M*M^{*(k-1)}$ is the k-fold convolution of M with itself. One then has that

$$\widehat{\Pi_M}(\boldsymbol{\xi}) = e^{-M(\mathbb{R}^N)} \sum_{k=0}^{\infty} \frac{\widehat{M}(\boldsymbol{\xi})^k}{k!} = \exp\left(\int_{\mathbb{R}^N} \left(e^{i(\boldsymbol{\xi},\mathbf{y})} - 1\right) M(d\mathbf{y})\right).$$

To handle general Lévy measures M, for each $r \in (0,1)$, define $M_r(d\mathbf{y}) = \mathbf{1}_{(r,\infty)}(|\mathbf{y}|)M(d\mathbf{y})$ and $\mathbf{a}_r = \int_{B(\mathbf{0},1)} \mathbf{y}\, M_r(d\mathbf{y})$. Then

$$\begin{aligned}\widehat{\delta_{-\mathbf{a}_r} * \Pi_{M_r}}(\boldsymbol{\xi}) &= \exp\left(\int_{B(\mathbf{0},r)\complement} \left(e^{i(\boldsymbol{\xi},\mathbf{y})_{\mathbb{R}^N}} - 1 - i\mathbf{1}_{B(\mathbf{0},1)}(\mathbf{y})(\boldsymbol{\xi},\mathbf{y})_{\mathbb{R}^N}\right) M(d\mathbf{y})\right)\\ &\longrightarrow \exp\left(\int_{\mathbb{R}^N} \left(e^{i(\boldsymbol{\xi},\mathbf{y})_{\mathbb{R}^N}} - 1 - i\mathbf{1}_{B(\mathbf{0},1)}(\mathbf{y})(\boldsymbol{\xi},\mathbf{y})_{\mathbb{R}^N}\right) M(d\mathbf{y})\right)\end{aligned}$$

uniformly for $\boldsymbol{\xi}$ in compact subsets of $\mathbb{R}^N$. To complete the proof, we need Lévy's continuity theorem (see Theorem 3.1.8 in [20]), which states that if $\{\mu_n : n \geq 0\} \subseteq \mathbf{M}_1(\mathbb{R}^N)$ and $\hat{\mu}_n$ is the characteristic function (i.e., the Fourier transform) of μ_n, then $\mu = \lim_{n\to\infty} \mu_n$ exists in $\mathbf{M}_1(\mathbb{R}^N)$ with the weak topology if and only if $\hat{\mu}_n(\boldsymbol{\xi})$ converges for each $\boldsymbol{\xi}$ and uniformly in a neighborhood of $\mathbf{0}$, in which case $\mu_n \longrightarrow \mu$ in $\mathbf{M}_1(\mathbb{R}^N)$ where $\hat{\mu}(\boldsymbol{\xi}) = \lim_{n\to\infty} \hat{\mu}_n(\boldsymbol{\xi})$. In particular, there is an element μ_M of $\mathbf{M}_1(\mathbb{R}^N)$ whose Fourier transform is

$$\exp\left(\int_{\mathbb{R}^N} \left(e^{i(\boldsymbol{\xi},\mathbf{y})_{\mathbb{R}^N}} - 1 - i\mathbf{1}_{B(\mathbf{0},1)}(\mathbf{y})(\boldsymbol{\xi},\mathbf{y})_{\mathbb{R}^N}\right) M(d\mathbf{y})\right).$$

Hence, if $\lambda = \delta_{\mathbf{m}} * \mu_M * \gamma_{\mathbf{0},C}$, then $\hat{\lambda}(\boldsymbol{\xi}) = e^{\ell(\boldsymbol{\xi})}$.

1.2 Kolmogorov's equations

The reason why the measures λ_t constructed in the preceding are *canonical* is that they are the analog in $\mathbf{M}_1(\mathbb{R}^N)$ of rays $\mathbb{R}^N$. This analogy is based on the equation $\lambda_{s+t} = \lambda_s * \lambda_t$. If one thinks of convolution as the analog in $\mathbf{M}_1(\mathbb{R}^N)$ of addition in $\mathbb{R}^N$, then this equation is the analog of the equation $F(s+t) = F(s) + F(t)$, which is the equation for a ray in $\mathbb{R}^N$.

There is an alternative way to think about this analogy. Namely, a ray is the integral curve starting at $\mathbf{0}$ of a vector field $\mathbf{V}$ that is constant in the sense that, for all $\mathbf{x} \in \mathbb{R}^N$, $\tau_{\mathbf{x}} \circ \mathbf{V} = \mathbf{V} \circ \tau_{\mathbf{x}}$, where $\tau_{\mathbf{x}}$ is the translation operator given by $\tau_{\mathbf{x}} f(\mathbf{y}) = f(\mathbf{x}+\mathbf{y})$ for $f\colon \mathbb{R}^N \longrightarrow \mathbb{R}$. Since $t \rightsquigarrow \lambda_t$ starts at $\delta_{\mathbf{0}}$, which is the analog in $\mathbf{M}_1(\mathbb{R}^N)$ of $\mathbf{0}$ in $\mathbb{R}^N$, checking that $t \rightsquigarrow \lambda_t$ is the analog of a ray reduces to showing that its tangent field is constant. To this end, let A be the linear functional determined by the Lévy system corresponding to $\{\lambda_t : t > 0\}$, and define the operator L so that $L\varphi(\mathbf{x}) = A \circ \tau_{\mathbf{x}}\varphi$. Then

$$
\begin{aligned}
t^{-1}\big(\langle\varphi,\lambda_{s+t}\rangle-\langle\varphi,\lambda_s\rangle\big)&=t^{-1}\int\left(\int\big(\varphi(\mathbf{x}+\mathbf{y})-\varphi(\mathbf{x})\big)\,\lambda_t(d\mathbf{y})\right)\lambda_s(d\mathbf{x})\\
&=\int t^{-1}\big(\langle\tau_{\mathbf{x}}\varphi,\mu_t\rangle-\tau_x\varphi(\mathbf{0})\big)\lambda_s(d\mathbf{x})\longrightarrow\langle A\circ\tau_{\mathbf{x}}\varphi,\lambda_s\rangle,
\end{aligned}
$$

as $t\searrow 0$. Hence,

$$
\lim_{t\searrow} t^{-1}\big(\langle\varphi,\lambda_{s+t}\rangle-\langle\varphi,\lambda_s\rangle\big)=\langle L\varphi,\lambda_s\rangle,
$$

and so

$$
\frac{d}{dt}\langle\varphi,\lambda_t\rangle=\langle L\varphi,\lambda_t\rangle. \tag{1.2.1}
$$

Thus L can be thought of as the tangent field along $t\rightsquigarrow\lambda_t$, and it is clearly constant in the sense that $\tau_{\mathbf{x}}\circ L=L\circ\tau_{\mathbf{x}}$.

As we will see, there are advantages to the second line of reasoning. For example, it gives us another characterization of $\{\lambda_t : t>0\}$. Namely, we know that (1.2.1) holds for $\varphi\in\mathbb{D}$, but one can easily show that it holds for all $\varphi\in C_{\mathrm{b}}(\mathbb{R}^N;\mathbb{C})$. Indeed, first note that

$$
\begin{aligned}
|L\varphi(\mathbf{x})|&\le\|C\|_{\mathrm{op}}\|\nabla^2\varphi(\mathbf{x})\|_{\mathrm{op}}+|\mathbf{m}||\nabla\varphi(\mathbf{x})|\\
&\quad+\frac{1}{2}\sup_{\mathbf{y}\in B(\mathbf{x},1)}\|\nabla^2\varphi(\mathbf{y})\|_{\mathrm{op}}\int_{B(\mathbf{0},1)}|\mathbf{y}|^2\,M(d\mathbf{y})+2M\big(B(\mathbf{0},1)\complement\big)\|\varphi\|_{\mathrm{u}}.
\end{aligned}
$$

Now let $\varphi\in C^2_{\mathrm{b}}(\mathbb{R}^N;\mathbb{R})$ be given. Then we can choose $\{\varphi_n : n\ge 1\}\subseteq\mathscr{S}(\mathbb{R}^N;\mathbb{R})$ to be a sequence of functions that are bounded in $C^2_{\mathrm{b}}(\mathbb{R}^N;\mathbb{R})$ and for which $\varphi_n\longrightarrow\varphi$, $\nabla\varphi_n\longrightarrow\nabla\varphi$, and $\nabla^2\varphi_n\longrightarrow\nabla^2\varphi$ uniformly on compacts, in which case the preceding says that $\sup_{n\ge1}\|L\varphi_n\|_{\mathrm{u}}<\infty$ and $L\varphi_n(\mathbf{x})\longrightarrow L\varphi(\mathbf{x})$ uniformly for $\mathbf{x}$ in compacts. Hence, from (1.2.1) for $\varphi\in\mathscr{S}(\mathbb{R}^N;\mathbb{R})$, one has that

$$
\begin{aligned}
\langle\varphi,\lambda_t\rangle-\varphi(\mathbf{0})&=\lim_{n\to\infty}\big(\langle\varphi_n,\lambda_t\rangle-\varphi_n(\mathbf{0})\big)\\
&=\lim_{n\to\infty}\int_0^t\langle L\varphi_n,\lambda_\tau\rangle\,d\tau=\int_0^t\langle L\varphi,\lambda_\tau\rangle\,d\tau,
\end{aligned}
$$

and so (1.2.1) continues to hold for $\varphi\in C^2_{\mathrm{b}}(\mathbb{R}^N;\mathbb{R})$. Knowing (1.2.1) for $\varphi\in C^2_{\mathrm{b}}(\mathbb{R}^N;\mathbb{R})$, one knows that it holds for $\varphi\in C^2_{\mathrm{b}}(\mathbb{R}^N;\mathbb{C})$, and applying it to $e^{i(\boldsymbol{\xi},\mathbf{y})_{\mathbb{R}^N}}$, one sees that it implies

$$
\frac{d}{dt}\widehat{\lambda_t}(\boldsymbol{\xi})=\ell(\boldsymbol{\xi})\widehat{\lambda_t}(\boldsymbol{\xi}).
$$

Hence (1.2.1) together with the initial condition $\lambda_0=\delta_0$ implies that $\widehat{\lambda_t}(\boldsymbol{\xi})=e^{t\ell(\boldsymbol{\xi})}$. That is, (1.2.1) together with $\lambda_0=\delta_0$ uniquely determine λ_t, and so *$t\rightsquigarrow\lambda_t$ is the unique integral curve of L starting at δ_0.*

Up to now we have looked only at solutions to (1.2.1) that satisfy the initial condition $\lambda_0 = \delta_{\mathbf{0}}$, but there are good reasons to look at ones corresponding to a general initial condition. Thus, let $\nu \in \mathbf{M}_1(\mathbb{R}^N)$ be given. Then using Fourier transforms, one sees that $t \rightsquigarrow \omega_t$ satisfies (1.2.1) with $\omega_0 = \nu$ if and only if $\widehat{\omega}_t = \hat{\nu}e^{t\ell}$, and so $t \rightsquigarrow \nu * \lambda_t$ is the one and only solution. In particular, consider $P(t, \mathbf{x}, \cdot\,) = \delta_{\mathbf{x}} * \lambda_t$. Then

$$\begin{aligned} P(s+t, \mathbf{x}, \Gamma) &= \int_{\mathbb{R}^N} \lambda_t(\Gamma - \mathbf{x} - \mathbf{y})\, \lambda_s(d\mathbf{y}) = \int_{\mathbb{R}^N} \lambda_t(\Gamma - \mathbf{y})\, P(s, \mathbf{x}, d\mathbf{y}) \\ &= \int_{\mathbb{R}^N} P(t, \mathbf{y}, \Gamma)\, P(s, \mathbf{x}, d\mathbf{y}), \end{aligned}$$

and so $P(t, \mathbf{x}, \cdot\,)$ satisfies the **Chapman–Kolmogorov equation**

$$P(s+t, \mathbf{x}, \Gamma) = \int_{\mathbb{R}^N} P(t, \mathbf{y}, \Gamma)\, P(s, \mathbf{x}, d\mathbf{y}). \tag{1.2.2}$$

Hence $(t, \mathbf{x}) \rightsquigarrow P(t, \mathbf{x}, \cdot\,)$ is a **transition probability function** which, for $\varphi \in C^2_{\mathrm{b}}(\mathbb{R}^N; \mathbb{R})$, satisfies

$$\frac{d}{dt}\int_{\mathbb{R}^N} \varphi(\mathbf{y})\, P(t, \mathbf{x}, dy) = \int_{\mathbb{R}^N} L\varphi(\mathbf{y})\, P(t, \mathbf{x}, d\mathbf{y}) \text{ and } P(0, \mathbf{x}, \cdot\,) = \delta_{\mathbf{x}}. \tag{1.2.3}$$

This equation is **Kolmogorov's forward equation for a transition probability function.**

Next observe that, because

$$\int_{\mathbb{R}^N} \varphi(\mathbf{y})\, P(t, \mathbf{x}, d\mathbf{y}) = \int_{\mathbb{R}^N} \varphi(\mathbf{x} + \mathbf{y})\, \lambda_t(d\mathbf{y}),$$

$\int_{\mathbb{R}^N} \varphi(\mathbf{y})\, P(t, \cdot\,, d\mathbf{y}) \in C^2_{\mathrm{b}}(\mathbb{R}^N; \mathbb{R})$ if $\varphi \in C^2_{\mathrm{b}}(\mathbb{R}^N; \mathbb{R})$, and therefore, as $h \searrow 0$,

$$\begin{aligned} &\frac{1}{h}\left(\int_{\mathbb{R}^N} \varphi(\mathbf{y})\, P(t+h, \cdot\,, d\mathbf{y}) - \int_{\mathbb{R}^N} \varphi(\mathbf{y})\, P(t, \cdot\,, d\mathbf{y})\right) \\ &\quad = \frac{1}{h}\left(\int_{\mathbb{R}^N}\left(\int_{\mathbb{R}^N} \varphi(\mathbf{y}')\, P(t, \mathbf{y}, d\mathbf{y}')\right) P(h, \cdot\,, d\mathbf{y}) - \int_{\mathbb{R}^N} \varphi(\mathbf{y}')\, P(t, \cdot\,, d\mathbf{y}')\right) \\ &\quad \longrightarrow L\int_{\mathbb{R}^N} \varphi(\mathbf{y})\, P(t, \cdot\,, d\mathbf{y}). \end{aligned}$$

Hence, if $\varphi \in C^2_{\mathrm{b}}(\mathbb{R}^N; \mathbb{R})$, then

$$\frac{d}{dt}\int_{\mathbb{R}^N} \varphi(\mathbf{y})\, P(t, \cdot\,, d\mathbf{y}) = L\int_{\mathbb{R}^N} \varphi(\mathbf{y})\, P(t, \cdot\,, d\mathbf{y})$$
$$\text{and} \quad \lim_{t\searrow 0}\int_{\mathbb{R}^N} \varphi(\mathbf{y})\, P(t, \cdot\,, d\mathbf{y}) = \varphi,$$

which means that $u(t,\mathbf{x}) = \int_{\mathbb{R}^N} \varphi(\mathbf{y})\, P(t,\mathbf{x},d\mathbf{y})$ is a solution to Kolmogorov's **backward equation.**

$$\partial_t u = Lu \quad \text{and} \quad \lim_{t\searrow 0} u(t,\,\cdot\,) = \varphi. \tag{1.2.4}$$

In fact, $(t,\mathbf{x}) \rightsquigarrow \int_{\mathbb{R}^N} \varphi(\mathbf{y})\, P(t,\mathbf{x},d\mathbf{y})$ is the one and only solution $u \in C_{\mathrm{b}}^{1,2}\big([0,\infty)\times\mathbb{R}^N;\mathbb{R}\big)$ to (1.2.4). To see that there is no other solution u, use (1.2.3) and the chain rule to show that

$$\frac{d}{d\tau}\int_{\mathbb{R}^N} u(t-\tau,\mathbf{y})\, P(\tau,\mathbf{x},d\mathbf{y}) = 0 \quad \text{for } \tau \in (0,t),$$

and conclude that $u(t,\mathbf{x}) = \int_{\mathbb{R}^N} \varphi(\mathbf{y})\, P(t,\mathbf{x},d\mathbf{y})$.

The reason for the names "forward" and "backward" is best understood when one thinks about the Markov process for which $P(t,\mathbf{x},d\mathbf{y})$ is the transition function: the $\mathbf{y}$ is where the path will be after time t and $\mathbf{x}$ is where it was initially. Thus, $\mathbf{y}$ is the relevant variable for an observer looking forward in time, and $\mathbf{x}$ is the relevant variable when he looks backward. Therefore the equation (1.2.3) describes the evolution of $P(t,\mathbf{x},\,\cdot\,)$ as a function of time and the forward variable, whereas (1.2.4) describes that evolution in terms of time and the backward variable. These ideas are easiest to understand when, as was the case in [9], $P(t,\mathbf{x},d\mathbf{y})$ admits a density $p(t,\mathbf{x},\mathbf{y})$ with respect to some reference measure. In that case, (1.2.3) can be written as

$$\partial_t p(t,\mathbf{x},\,\cdot\,) = L^* p(t,\mathbf{x},\,\cdot\,) \quad \text{and} \quad \lim_{t\searrow 0} p(t,\mathbf{x},\,\cdot\,) = \delta_{\mathbf{x}},$$

where L^* is the adjoint of L with respect to the reference measure, and (1.2.4) becomes

$$\partial_t p(t,\,\cdot\,,\mathbf{y}) = Lp(t,\,\cdot\,,\mathbf{y}) \quad \text{and} \quad \lim_{t\searrow 0} p(t,\,\cdot\,,\mathbf{y}) = \delta_{\mathbf{y}}.$$

1.2.1 *The forward equation with variable coefficients*

In the preceding, we showed that, for any $\nu \in \mathbf{M}_1(\mathbb{R}^N)$, $\{\nu * \lambda_t : t > 0\}$ can be described as the unique integral curve of the vector field L determined by a Lévy system $(\mathbf{m}, C, M)$. As we pointed out, L is a constant vector field. Here we will show how one can go about solving Kolmogorov's forward equation for L's having variable coefficients, although, because it is the case dealt with in the rest of this book, we will restrict our attention to local operators. That is, until further notice, we will be dealing with operators

$$L\varphi(\mathbf{x}) = \frac{1}{2}\sum_{i,j=1}^N a_{ij}(\mathbf{x})\partial_{x_i}\partial_{x_j}\varphi(\mathbf{x}) + \sum_{i=1}^N b_i(\mathbf{x})\partial_{x_i}\varphi(\mathbf{x}), \tag{1.2.5}$$

where $a(\mathbf{x}) = \big((a_{ij}(\mathbf{x}))\big)_{1\le i,j\le N}$ is a non-negative definite, symmetric matrix for each $x \in \mathbb{R}^N$. In the probability literature, a is called the **diffusion coefficient** and b is called the **drift coefficient**. (Cf. the discussion at the beginning of §4.5 regarding these designations.)

The goal is to find a family $\{\mu_t : t \ge 0\} \subseteq \mathbf{M}_1(\mathbb{R}^N)$ which satisfies Kolmogorov's forward equation

$$\partial_t\langle\varphi,\mu_t\rangle = \langle L\varphi,\mu_t\rangle \quad \text{with}\, \mu_0 = \nu. \tag{1.2.6}$$

Although Kolmogorov interpreted this problem from a purely analytic standpoint and wrote (1.2.6) as $\partial_t f_t = L^* f_t$, where f_t is the density of μ_t with respect to Lebesgue measure and

$$L^*\varphi = \frac{1}{2}\sum_{i,j=1}^N \partial_{x_i}\partial_{x_j}(a_{ij}\varphi) - \sum_{i=1}^N \partial_{x_i}(b_i\varphi)$$

is the formal adjoint of L, K. Itô chose (cf. [18]) an interpretation based on the idea that L is a vector field on $\mathbf{M}_1(\mathbb{R}^N)$ and that (1.2.6) is the equation that describes its integral curves starting at ν. One of the many advantages to adopting Itô's interpretation is that it leads to the following general existence theorem and explains how he arrived at the ideas developed in Chapters 2 and 3.

Theorem 1.2.1. *Assume that a and b are continuous and that*

$$\Lambda := \sup_{x\in\mathbb{R}^N} \frac{\mathrm{Trace}\big(a(\mathbf{x})\big) + 2\big(\mathbf{x}, b(\mathbf{x})\big)^+_{\mathbb{R}^N}}{1+|\mathbf{x}|^2} < \infty. \tag{1.2.7}$$

Then, for each $\nu \in \mathbf{M}_1(\mathbb{R}^N)$, there is a continuous $t \in [0,\infty) \longmapsto \mu_t \in \mathbf{M}_1(\mathbb{R}^N)$ which satisfies

$$\langle\varphi,\mu_t\rangle - \langle\varphi,\nu\rangle = \int_0^t \langle L\varphi,\mu_\tau\rangle\, d\tau, \tag{1.2.8}$$

for all $\varphi \in C^2_c(\mathbb{R}^N;\mathbb{C})$, where L is the operator in (1.2.5). *Moreover,*

$$\int (1+|\mathbf{y}|^2)\,\mu_t(d\mathbf{y}) \le e^{\Lambda t}\int (1+|\mathbf{x}|^2)\,\nu(d\mathbf{x}), \quad t \ge 0. \tag{1.2.9}$$

Before giving the proof, it may be helpful to review how one goes about integrating vector fields on $\mathbb{R}^N$. Indeed, when applied to the case when $a = 0$, our proof is exactly the same as the Euler approximation scheme used to solve first order ordinary differential equations. Namely, when $a = 0$, except for the initial condition, there should be no randomness, and so, when we remove the randomness from the initial condition by taking $\nu = \delta_{\mathbf{x}}$, we should expect (cf. Exercise 5.1) that $\mu_t = \delta_{X(t,\mathbf{x})}$, where $t \in [0,\infty) \longmapsto X(t,\mathbf{x}) \in \mathbb{R}^N$ satisfies

$$\varphi\big(X(t,\mathbf{x})\big) - \varphi(\mathbf{x}) = \int_0^t \big(b(X(\tau,\mathbf{x})), \nabla\varphi(X(\tau))\big)_{\mathbb{R}^N}\, d\tau.$$

Equivalently, $t \rightsquigarrow X(t,\mathbf{x})$ is an integral curve of the vector field b starting at $\mathbf{x}$. That is,

$$X(t,\mathbf{x}) = \mathbf{x} + \int_0^t b\big(X(\tau,\mathbf{x})\big)\, d\tau.$$

To show that such an integral curve exists, Euler used the following approximation scheme. For each $n \geq 0$, define $t \rightsquigarrow X_n(t,\mathbf{x})$ so that $X_n(0,\mathbf{x}) = \mathbf{x}$ and

$$X_n(t,\mathbf{x}) = X_n(m2^{-n},\mathbf{x}) + (t - m2^{-n})b\big(X_n(m2^{-n},\mathbf{x})\big)$$
$$\text{for } m2^{-n} < t \leq (m+1)2^{-n}.$$

Clearly,

$$X_n(t,\mathbf{x}) = \mathbf{x} + \int_0^t b\big(X_n(\lfloor\tau\rfloor_n, \mathbf{x}\big)\, d\tau,$$

where[4] $\lfloor\tau\rfloor_n = 2^{-n}\lfloor 2^n\tau\rfloor$ is the largest diadic number $m2^{-n}$ dominated by τ. Hence, if we can show that $\{X_n(\,\cdot\,,\mathbf{x}) : n \geq 0\}$ is relatively compact in the space $C\big([0,\infty);\mathbb{R}^N\big)$ with the topology of uniform convergence on compacts, then we can take $t \rightsquigarrow X(t,\mathbf{x})$ to be any limit of the $X_n(\,\cdot\,,\mathbf{x})$'s.

To simplify matters, assume for the moment that b is bounded. In that case it is clear that $|X_n(t,\mathbf{x}) - X_n(s,\mathbf{x})| \leq \|b\|_{\mathrm{u}}|t-s|$, and so the Ascoli–Arzela theorem guarantees the required compactness. To remove the boundedness assumption, choose an $\eta \in C^\infty_{\mathrm{c}}\big(\mathbb{R}^N;[0,1]\big)$ so that $\eta = 1$ on $\overline{B(0,1)}$ and 0 off of $B(\mathbf{0},2)$, and, for each $k \geq 1$, replace b by b_k, where $b_k(\mathbf{y}) = \eta(k^{-1}\mathbf{x})b(\mathbf{x})$. Next, let $t \rightsquigarrow X_k(t,\mathbf{x})$ be an integral curve of b_k starting at $\mathbf{x}$, and observe that

$$\frac{d}{dt}|X_k(t,\mathbf{x})|^2 = 2\big(X_k(t,\mathbf{x}), b_k(X_k(t,\mathbf{x}))\big)_{\mathbb{R}^N} \leq \Lambda\big(1 + |X_k(t,\mathbf{x})|^2\big),$$

from which, even without Lemma 1.2.4 below, it is an easy step to the conclusion that

$$1 + |X_k(t,\mathbf{x})|^2 \leq (1+|\mathbf{x}|^2)e^{t\Lambda}.$$

But this means that, for each $T > 0$, $|X_k(t,\mathbf{x}) - X_k(s,\mathbf{x})| \leq C(T)|t-s|$ for $s,t \in [0,T]$, where $C(T)$ is the maximum value of $|b|$ on the closed ball of radius $(1 + \|\mathbf{x}|)^{\frac{\Lambda t}{2}}$ centered at the origin, and so we again can invoke the Ascoli–Arzela theorem to see that $\{X_k(\,\cdot\,,\mathbf{x}) : k \geq 1\}$ is relatively compact and therefore has a limit which is an integral curve of b.

We will now show how to construct a solution to (1.2.6) when $a \neq 0$. In view of the preceding, it should be clear that our first task is to find an appropriate replacement for the Ascoli–Arzela theorem, and the one that we will choose is a variant of P. Lévy's continuity theorem.

[4] We will use $\lfloor\tau\rfloor$ to denote the integer part of a number $\tau \in \mathbb{R}$.

In the following, and elsewhere, say that $\{\varphi_k : k \geq 1\} \subseteq C_b(\mathbb{R}^N;\mathbb{C})$ converges to φ in $C_b(\mathbb{R}^N;\mathbb{C})$ and write $\varphi_k \longrightarrow \varphi$ in $C_b(\mathbb{R}^N;\mathbb{C})$ if $\sup_k \|\varphi_k\|_u < \infty$ and $\varphi_k(\mathbf{x}) \longrightarrow \varphi(\mathbf{x})$ uniformly for $\mathbf{x}$ in compact subsets of $\mathbb{R}^N$. Also, given a σ-compact metric space E, say that $\{\mu_k : k \geq 1\} \subseteq C(E;\mathbf{M}_1(\mathbb{R}^N))$ **converges to μ in $C(E;\mathbf{M}_1(\mathbb{R}^N))$** and write $\mu_k \longrightarrow \mu$ in $C(E;\mathbf{M}_1(\mathbb{R}^N))$ if, for each $\varphi \in C_b(\mathbb{R}^N;\mathbb{C})$, $\langle\varphi, \mu_k(\mathbf{z})\rangle \longrightarrow \langle\varphi, \mu(\mathbf{z})\rangle$ uniformly for $\mathbf{z}$ in compact subsets of E.

Theorem 1.2.2. *If $\mu_k \longrightarrow \mu$ in $C(E;\mathbf{M}_1(\mathbb{R}^N))$, then*

$$\langle\varphi_k, \mu_k(\mathbf{z}_k)\rangle \longrightarrow \langle\varphi, \mu(\mathbf{z})\rangle$$

whenever $\mathbf{z}_k \longrightarrow \mathbf{z}$ in E and $\varphi_k \longrightarrow \varphi$ in $C_b(\mathbb{R}^N;\mathbb{C})$. Moreover, if $\{\mu_n : n \geq 0\} \subseteq C(E;\mathbf{M}_1(\mathbb{R}^N))$ and $f_n(\mathbf{z},\boldsymbol{\xi}) = \widehat{\mu_n(\mathbf{z})}(\boldsymbol{\xi})$, then $\{\mu_n : n \geq 0\}$ is relatively compact in $C(E;\mathbf{M}_1(\mathbb{R}^N))$ if $\{f_n : n \geq 0\}$ is equicontinuous at each $(\mathbf{z},\boldsymbol{\xi}) \in E \times \mathbb{R}^N$. In particular, $\{\mu_n : n \geq 0\}$ is relatively compact if, for each $\boldsymbol{\xi} \in \mathbb{R}^N$, $\{f_n(\,\cdot\,,\boldsymbol{\xi}) : n \geq 0\}$ is equicontinuous at each $\mathbf{z} \in E$ and, for each compact $K \subseteq E$,

$$\lim_{R\to\infty} \sup_{n\geq 0} \sup_{\mathbf{z}\in K} \mu_n\big(\mathbf{z}, \mathbb{R}^N \setminus B(0,R)\big) = 0.$$

Proof. To prove the first assertion, suppose $\mu_k \longrightarrow \mu$ in $C(E;\mathbf{M}_1(\mathbb{R}^N))$, $\mathbf{z}_k \longrightarrow \mathbf{z}$ in E, and $\varphi_k \longrightarrow \varphi$ in $C_b(\mathbb{R}^N;\mathbb{C})$. Then, for every $R > 0$,

$$\begin{aligned}
&\overline{\lim_{k\to\infty}} \big|\langle\varphi_k, \mu_k(\mathbf{z}_k)\rangle - \langle\varphi, \mu(\mathbf{z})\rangle\big| \\
&\leq \overline{\lim_{k\to\infty}} \Big(\big|\langle\varphi_k - \varphi, \mu_k(\mathbf{z}_k)\rangle\big| + \big|\langle\varphi, \mu_k(\mathbf{z}_k)\rangle - \langle\varphi, \mu(\mathbf{z})\rangle\big|\Big) \\
&\leq \overline{\lim_{k\to\infty}} \sup_{y\in B(0,R)} |\varphi_k(y) - \varphi(y)| + 2\sup_k \|\varphi_k\|_u \overline{\lim_{k\to\infty}} \mu_k\big(\mathbf{z}_k, B(0,R)\complement\big) \\
&\leq 2\sup_k \|\varphi_k\|_u \mu\big(\mathbf{z}, B(0,R)\complement\big)
\end{aligned}$$

since (cf. Theorem 9.1.5 in [20]) $\overline{\lim}_{k\to\infty} \mu_k(\mathbf{z}_k, F) \leq \mu(\mathbf{z}, F)$ for any closed $F \subseteq \mathbb{R}^N$. Hence, the required conclusion follows after one lets $R \to \infty$.

Turning to the second assertion, apply the Arzela–Ascoli theorem to produce an $f \in C_b(E \times \mathbb{R}^N;\mathbb{C})$ and a subsequence $\{n_k : k \geq 0\}$ such that $f_{n_k} \longrightarrow f$ uniformly on compacts. By Lévy's continuity theorem, there is, for each $\mathbf{z} \in E$, a $\mu(\mathbf{z}) \in \mathbf{M}_1(\mathbb{R}^N)$ for which $f(\mathbf{z},\,\cdot\,) = \widehat{\mu(\mathbf{z})}$. Moreover, if $\mathbf{z}_k \longrightarrow \mathbf{z}$ in E, then, because $f_{n_k}(\mathbf{z}_k,\,\cdot\,) \longrightarrow f(\mathbf{z},\,\cdot\,)$ uniformly on compact subsets of $\mathbb{R}^N$, another application of Lévy's theorem shows that $\mu_{n_k}(\mathbf{z}_k) \longrightarrow \mu(\mathbf{z})$ in $\mathbf{M}_1(\mathbb{R}^N)$, and from this it is clear that $\mu_{n_k} \longrightarrow \mu$ in $C(E;\mathbf{M}_1(\mathbb{R}^N))$.

It remains to show that, under the conditions in the final assertion, $\{f_n : n \geq 0\}$ is equicontinuous at each $(\mathbf{z},\boldsymbol{\xi})$. But, by assumption, for each $\boldsymbol{\xi} \in \mathbb{R}^N$, $\{f_n(\,\cdot\,,\boldsymbol{\xi}) : n \geq 0\}$ is equicontinuous at every $\mathbf{z} \in E$. Thus, it suffices to show that if $\boldsymbol{\xi}_k \longrightarrow \boldsymbol{\xi}$ in $\mathbb{R}^N$, then, for each compact $K \subseteq E$,

$$\lim_{k\to\infty}\sup_{n\geq 0}\sup_{\mathbf{z}\in K}\left|f_n(\mathbf{z},\boldsymbol{\xi}_k)-f_n(\mathbf{z},\boldsymbol{\xi})\right|=0.$$

To this end, note that, for any $R>0$,

$$\begin{aligned}\left|f_n(\mathbf{z},\boldsymbol{\xi}_k)-f_n(\mathbf{z},\boldsymbol{\xi})\right| &\leq \int_{B(\mathbf{0},R)}\left|e^{i(\boldsymbol{\xi}_k-\boldsymbol{\xi},\mathbf{y})_{\mathbb{R}^N}}-1\right|\mu_n(\mathbf{z},d\mathbf{y})+2\mu_n\big(\mathbf{z},B(0,R)\complement\big)\\ &\leq R|\boldsymbol{\xi}_k-\boldsymbol{\xi}|+2\mu_n\big(\mathbf{z},B(0,R)\complement\big),\end{aligned}$$

and therefore

$$\varlimsup_{k\to\infty}\sup_{n\geq 0}\sup_{\mathbf{z}\in K}\left|f_n(\mathbf{z},\boldsymbol{\xi}_k)-f_n(\mathbf{z},\boldsymbol{\xi})\right|\leq 2\sup_{n\geq 0}\sup_{\mathbf{z}\in K}\mu_n\big(\mathbf{z},B(0,R)\complement\big)\longrightarrow 0$$

when $R\to\infty$. □

Now that we have a suitable compactness criterion, the next step is to develop an Euler approximation scheme. To do so, we will adopt the idea that convulsion plays the role in $\mathbf{M}_1(\mathbb{R}^N)$ that linear translation plays in $\mathbb{R}^N$. Thus, "linear translation" in $\mathbf{M}_1(\mathbb{R}^N)$ should be a path $t\in[0,\infty)\longmapsto\mu_t\in\mathbf{M}_1(\mathbb{R}^N)$ given by $\mu_t=\nu*\lambda_t$, where $t\rightsquigarrow\lambda_t$ satisfies $\lambda_0=\delta_{\mathbf{0}}$ and $\lambda_{s+t}=\lambda_s*\lambda_t$. That is, $\mu_t=\nu*\lambda_t$, where λ_t is infinitely divisible (cf. Exercises 1.2 and 1.3 below). Moreover, because L is local, the only infinitely divisible laws that can appear here must be Gaussian. Given these hints, we will now take $Q(t,\mathbf{x})=\gamma_{\mathbf{x}+tb(\mathbf{x}),ta(\mathbf{x})}$, the distribution of

$$\mathbf{y}\rightsquigarrow\mathbf{x}+tb(\mathbf{x})+t^{\frac{1}{2}}\sigma(\mathbf{x})\mathbf{y}\quad\text{under }\gamma_{\mathbf{0},\mathbf{I}},$$

where $\sigma:\mathbb{R}^N\longrightarrow\mathrm{Hom}(\mathbb{R}^M;\mathbb{R}^N)$ is a square root[5] of a in the sense that $a(\mathbf{x})=\sigma(\mathbf{x})\sigma(\mathbf{x})^\top$. To check that $Q(t,\mathbf{x})$ will play the role that $\mathbf{x}+tb(\mathbf{x})$ played above, observe that if $\varphi\in C^2(\mathbb{R}^N;\mathbb{C})$ and φ together with its derivatives have at most exponential growth, then

$$\begin{aligned}&\langle\varphi,Q(t,\mathbf{x})\rangle-\varphi(\mathbf{x})=\int_0^t\langle L^{\mathbf{x}}\varphi,Q(\tau,\mathbf{x})\rangle\,d\tau,\\ &\text{where }L^{\mathbf{x}}\varphi(\mathbf{y})=\frac{1}{2}\sum_{i,j}^N a_{ij}(\mathbf{x})\partial_{\mathbf{y}_i}\partial_{\mathbf{y}_j}\varphi(\mathbf{y})+\sum_{i=1}^N b_i(\mathbf{x})\partial_{\mathbf{y}_i}\varphi(\mathbf{y}).\end{aligned}\tag{1.2.10}$$

To verify (1.2.10), simply note that

---

[5] At the moment, it makes no difference which square root of a one chooses. Thus, one might as well assume here that $\sigma(\mathbf{x})=a(\mathbf{x})^{\frac{1}{2}}$, the non-negative definite, symmetric square root $a(\mathbf{x})$. However, later on it will be useful to have kept our options open.

$$
\begin{aligned}
\frac{d}{dt}\langle\varphi, Q(t,\mathbf{x})\rangle &= \frac{d}{dt}\int_{\mathbb{R}^M} \varphi\big(\mathbf{x}+\sigma(\mathbf{x})\mathbf{y}+tb(\mathbf{x})\big)\,\gamma_{\mathbf{0},t\mathbf{I}}(d\mathbf{y}) \\
&= \int_{\mathbb{R}^M} \Big(b(\mathbf{x}), \nabla\varphi\big(\mathbf{x}+\sigma(\mathbf{x})\mathbf{y}+tb(\mathbf{x})\big)\Big)_{\mathbb{R}^N}\,\gamma_{\mathbf{0},t\mathbf{I}}(d\mathbf{y}) \\
&\quad + \frac{1}{2}\int_{\mathbb{R}^M} \varphi\big(\mathbf{x}+\sigma(\mathbf{x})\mathbf{y}+tb(\mathbf{x})\big)\,\Delta_{\mathbf{y}}\left(\frac{1}{(2\pi t)^{\frac{M}{2}}}e^{-\frac{|\mathbf{y}|^2}{2t}}\right)d\mathbf{y},
\end{aligned}
$$

and integrate twice by parts to move the derivatives in $\Delta_{\mathbf{y}}$ over to φ. As a consequence of either (1.2.10) or direct computation, we have

$$
\int |y|^2\,Q(t,\mathbf{x},d\mathbf{y}) = \big|\mathbf{x}+tb(\mathbf{x})\big|^2 + t\,\mathrm{Trace}\big(a(\mathbf{x})\big). \tag{1.2.11}
$$

Now, for each $n \geq 0$, define the Euler approximation $t \in [0,\infty) \longmapsto \mu_{t,n} \in \mathbf{M}_1(\mathbb{R}^N)$ so that

$$
\begin{aligned}
\mu_{0,n} = \nu \quad \text{and} \quad \mu_{t,n} &= \int Q\big(t-m2^{-n},\mathbf{y}\big)\,\mu_{m2^{-n},n}(d\mathbf{y}) \\
&\text{for } m2^{-n} < t \leq (m+1)2^{-n}.
\end{aligned} \tag{1.2.12}
$$

By (1.2.11), we know that

$$
\begin{aligned}
&\int_{\mathbb{R}^N} |\mathbf{y}|^2\,\mu_{t,n}(d\mathbf{y}) \\
&= \int_{\mathbb{R}^N} \Big[\big|\mathbf{y}+(t-m2^{-n})b(\mathbf{y})\big|^2 \\
&\qquad + (t-m2^{-n})\,\mathrm{Trace}\big(a(\mathbf{y})\big)\Big]\,\mu_{m2^{-n},n}(d\mathbf{y})
\end{aligned} \tag{1.2.13}
$$

for $m2^{-n} \leq t \leq (m+1)2^{-n}$.

Lemma 1.2.3. *Assume that*

$$
\lambda := \sup_{x\in\mathbb{R}^N} \frac{\mathrm{Trace}\big(a(\mathbf{x})\big)+2|b(\mathbf{x})|^2}{1+|\mathbf{x}|^2} < \infty.
$$

Then

$$
\sup_{n\geq 0}\int_{\mathbb{R}^N} (1+|\mathbf{y}|^2)\,\mu_{t,n}(d\mathbf{y}) \leq e^{(1+\lambda)t}\int_{\mathbb{R}^N}(1+|\mathbf{x}|^2)\,\nu(d\mathbf{x}). \tag{1.2.14}
$$

In particular, if $\int |\mathbf{x}|^2\,\nu(d\mathbf{x}) < \infty$, *then* $\{\mu_{\cdot,n} : n \geq 0\}$ *is a relatively compact subset of* $C\big([0,\infty);\mathbf{M}_1(\mathbb{R}^N)\big)$.

Proof. Suppose that $m2^{-n} \leq t \leq (m+1)2^{-n}$, and set $\tau = t - m2^{-n}$. First note that

$$\begin{aligned}&\big|\mathbf{y}+\tau b(\mathbf{y})\big|^2+\tau\,\mathrm{Trace}\big(a(\mathbf{y})\big)\\&=|\mathbf{y}|^2+2\tau\big(\mathbf{y},b(\mathbf{y})\big)_{\mathbb{R}^N}+\tau^2|b(\mathbf{y})|^2+\tau\,\mathrm{Trace}\big(a(\mathbf{y})\big)\\&\le|\mathbf{y}|^2+\tau\big[|\mathbf{y}|^2+2|b(\mathbf{y})|^2+\mathrm{Trace}\big(a(\mathbf{y})\big)\big]\le|\mathbf{y}|^2+(1+\lambda)\tau(1+|\mathbf{y}|^2),\end{aligned}$$

and therefore, by (1.2.13),

$$\int(1+|\mathbf{y}|^2)\,\mu_{t,n}(d\mathbf{y})\le\big(1+(1+\lambda)\tau\big)\int(1+|\mathbf{y}|^2)\,\mu_{m2^{-n},n}(d\mathbf{y}).$$

Hence, by induction on $m\ge 1$,

$$\begin{aligned}&\int(1+|\mathbf{y}|^2)\,\mu_{t,n}(d\mathbf{y})\\&\quad\le\big(1+(1+\lambda)2^{-n}\big)^m\big(1+(1+\lambda)\tau\big)\int[t]\int(1+|\mathbf{x}|^2)\,\nu(d\mathbf{x})\\&\quad\le e^{(1+\lambda)t}\int(1+|\mathbf{x}|^2)\,\nu(d\mathbf{x}).\end{aligned}$$

Next, set $f_n(t,\boldsymbol{\xi})=\widehat{\mu_{t,n}}(\boldsymbol{\xi})$. Under the assumption that the second moment $S=\int|\mathbf{x}|^2\,\nu(d\mathbf{x})<\infty$, we want to show that $\{f_n:n\ge 0\}$ is equicontinuous at each $(t,\boldsymbol{\xi})\in[0,\infty)\times\mathbb{R}^N$. Since, by (1.2.14),

$$\mu_{t,n}\big(\overline{B(0,R)}\complement\big)\le(1+S)(1+R^2)^{-1}e^{(1+\lambda)t},$$

the last part of Theorem 1.2.2 says that it suffices to show that, for each $\boldsymbol{\xi}\in\mathbb{R}^N$, $\{f_n(\,\cdot\,,\boldsymbol{\xi}):n\ge 0\}$ is equicontinuous at every $t\in[0,\infty)$. To this end, first observe that, for $m2^{-n}\le s<t\le(m+1)2^{-n}$,

$$\big|f_n(t,\boldsymbol{\xi})-f_n(s,\boldsymbol{\xi})\big|\le\int\big|\widehat{Q(t,\mathbf{y})}(\boldsymbol{\xi})-\widehat{Q(s,\mathbf{y})}(\boldsymbol{\xi})\big|\,\mu_{m2^{-n},n}(d\mathbf{y})$$

and, by (1.2.10),

$$\begin{aligned}&\big|\widehat{Q(t,\mathbf{y})}(\boldsymbol{\xi})-\widehat{Q(s,\mathbf{y})}(\boldsymbol{\xi})]\big|=\left|\int_s^t\left(\int L^{\mathbf{y}}e^{i(\boldsymbol{\xi},\mathbf{y}')_{\mathbb{R}^N}}\,Q(\tau,\mathbf{y},d\mathbf{y}')\right)d\tau\right|\\&\le(t-s)\Big(\tfrac12\big(\boldsymbol{\xi},a(\mathbf{y})\boldsymbol{\xi}\big)_{\mathbb{R}^N}+|\boldsymbol{\xi}||b(\mathbf{y})|\Big)\le\tfrac12(1+\lambda)(1+|\mathbf{y}|^2)(1+|\boldsymbol{\xi}|^2)(t-s).\end{aligned}$$

Hence, by (1.2.14),

$$\big|f_n(t,\boldsymbol{\xi})-f_n(s,\boldsymbol{\xi})\big|\le\frac{(1+\lambda)(1+|\boldsymbol{\xi}|^2)}{2}e^{(1+\lambda)t}\int(1+|\mathbf{x}|^2)\,\nu(d\mathbf{x})(t-s),$$

first for $s<t$ in the same diadic interval and then for all $s<t$. □

Having Lemma 1.2.3, we can now prove Theorem 1.2.1 under the assumptions that a and b are bounded and that $\int|\mathbf{x}|^2\,\nu(d\mathbf{x})<\infty$. Indeed, because we know then that $\{\mu_{\cdot,n} : n \geq 0\}$ is relatively compact in $C\big([0,\infty);\mathbf{M}_1(\mathbb{R}^N)\big)$, all that we have to do is show that every limit satisfies (1.2.8). For this purpose, first note that, by (1.2.10),

$$\langle \varphi, \mu_{t,n}\rangle - \langle \varphi, \nu\rangle = \int_0^t \left(\int \langle L^{\mathbf{y}}\varphi, Q(\tau - \lfloor \tau \rfloor_n, \mathbf{y})\rangle \mu_{\lfloor \tau \rfloor_n, n}(d\mathbf{y}) \right) d\tau$$

for any $\varphi \in C^2_{\mathrm{b}}(\mathbb{R}^N;\mathbb{C})$. Next, observe that, as $n \to \infty$,

$$\langle L^{\mathbf{y}}\varphi, Q(\tau - \lfloor \tau \rfloor_n, \mathbf{y})\rangle \longrightarrow L\varphi(\mathbf{y})$$

boundedly and uniformly for $(\tau, \mathbf{y})$ in compacts. Hence, if

$$\mu_{\cdot,n_k} \longrightarrow \mu_{\cdot} \text{ in } C\big([0,\infty);\mathbf{M}_1(\mathbb{R}^N)\big),$$

then, by Theorem 1.2.2,

$$\langle \varphi, \mu_{t,n_k}\rangle \longrightarrow \langle \varphi, \mu_t\rangle \quad \text{and}$$
$$\int_0^t \left(\int \langle L^{\mathbf{y}}\varphi, Q(\tau - \lfloor \tau \rfloor_n, \mathbf{y})\rangle \mu_{\lfloor \tau \rfloor_n, n}(d\mathbf{y}) \right) d\tau \longrightarrow \int_0^t \langle L\varphi, \mu_\tau\rangle\, d\tau.$$

Before removing the boundedness assumptions on a and b, we want to show that $\int |\mathbf{x}|^2\,\nu(d\mathbf{x}) < \infty$ implies that (1.2.8) continues to hold for $\varphi \in C^2(\mathbb{R}^N;\mathbb{C})$ with bounded second order derivatives. First, from (1.2.14), we know that

$$(*) \qquad \int (1+|\mathbf{y}|^2)\,\mu_t(d\mathbf{y}) \leq e^{(1+\lambda)t} \int (1+|\mathbf{y}|^2)\,\nu(d\mathbf{y}).$$

Now choose $\eta \in C^\infty_{\mathrm{c}}\big(\mathbb{R}^N;[0,1]\big)$ so that $\eta = 1$ on $\overline{B(0,1)}$ and $\eta = 0$ off of $B(0,2)$, define η_R by $\eta_R(\mathbf{y}) = \eta(R^{-1}\mathbf{y})$ for $R \geq 1$, and set $\varphi_R = \eta_R\varphi$. Observe that

$$\frac{|\varphi(\mathbf{y})|}{1+|\mathbf{y}|^2} \vee \frac{|\nabla\varphi(\mathbf{y})|}{1+|\mathbf{y}|} \vee \|\nabla^2\varphi(\mathbf{y})\|_{\mathrm{H.S.}}$$

is bounded independent of $\mathbf{y} \in \mathbb{R}^N$, and therefore so is $\frac{|L\varphi(\mathbf{y})|}{1+|\mathbf{y}|^2}$. Thus, by $(*)$, there is no problem about integrability of the expressions in (1.2.8). Moreover, because (1.2.8) holds for each φ_R, all that we have to do is check that

$$\langle \varphi, \mu_t\rangle = \lim_{R\to\infty} \langle \varphi_R, \mu_t\rangle$$
$$\int_0^t \langle L\varphi, \mu_\tau\rangle\, d\tau = \lim_{R\to\infty} \int_0^t \langle L\varphi_R, \mu_\tau\rangle\, d\tau.$$

The first of these is an immediate application of Lebesgue's dominated convergence theorem. To prove the second, observe that

$$L\varphi_R(\mathbf{y}) = \eta_R(\mathbf{y})L\varphi(\mathbf{y}) + \big(\nabla\eta_R(\mathbf{y}), a(\mathbf{y})\nabla\varphi\big)_{\mathbb{R}^N} + \varphi(\mathbf{y})L\eta_R(\mathbf{y}).$$

Again the first term on the right causes no problem. To handle the other two terms, note that, because η_R is constant off of $\overline{B(0,2R)}\setminus B(0,R)$ and because $\nabla\eta_R(\mathbf{y}) = R^{-1}\nabla\eta(R^{-1}\mathbf{y})$ while $\nabla^2\eta_R(\mathbf{y}) = R^{-2}\nabla^2\eta(R^{-1}\mathbf{y})$, one can easily check that they are dominated by a constant, which is independent of R, times $(1+|\mathbf{y}|^2)\mathbf{1}_{[R,2R]}(|\mathbf{y}|)$. Hence, once again $(*)$ plus Lebesgue's dominated convergence theorem gives the desired result.

Knowing that (1.2.8) holds for $\varphi \in C^2(\mathbb{R}^N;\mathbb{C})$ with bounded second order derivatives, we can prove (1.2.9) by taking $\varphi(\mathbf{y}) = 1+|\mathbf{y}|^2$ and thereby obtain

$$
\begin{aligned}
&\int (1+|\mathbf{y}|^2)\,\mu_t(d\mathbf{y}) \\
&\quad = \int (1+|\mathbf{y}|^2)\,\nu(d\mathbf{y}) + \int_0^t \left(\int \Big[\mathrm{Trace}\big(a(\mathbf{y})\big) + 2\big(\mathbf{y},b(\mathbf{y})\big)_{\mathbb{R}^N}\Big]\,\mu_\tau(d\mathbf{y})\right) d\tau \\
&\quad \le \int (1+|\mathbf{y}|^2)\,\nu(d\mathbf{y}) + \Lambda\int_0^t \left(\int (1+|\mathbf{y}|^2)\,\mu_\tau(d\mathbf{y})\right) d\tau.
\end{aligned}
$$

Given the preceding, (1.2.9) becomes an easy application of the following simple but useful lemma of T. Gronwall.

Lemma 1.2.4. *Let α and β be continuous non-negative functions on $\mathbb{R}$, assume that α is non-decreasing, and set $B(t) = \int_0^t \beta(\tau)\,d\tau$. If $u \in C\big([0,T];\mathbb{R}\big)$ satisfies*

$$u(t) \le \alpha(t) + \int_0^t \beta(\tau)u(\tau)\,d\tau \quad \text{for } t\in[0,T],$$

then

$$u(T) \le \alpha(0)e^{B(T)} + \int_0^T e^{B(T)-B(\tau)}\,d\alpha(\tau).$$

In particular, if α and β are constant on $[0,T]$ and

$$u(t) \le \alpha + \beta\int_0^t \big(1+u(\tau)\big)\,d\tau \quad \text{for } t\in[0,T],$$

then

$$1+u(T) \le (1+\alpha)e^{\beta T}.$$

Proof. Set $U(t) = \int_0^t \beta(\tau)u(\tau)\,d\tau$ for $t\in[0,T]$. Then[6]

$$\dot U(t) = \beta(t)u(t) \le \alpha(t)\beta(t) + \beta(t)U(t),$$

and so $\partial_t\big(e^{-B(t)}U(t)\big) \le \alpha(t)\beta(t)e^{-B(t)}$, which means that

$$e^{-B(T)}U(T) \le \int_0^T \alpha(\tau)\beta(\tau)e^{-B(\tau)}\,d\tau = \alpha(0) - \alpha(T)e^{-B(T)} + \int_0^T e^{-B(\tau)}\,d\alpha(\tau).$$

[6] When f is a function of time, we will sometimes use $\dot f$ to denote its derivative.

Finally, remember that $u(T) \leq \alpha(T) + U(T)$. □

Continuing with the assumption that $\int |\mathbf{x}|^2\,\nu(d\mathbf{x}) < \infty$, we want to remove the boundedness assumption on a and b and replace it by (1.2.7). To do this, take η_R as we did before, set $a_k = \eta_k a$, $b_k = \eta_k b$, define L_k accordingly for a_k and b_k, and choose $t \rightsquigarrow \mu_{t,k}$ so that (1.2.9) is satisfied and (1.2.8) holds when $\mu_{\cdot}$ and L are replaced there by $\mu_{\cdot,k}$ and L_k. Because of (1.2.9), the argument which was used earlier can be repeated to show that $\{\mu_{\cdot,k} : k \geq 1\}$ is relatively compact in $C\big([0,\infty);\mathbf{M}_1(\mathbb{R}^N)\big)$. Moreover, if $\mu_{\cdot}$ is any limit of $\{\mu_{\cdot,k} : k \geq 1\}$, then it satisfies (1.2.9) and, just as we did above, one can check that (1.2.8) holds, first for $\varphi \in C^2_{\mathrm{c}}(\mathbb{R}^N;\mathbb{C})$ and then for all $\varphi \in C^2(\mathbb{R}^N;\mathbb{C})$ with bounded second order derivatives.

To remove the second moment condition on ν, assume that it fails, and choose $r_k \nearrow \infty$ so that

$$\alpha_1 = \nu\big(B(0,r_1)\big) > 0 \quad \text{and} \quad \alpha_k = \nu\big(B(0,r_k) \setminus B(0,r_{k-1})\big) > 0 \text{ for each } k \geq 2,$$

and set $\nu_1 = \alpha_1^{-1}\nu \restriction B(0,r_1)$ and $\nu_k = \alpha_k^{-1}\nu \restriction B(0,r_k) \setminus B(0,r_{k-1})$ when $k \geq 2$. Finally, choose $t \rightsquigarrow \mu_{t,k}$ for L and ν_k, and define $\mu_t = \sum_{k=1}^{\infty} \alpha_k \mu_{t,k}$. It is an easy matter to check that this $t \rightsquigarrow \mu_t$ satisfies (1.2.8) for all $\varphi \in C^2_{\mathrm{c}}(\mathbb{R}^N;\mathbb{C})$.

Although Theorem 1.2.1 is very general, it is less than satisfactory. In particular, by itself, it does not produce a transition probability function that solves (1.2.3). That is, we would like to construct a map

$$(t,\mathbf{x}) \in [0,\infty) \times \mathbb{R}^N \longmapsto P(t,\mathbf{x},\,\cdot\,) \in \mathbf{M}_1(\mathbb{R}^N)$$

such that, for each $\mathbf{x}$, $t \rightsquigarrow P(t,\mathbf{x})$ satisfies (1.2.6) with $\nu = \delta_{\mathbf{x}}$, $(t,\mathbf{x}) \rightsquigarrow P(t,\mathbf{x},\Gamma)$ is a Borel measurable for all $\Gamma \in \mathcal{B}_{\mathbb{R}^N}$, and (1.2.2) holds. To get such a map, one might proceed as follows. For each $\mathbf{x} \in \mathbb{R}^N$, set $P_n(t,\mathbf{x}) = \mu_{t,n}$, where $t \rightsquigarrow \mu_{t,n}$ is given by the prescription in (1.2.12) when $\nu = \delta_x$. Clearly $(t,\mathbf{x}) \rightsquigarrow P_n(t,\mathbf{x})$ is continuous, and therefore $(t,\mathbf{x}) \rightsquigarrow P_n(t,\mathbf{x},\Gamma)$ is Borel measurable for all $\Gamma \in \mathcal{B}_{\mathbb{R}^N}$. Further, by construction, one knows that

$$\langle \varphi, P_n(m2^{-n} + t,\mathbf{x})\rangle = \int \langle \varphi, P_n(t,\mathbf{y})\rangle P_n(m2^{-n},\mathbf{x},d\mathbf{y})$$

for all $m \in \mathbb{N}$, $t \geq 0$, and bounded, Borel measurable φ's. Thus, if we could find a subsequence $\{P_{n_k} : k \geq 1\}$ of $\{P_n : n \geq 1\}$ which converged in $C\big([0,\infty) \times \mathbb{R}^N;\mathbf{M}_1(\mathbb{R}^N)\big)$ to some P, then $(t,\mathbf{x}) \rightsquigarrow P(t,\mathbf{x})$ would be a transition probability function that satisfies (1.2.3). Of course, once we had such a transition probability function, we could produce a solution to (1.2.6) for any $\nu \in \mathbf{M}_1(\mathbb{R}^N)$ by taking $\mu_t = \int P(t,\mathbf{x})\nu(d\mathbf{x})$.

1.3 Exercises

Exercise 1.1. There are very few L's for which a closed form solution to (1.2.3) or (1.2.4) is known. Of course, when $L = \frac{1}{2}\Delta$,

$$P(t, \mathbf{x}, d\mathbf{y}) = g(t, \mathbf{y} - \mathbf{x})\, d\mathbf{y} \quad \text{where}\, g(t, \mathbf{y}) = (2\pi t)^{-\frac{N}{2}} e^{-\frac{|\mathbf{y}|^2}{2t}}.$$

Another case is when L is the **Ornstein–Uhlenbeck operator** given by

$$L\varphi(\mathbf{x}) = \tfrac{1}{2}\Delta\varphi(\mathbf{x}) - \big(\mathbf{x}, \nabla\varphi(\mathbf{x})\big)_{\mathbb{R}^N}.$$

Perhaps the most elementary way to find the associated transition probability function $P(t, \mathbf{x}, \cdot\,)$ is to use (1.2.4). Namely, suppose $u \in C_{\mathrm{b}}^{1,2}\big([0,\infty)\times\mathbb{R}^N; \mathbb{R}\big)$ satisfies the **heat equation** $\partial_t u = \frac{1}{2}\Delta u$, and set

$$v(t, \mathbf{x}) = u\left(\frac{1 - e^{-2t}}{2}, e^{-t}\mathbf{x}\right).$$

Show that $\partial_t v = Lv$, and conclude that

$$P(t, \mathbf{x}, d\mathbf{y}) = g\left(\frac{1 - e^{-2t}}{2}, \mathbf{y} - e^{-t}\mathbf{x}\right) d\mathbf{y}.$$

Exercise 1.2. Given the characterization of linear functionals that satisfy the minimum principle and are quasi-local, it is quite easy to derive the **Lévy–Khinchine formula** for **infinitely divisible laws.** A $\mu \in \mathbf{M}_1(\mathbb{R}^N)$ is said to be infinitely divisible if, for each $n \geq 1$, there is a $\mu_{\frac{1}{n}} \in \mathbf{M}_1(\mathbb{R}^N)$ such that $\mu = \mu_{\frac{1}{n}}^{*n}$, and the Lévy–Khinchine formula says that $\mu \in \mathbf{M}_1(\mathbb{R}^N)$ is infinitely divisible if and only if there is a Lévy system $\big(\mathbf{m}, C, M\big)$ such that $\hat{\mu} = e^{\ell}$ where

$$(*) \qquad \ell(\boldsymbol{\xi}) = i(\mathbf{m}, \boldsymbol{\xi})_{\mathbb{R}^N} - \tfrac{1}{2}(\boldsymbol{\xi}, C\boldsymbol{\xi})_{\mathbb{R}^N} + \int_{\mathbb{R}^N} \Big(e^{i(\boldsymbol{\xi},\mathbf{y})_{\mathbb{R}^N}} - 1 - i\mathbf{1}_{B(\mathbf{0},1)}(\mathbf{y})(\boldsymbol{\xi}, \mathbf{y})_{\mathbb{R}^N}\Big) M(d\mathbf{y}).$$

As a consequence, one sees that $\mu = \lambda_1$, where $\{\lambda_t : t > 0\}$ is the canonical family determined by $(\mathbf{m}, C, M)$. In this and the following exercise, you are to derive their formula.

Show that it suffices to know that

(i) There is a unique $\ell \in C(\mathbb{R}^N; \mathbb{C})$ satisfying $\ell(\mathbf{0}) = 0$ and $\hat{\mu}_{\frac{1}{n}} = e^{\frac{1}{n}\ell}$ for all $n \geq 1$.
(ii) There is a $C < \infty$ such $|\ell(\boldsymbol{\xi})| \leq C\big(1 + |\boldsymbol{\xi}|^2\big)$.

Assuming (i) and (ii), here are some hints for proving the Lévy–Khinchine formula.

First, using (i), (ii) and Parseval's identity, show that

$$(**)\qquad A\varphi := \lim_{n\to\infty} n\big(\langle\varphi,\mu_{\frac{1}{n}}\rangle - \varphi(\mathbf{0})\big) = (2\pi)^{-N}\int_{\mathbb{R}^N} \ell(-\boldsymbol{\xi})\hat{\varphi}(\boldsymbol{\xi})\,d\boldsymbol{\xi}$$

for $\varphi \in \mathscr{S}(\mathbb{R}^N;\mathbb{C})$. Observe that A satisfies the minimum principle. In addition, given $\varphi \in \mathscr{S}(\mathbb{R}^N;\mathbb{C})$, show that, as $R\to\infty$,

$$(2\pi)^N A\varphi_R = R^N \int_{\mathbb{R}^N} \ell(-\boldsymbol{\xi})\hat{\varphi}(R\boldsymbol{\xi})\,d\boldsymbol{\xi} = \int_{\mathbb{R}^N} \ell(-R^{-1}\boldsymbol{\xi})\hat{\varphi}(\boldsymbol{\xi})\,d\boldsymbol{\xi} \longrightarrow 0,$$

and therefore that A is quasi-local. Now apply Theorem 1.1.1 to show that there exists a Lévy system $(\mathbf{m},\mathbf{C},M)$ such that $A\varphi$ equals

$$(2\pi)^{-N}\int_{\mathbb{R}^N}\bigg(-i(\mathbf{m},\boldsymbol{\xi})_{\mathbb{R}^N} - \frac{1}{2}(\boldsymbol{\xi},\mathbf{C}\boldsymbol{\xi})_{\mathbb{R}^N} + \int_{\mathbb{R}^N}\Big(e^{-i(\boldsymbol{\xi},\mathbf{y})_{\mathbb{R}^N}} - 1 + i\mathbf{1}_{B(\mathbf{0},1)}(\mathbf{y})(\boldsymbol{\xi},\mathbf{y})_{\mathbb{R}^N}\Big)M(\mathbf{y})\bigg)\hat{\varphi}(\boldsymbol{\xi})\,d\boldsymbol{\xi}$$

for all $\varphi \in \mathscr{S}(\mathbb{R}^N;\mathbb{C})$. By combining this with $(**)$, one arrives at $(*)$.

Exercise 1.3. In this exercise you are to show that an infinitely divisible μ satisfies the conditions (i) and (ii) in Exercise 1.2. For this purpose, it is important to know (cf. Lemma 3.2.3 in [20]) that if $f \in C\big(\overline{B(\mathbf{0},R)};\mathbb{C}\setminus\{\mathbf{0}\}\big)$ and $f(\mathbf{0}) = 1$, then there is a unique $\ell \in C\big(\overline{B(\mathbf{0};R)};\mathbb{C}\big)$ such that $\ell(\mathbf{0}) = 0$ and $f = e^{\ell}$.

Choose $r > 0$ so that $|1-\hat{\mu}(\boldsymbol{\xi})| \le \frac{1}{2}$ for $|\boldsymbol{\xi}| \le r$, and set $\ell = \log\hat{\mu}$ on $\overline{B(\mathbf{0},r)}$, where

$$\log z = -\sum_{m=1}^{\infty}\frac{(1-z)^m}{m} \quad \text{for } |z-1|<1$$

is the principal branch of the logarithm function. Clearly $\hat{\mu} = e^{\ell}$ on $\overline{B(\mathbf{0},r)}$, and so $\mathfrak{Re}(\ell) \le 0$. In addition, because $\widehat{\mu_{\frac{1}{n}}}(\boldsymbol{\xi}) \ne 0$ for $|\boldsymbol{\xi}| \le r$, there is a unique continuous $\ell_{\frac{1}{n}}$ on $\overline{B(\mathbf{0},r)}$ such that $\ell_{\frac{1}{n}}(\mathbf{0}) = 0$ and $\widehat{\mu_{\frac{1}{n}}} = e^{\ell_{\frac{1}{n}}}$. Using $\hat{\mu} = (\widehat{\mu}_{\frac{1}{n}})^n$, conclude that $\ell_{\frac{1}{n}} = \frac{1}{n}\ell$.

Next show that $|\log z| \le 2|1-z|$ if $|1-z| \le \frac{1}{2}$, and use this and the preceding to show that

$$\big|1-\widehat{\mu_{\frac{1}{n}}}(\boldsymbol{\xi})\big| = \big|1-e^{\frac{1}{n}\ell(\boldsymbol{\xi})}\big| \le \frac{1}{n} \quad \text{for } |\boldsymbol{\xi}| \le r.$$

Starting from

$$|1-\widehat{\mu_{\frac{1}{n}}}(\boldsymbol{\xi})| \ge \mathfrak{Re}\big(1-\widehat{\mu_{\frac{1}{n}}}(\boldsymbol{\xi})\big) = \int_{\mathbb{R}^N}\big(1-\cos(\boldsymbol{\xi},\mathbf{y})\big)\,\mu_{\frac{1}{n}}(d\mathbf{y}),$$

show that, for any $\mathbf{e} \in \mathbb{S}^{N-1}$,

$$\frac{1}{n} \geq \frac{1}{r} \int_{\mathbb{R}^N} \left(\int_0^r \big(1 - \cos t(\mathbf{e}, \mathbf{y})\big)\, dt \right) \mu_{\frac{1}{n}}(d\mathbf{y})$$
$$= \int_{\mathbb{R}^N} \left(1 - \frac{\sin r(\mathbf{e}, \mathbf{y})_{\mathbb{R}^N}}{r(\mathbf{e}, \mathbf{y})_{\mathbb{R}^N}} \right) \mu_{\frac{1}{n}}(d\mathbf{y}).$$

Hence, if

$$s(T) := \inf_{|t| \geq T} \left(1 - \frac{\sin t}{t} \right) \quad \text{for } T > 0,$$

then, for any $R > 0$,

$$\frac{1}{n} \geq s(rR)\mu_{\frac{1}{n}}\big(\{\mathbf{y} : |(\mathbf{e}, \mathbf{y})_{\mathbb{R}^N}| \geq R\}\big).$$

Further, observe that because $\sin t = \int_0^t \cos\tau\, d\tau < t$ and $\frac{\sin t}{t} \longrightarrow 0$ as $t \to \infty$, $s(T) > 0$ for all $T > 0$ and so

$$(*) \qquad \mu_{\frac{1}{n}}\big(\{\mathbf{y} : |(\mathbf{e}, \mathbf{y})_{\mathbb{R}^N}| \geq R\}\big) \leq \frac{1}{ns(rR)}.$$

Since

$$|1 - \widehat{\mu_{\frac{1}{n}}}(\rho\mathbf{e})| \leq \rho R + 2\mu_{\frac{1}{n}}\big(\{\mathbf{y} : |(\mathbf{e}, \mathbf{y})_{\mathbb{R}^N}| \geq R\}\big),$$

$(*)$ implies that, for any $\rho > 0$,

$$\sup_{|\boldsymbol{\xi}| \leq \rho} |1 - \widehat{\mu_{\frac{1}{n}}}(\boldsymbol{\xi})| \leq \rho R + \frac{2}{ns(rR)} \quad \text{for all } (\rho, R) \in (0, \infty)^2,$$

by taking $R = \frac{1}{4\rho}$, one arrives first at

$$\sup_{|\boldsymbol{\xi}| \leq \rho} |1 - \widehat{\mu_{\frac{1}{n}}}(\boldsymbol{\xi})| \leq \frac{1}{4} + \frac{2}{ns\left(\frac{r}{4\rho}\right)},$$

and then at

$$(**) \qquad \sup_{|\boldsymbol{\xi}| \leq \rho} |1 - \widehat{\mu_{\frac{1}{n}}}(\boldsymbol{\xi})| \leq \frac{1}{2} \quad \text{if } n \geq \frac{8}{s\left(\frac{r}{4\rho}\right)}$$

for all $\rho > 0$. From $(**)$ it is clear that, for each $\rho > 0$, there is an n such that $|\widehat{\mu_{\frac{1}{n}}}(\boldsymbol{\xi})| \geq \frac{1}{2}$ and therefore $|\hat{\mu}(\boldsymbol{\xi})| \geq 2^{-n}$ for $|\boldsymbol{\xi}| \leq \rho$. Because this proves that $\hat{\mu}$ never vanishes, show that ℓ admits a unique continuous extention to $\mathbb{R}^N$ such that $\hat{\mu} = e^{\ell}$ and $\widehat{\mu_{\frac{1}{n}}} = e^{\frac{1}{n}\ell}$ on $\mathbb{R}^N$.

Using $(**)$, show that $|\ell(\boldsymbol{\xi})| \leq n$ if $|\boldsymbol{\xi}| \leq \rho$ and $n \geq \frac{8}{s(\frac{r}{4\rho})}$. Check that $\lim_{t \searrow 0} t^{-2}\big(1 - \frac{\sin t}{t}\big) = \frac{1}{6}$ and therefore that there exists an $\epsilon > 0$ such that

$s(T) \geq \epsilon(T \wedge 1)^2$. Finally, from these, show that there is a $C < \infty$ such that $|\ell(\boldsymbol{\xi})| \leq C(1 + |\boldsymbol{\xi}|^2)$.

Exercise 1.4. Show that

$$\int_{\mathbb{R}^N\setminus\{\mathbf{0}\}} \left(e^{i(\boldsymbol{\xi},\mathbf{y})_{\mathbb{R}^N}} - 1 - i\mathbf{1}_{B(\mathbf{0},1)}(\mathbf{y})(\boldsymbol{\xi},\mathbf{y})_{\mathbb{R}^N} \right) \frac{d\mathbf{y}}{|\mathbf{y}|^{N+1}}$$
$$= |\boldsymbol{\xi}| \int_{\mathbb{R}^N\setminus\{\mathbf{0}\}} \left(\cos(\mathbf{e},\mathbf{y})_{\mathbb{R}^N} - 1 \right) \frac{d\mathbf{y}}{|\mathbf{y}|^{N+1}}$$

for any $\mathbf{e} \in \mathbb{S}^{N-1}$, and conclude that there is a $c > 0$ such that $\ell(\boldsymbol{\xi})$ corresponding to $\mathbf{m} = \mathbf{0}$, $C = 0$, and $M(d\mathbf{y}) = c\mathbf{1}_{\mathbb{R}^N\setminus\{\mathbf{0}\}}(\mathbf{y})|\mathbf{y}|^{-N-1}\, d\mathbf{y}$ is equal to $-|\boldsymbol{\xi}|$. The associated infinitely divisible laws are called **Cauchy distributions**. To see what they look like, begin by showing that

$$\int_0^\infty t^{-\frac{1}{2}} e^{-\frac{a^2}{2t} - \frac{b^2 t}{2}}\, dt = \frac{\sqrt{2\pi} e^{-ab}}{b}$$

and then that

$$\int_0^\infty t^{-\frac{3}{2}} e^{-\frac{a^2}{2t} - \frac{b^2 t}{2}}\, dt = \frac{\sqrt{2\pi} e^{-ab}}{a}$$

for $a, b > 0$. To do the first of these, try the change of variable $\tau = bt^{\frac{1}{2}} - at^{-\frac{1}{2}}$, and get the second by differentiating the first with respect to a. Now apply the second one to see that

$$t \int_0^\infty \tau^{-\frac{3}{2}} e^{-\frac{t^2}{2\tau}} \widehat{\gamma_{\mathbf{0},\tau\mathbf{I}}}(\boldsymbol{\xi})\, d\tau = \sqrt{2\pi} e^{-t|\boldsymbol{\xi}|}.$$

and conclude from this that if

$$P_t(d\mathbf{y}) = \left(\frac{t}{\sqrt{2\pi}} \int_0^\infty \tau^{-\frac{3}{2}} e^{-\frac{t^2}{2\tau}} \gamma_{\mathbf{0},\tau\mathbf{I}}(\mathbf{y})\, d\tau \right) d\mathbf{y} = \frac{2}{\omega_N} \frac{t}{(t^2 + |\mathbf{y}|^2)^{\frac{N+1}{2}}}\, d\mathbf{y},$$

where

$$\omega_N = \frac{2\pi^{\frac{N+1}{2}}}{\Gamma\left(\frac{N+1}{2}\right)}$$

is the surface area of the unit N-sphere $\mathbb{S}^N = \{\mathbf{x} \in \mathbb{R}^{N+1} : |\mathbf{x}| = 1\}$, then

$$\widehat{P_t}(\boldsymbol{\xi}) = e^{-t|\boldsymbol{\xi}|}.$$

Finally, use this to show that the constant c above is $\frac{2}{\omega_N}$.

Chapter 2
Itô's Approach

To address the problem of convergence raised at the end of Chapter 1, K. Itô used a technique known as **coupling**. Given a pair of Borel probability measures μ_1 and μ_2 on some metric space (E, ρ), a coupling of μ_1 to μ_2 is a pair of E-valued random variables X_1 on X_2 on some probability space $(\Omega, \mathcal{B}, \mathbb{P})$ such that μ_1 is the distribution of X_1 and μ_2 is the distribution of X_2. Given such a coupling, one can compare μ_1 to μ_2 by looking at

$$\mathbb{E}^{\mathbb{P}}\big[\rho(X_1, X_2)^p\big]^{\frac{1}{p}}.$$

To yield useful information, the coupling technique requires one to make a judicious choice of the random variables. On the one hand, the choice should be good enough to give a reasonably accurate assessment of the difference between the measures. On the other hand, unless the choice is one for which calculations are possible, it has no value. The choice that Itô made was a very clever compromise between accuracy and practicality. Namely, he lifted everything to pathspace and performed his coupling there. If one thinks, as Itô did, of Kolmogorov's equations as describing the evolution of measures in $\mathbf{M}_1(\mathbb{R}^N)$, moving to pathspace is a natural idea. Indeed, the measure μ_t should be the distribution at time t of a randomly diffusing particle, and so the position of that particle should be a good candidate for ones coupling procedure. However, in order to fully appreciate just how clever Itô's coupling procedure is, it may be helpful to start by using a less clever one.

Let coefficients a and b be given, and, for each $\mathbf{x} \in \mathbb{R}^N$ and $n \geq 1$, determine $t \rightsquigarrow P_n(t, \mathbf{x})$ by (1.2.12) with $\nu = \delta_{\mathbf{x}}$. Suppose that $\sigma \colon \mathbb{R}^N \longrightarrow \mathrm{Hom}(\mathbb{R}^M; \mathbb{R}^N)$ is a Borel measurable function for which $a = \sigma\sigma^\top$. Next, let $\{Y_m : m \geq 1\}$ be a sequence of mutually independent, $\mathbb{R}^M$-valued Gaussian random variables with mean $\mathbf{0}$ and covariance $\mathbf{I}$ on some probability space $(\Omega, \mathcal{F}, \mathbb{P})$, and define the random variable $X_n(t, \mathbf{x})$ for $(t, \mathbf{x}) \in [0, \infty) \times \mathbb{R}^N$ by $X_n(0, \mathbf{x}) = \mathbf{x}$ and

D. W. Stroock, *Elements of Stochastic Calculus and Analysis*,
CRM Short Courses, https://doi.org/10.1007/978-3-319-77038-3_2

$$X_n(t,\mathbf{x}) = X_n(m2^{-n},\mathbf{x}) + (t-m2^{-n})b\big(X_n(m2^{-n},\mathbf{x})\big) + (t-m2^{-n})^{\frac{1}{2}}\sigma\big(X_n(m2^{-n},\mathbf{x})\big)Y_{m+1}$$

for $m2^{-n} < t \le (m+1)2^{-n}$. Using induction on $n \ge 0$, one can check that $P_n(t,\mathbf{x})$ is the distribution of $X_n(t,\mathbf{x})$ and therefore that $X_n(t,\mathbf{x})$ and $X_n(t,\mathbf{y})$ provide a coupling of $P_n(t,\mathbf{x})$ to $P_n(t,\mathbf{y})$. Further,

$$\begin{aligned} &X_n(t,\mathbf{y}) - X_n(t,\mathbf{x}) \\ &= X_n(m2^{-n},\mathbf{y}) - X_n(m2^{-n},\mathbf{x}) \\ &\quad + (t-m2^{-n})\Big(b\big(X_n(m2^{-n},\mathbf{y})\big) - b\big(X_n(m2^{-n},\mathbf{x})\big)\Big) \\ &\quad + (t-m2^{-n})^{\frac{1}{2}}\Big(\sigma\big(X_n(m2^{-n},\mathbf{y})\big) - \sigma\big(X_n(m2^{-n},\mathbf{x})\big)\Big)Y_{m+1} \end{aligned}$$

for $m2^{-n} \le t \le (m+1)2^{-n}$. Thus, because Y_{m+1} is independent of $X_n(m2^{-n},\,\cdot\,)$,

$$\begin{aligned} &\mathbb{E}^{\mathbb{P}}\big[|X_n(t,\mathbf{y}) - X_n(t,\mathbf{x})|^2\big] \\ &= \mathbb{E}^{\mathbb{P}}\big[|X_n(m2^{-n},\mathbf{y}) - X_n(m2^{-n},\mathbf{x})|^2\big] \\ &\quad + 2(t-m2^{-n})\mathbb{E}^{\mathbb{P}}\Big[\Big(X_n(m2^{-n},\mathbf{y}) - X_n(m2^{-n},\mathbf{x}), \\ &\qquad\qquad b\big(X_n(m2^{-n},\mathbf{y})\big) - b\big(X_n(m2^{-n},\mathbf{x})\big)\Big)_{\mathbb{R}^N}\Big] \\ &\quad + (t-m2^{-n})^2\mathbb{E}^{\mathbb{P}}\Big[\big|b\big(X_n(m2^{-n},\mathbf{y})\big) - b\big(X_n(m2^{-n},\mathbf{x})\big)\big|^2\Big] \\ &\quad + (t-m2^{-n})\mathbb{E}^{\mathbb{P}}\Big[\big\|\sigma\big(X_n(m2^{-n},\mathbf{y})\big) - \sigma\big(X_n(m2^{-n},\mathbf{x})\big)\big\|_{\mathrm{H.S.}}^2\Big]. \end{aligned}$$

Now assume that σ and b are uniformly Lipschitz continuous. Then, from the preceding, we see that there would exist a $C < \infty$ such that

$$\mathbb{E}^{\mathbb{P}}\big[|X_n(t,\mathbf{y}) - X_n(t,\mathbf{x})|^2\big] \le \big(1 + C(t-m2^{-n})\big)\mathbb{E}^{\mathbb{P}}\big[|X_n(m2^{-n},\mathbf{y}) - X_n(m2^{-n},\mathbf{x})|^2\big]$$

for $m2^{-n} \le t \le (m+1)2^{-n}$, and so

$$\mathbb{E}^{\mathbb{P}}\big[|X_n(t,\mathbf{y}) - X_n(t,\mathbf{x})|^2\big] \le (1+C2^{-n})^m\big(1+C(t-m2^{-n})\big)|\mathbf{y}-\mathbf{x}| \le e^{Ct}|\mathbf{y}-\mathbf{x}|^2.$$

From this it follows that, for each $T > 0$, there is a $C(T) < \infty$ such that

$$\begin{aligned} \sup_{t\in[0,T]}\big|\widehat{P_n(t,\mathbf{y})}(\boldsymbol{\xi}) - \widehat{P_n(t,\mathbf{x})}(\boldsymbol{\xi})\big| &\le \mathbb{E}^{\mathbb{P}}\Big[\big|e^{i(\boldsymbol{\xi},X_n(t,\mathbf{y}))_{\mathbb{R}^N}} - e^{i(\boldsymbol{\xi},X_n(t,\mathbf{x}))_{\mathbb{R}^N}}\big|\Big] \\ &\le C(T)|\boldsymbol{\xi}||\mathbf{y}-\mathbf{x}|, \end{aligned}$$

and so, after this is combined with our earlier results, Theorem 1.2.2 says that $\{P_n : n \ge 1\}$ is relatively compact in $C\big([0,\infty) \times \mathbb{R}^N; \mathbf{M}_1(\mathbb{R}^N)\big)$. Hence, by the reasoning given at the end of Chapter 1, we have shown that, when

σ and b are uniformly Lipschitz continuous, there is a transition probability function that satisfies Kolmogorov's forward equation.

There is a similar way to couple $P_{n_1}(t,\mathbf{x})$ to $P_{n_2}(t,\mathbf{x})$ for $n_2 \neq n_1$ and thereby prove that $\{P_n : n \geq 1\}$ converges in $C([0,\infty) \times \mathbb{R}^N; \mathbf{M}_1(\mathbb{R}^N))$ when σ and b are uniformly Lipschitz continuous. However, the argument is cumbersome, and, as Itô understood and we will see below, it is much smarter to use a coupling in which increments of Brownian motion are one's source of independent Gaussian random variables.

2.1 Brownian motion and Wiener measure

Given the family $\{\lambda_t : t \geq 0\}$ associated with a Lévy system $(\mathbf{m}, C, M)$, Kolmogorov's consistency theorem (cf. Exercise 9.1.17 in [20]) guarantees that there is a family $\{X(t) : t \geq 0\}$ of random variables on a probability space $(\Omega, \mathcal{F}, \mathbb{P})$ with the properties that $\mathbb{P}(X(0) = \mathbf{0}) = 1$, and, for each $n \geq 1$, $0 = t_0 < t_1 \cdots < t_n$, and $\Gamma_0, \ldots, \Gamma_n \in \mathcal{B}_{\mathbb{R}^N}$,

$$\mathbb{P}\big(X(t_m) - X(t_{m-1}) \in \Gamma_m \quad \text{for } 0 \leq j \leq n\big) = \prod_{m=1}^{n} \lambda_{t_m - t_{m-1}}(\Gamma_m).$$

In fact (cf. Chapter 4 in [20]), one can always choose these random variables so that the paths $t \rightsquigarrow X(t)$ are right-continuous and have a left limit at each $t \in (0,\infty)$. That is, although they may have discontinuities, their only discontinuities are simple jumps and not oscillatory ones. Furthermore, a major goal of this section is to prove that the paths can be chosen to be continuous when $M = 0$.

2.1.1 *Lévy's construction of Brownian Motion*

The family of measures corresponding to the Lévy system $(\mathbf{0}, \mathbf{I}, 0)$ are the Gaussian measures $\gamma_{\mathbf{0},t\mathbf{I}}$, and a family $\{B(t) : t \geq 0\}$ of random variables on a probability space $(\Omega, \mathcal{F}, \mathbb{P})$ satisfying

$$\begin{aligned} \mathbb{P}\big(B(0) \in & \Gamma_0 \text{ and } B(t_m) - B(t_{m-1}) \in \Gamma_m \text{ for } 0 \leq j \leq n\big) \\ &= \mathbf{1}_{\Gamma_0}(\mathbf{0}) \prod_{m=1}^{n} \gamma_{\mathbf{0},(t_m - t_{m-1})\mathbf{I}}(\Gamma_m) \end{aligned} \tag{2.1.1}$$

is called an $\mathbb{R}^N$-valued **Brownian motion** if, $\mathbb{P}$-almost surely, $t \rightsquigarrow B(t)$ is continuous. The first person to prove the existence of such random variables was N. Wiener, but, because it is more transparent than Wiener's, we will use a proof devised by P. Lévy.

To understand Lévy's idea, it is best to begin by assuming that a Brownian motion exists and examine its polygonal approximations. Thus, suppose that $\{B(t) : t \geq 0\}$ is a Brownian motion, and, for $n \geq 0$, let $t \rightsquigarrow B_n(t)$ be the polygonal path that linearly interpolates $t \rightsquigarrow B(t)$ between times $m2^{-n}$. In other words, $B_n(m2^{-n}) = B(m2^{-n})$ and

$$B_n(t) = (m+1-2^n t)B(m2^{-n}) + (2^n t - m)B\big((m+1)2^{-n}\big)$$

for $m \geq 0$ and $t \in I_{m,n} := \big[m2^{-n}, (m+1)2^{-n}\big]$. The distribution of each individual family $\{B_n(t) : t \geq 0\}$ is very easy to understand, but what we need to understand is the relationship between successive families. Obviously, since $B_{n+1}(m2^{-n}) = B_n(m2^{-n})$ and $t \rightsquigarrow B_{n+1}(t) - B_n(t)$ is linear on the intervals $I_{2m-2,n+1}$ and $I_{2m-1,n+1}$, the maximum difference between $B_{n+1}(\,\cdot\,)$ and $B_n(\,\cdot\,)$ occurs at times $(2m-1)2^{-n-1}$. With this in mind, set $X_{m,0} = B_0(m) - B_0(m-1)$ and, for $m \geq 1$ and $n \geq 0$,

$$\begin{aligned}
X_{m,n+1} &= 2^{\frac{n}{2}+1}\Big(B_{n+1}\big((2m-1)2^{-n-1}\big) - B_n\big((2m-1)2^{-n-1}\big)\Big) \\
&= 2^{\frac{n}{2}+1}\left(B\big((2m-1)2^{-n-1}\big) - \frac{B\big((m-1)2^{-n}\big) + B(m2^{-n})}{2}\right) \\
&= 2^{\frac{n}{2}}\bigg[\Big(B\big((2m-1)2^{-n-1}\big) - B\big((m-1)2^{-n}\big)\Big) \\
&\qquad - \Big(B(m2^{-n}) - B\big((2m-1)2^{-n-1}\big)\Big)\bigg].
\end{aligned}$$

It is clear that, for each $n \geq 0$, $\{X_{m,n} : m \geq 1\}$ is a sequence of mutually independent, Gaussian random variables with mean $\mathbf{0}$ and covariance $\mathbf{I}$. However, what is less evident is that these sequences are independent of one another. To prove that they are, we will need the following lemma about spaces of Gaussian random variables. A linear subspace $\mathfrak{G}$ of $L^2(\mathbb{P};\mathbb{R})$ is said to be a **centered Gaussian family** if all its elements are centered (i.e., mean 0) Gaussian random variables. That is, for all $X \in \mathfrak{G}$ and $\zeta \in \mathbb{C}$,

$$\mathbb{E}^{\mathbb{P}}\big[e^{\zeta X}\big] = \exp\left(\frac{\mathbb{E}^{\mathbb{P}}[X^2]\zeta^2}{2}\right).$$

Notice that if $\{B(t) : t \geq 0\}$ is a Brownian motion, then the span of

$$\big\{(\boldsymbol{\xi}, B(t))_{\mathbb{R}^N} : t \geq 0 \text{ and } \boldsymbol{\xi} \in \mathbb{R}^N\big\}$$

is a Gaussian family. To check this, for $n \geq 1$, $0 = t_0 < t_2 < \cdots < t_n$ and $\boldsymbol{\xi}_1, \ldots, \boldsymbol{\xi}_n \in \mathbb{R}^N$, write

$$\sum_{m=1}^{n} (\boldsymbol{\xi}_m, B(t_m))_{\mathbb{R}^N} = \sum_{\ell=1}^{n} \left(\sum_{m=\ell}^{n} \boldsymbol{\xi}_m, B(t_\ell) - B(t_{\ell-1}) \right)_{\mathbb{R}^N},$$

which, as the sum of mutually independent, centered Gaussian random variables, is also a centered Gaussian random variable.

Lemma 2.1.1. *Let $\mathfrak{G} \subseteq L^2(\mathbb{P};\mathbb{R})$ be a centered Gaussian family. Then its $L^2(\mathbb{P};\mathbb{R})$ closure $\overline{\mathfrak{G}}$ is also a centered Gaussian family. In addition, if $S \subseteq \mathfrak{G}$, then S is independent of $S^\perp \cap \mathfrak{G}$, where $S^\perp$ is the perpendicular complement of S in $L^2(\mathbb{P};\mathbb{R})$.*

Proof. To prove that $\overline{\mathfrak{G}}$ is a centered Gaussian family, suppose that $\mathfrak{G} \ni X_k \longrightarrow X$ in $L^2(\mathbb{P};\mathbb{R})$. Then, for all $\xi \in \mathbb{R}$,

$$\mathbb{E}^{\mathbb{P}}\left[e^{i\xi X}\right] = \lim_{n\to\infty} \mathbb{E}^{\mathbb{P}}\left[e^{i\xi X_n}\right] = \lim_{n\to\infty} e^{-\frac{\xi^2}{2}\mathbb{E}^{\mathbb{P}}[X_n]^2} = e^{-\frac{\xi^2}{2}\mathbb{E}^{\mathbb{P}}[X]^2},$$

and so X is a centered Gaussian random variable.

Turning to the second assertion, what we must show is that if $X_1, \dots, X_n \in S$ and $X_1', \dots, X_n' \in S^\perp \cap \mathfrak{G}$, then

$$\begin{aligned}
&\mathbb{E}^{\mathbb{P}}\left[\prod_{m=1}^{n} e^{i\xi_m X_m} \prod_{m=1}^{n} e^{i\xi_m' X_m'}\right] \\
&\quad = \mathbb{E}^{\mathbb{P}}\left[\prod_{m=1}^{n} e^{i\xi_m X_m}\right] \mathbb{E}^{\mathbb{P}}\left[\prod_{m=1}^{n} e^{i\xi_m' X_m'}\right]
\end{aligned}$$

for any choice of $\{\xi_m : 1 \le m \le n\} \cup \{\xi_m' : 1 \le m \le n\} \subseteq \mathbb{R}$. But the expectation value on the left is equal to

$$\begin{aligned}
&\exp\left(-\frac{1}{2}\mathbb{E}^{\mathbb{P}}\left[\left(\sum_{m=1}^{n} (\xi_m X_m + \xi_m' X_m')\right)^2\right]\right) \\
&= \exp\left(-\frac{1}{2}\mathbb{E}^{\mathbb{P}}\left[\left(\sum_{m=1}^{n} \xi_m X_m\right)^2\right] - \frac{1}{2}\mathbb{E}^{\mathbb{P}}\left[\left(\sum_{m=1}^{n} \xi_m' X_m'\right)^2\right]\right) \\
&= \mathbb{E}^{\mathbb{P}}\left[\prod_{m=1}^{n} e^{i\xi_m X_m}\right] \mathbb{E}^{\mathbb{P}}\left[\prod_{m=1}^{n} e^{i\xi_m' X_m'}\right],
\end{aligned}$$

since $\mathbb{E}^{\mathbb{P}}[X_m X_{m'}'] = 0$ for all $1 \le m, m' \le n$. □

Armed with Lemma 2.1.1, we can now check that the elements of $\{X_{m,n} : (m,n) \in \mathbb{Z}^+ \times \mathbb{N}\}$ are mutually independent. Indeed, since, for all $(m,n) \in \mathbb{Z}^+ \times \mathbb{N}$ and $\boldsymbol{\xi} \in \mathbb{R}^N$, $(\boldsymbol{\xi}, \mathbf{X}_{m,n})_{\mathbb{R}^N}$ is a member of the centered Gaussian family $\mathfrak{G}$ generated by $\{B(t) : t \ge 0\}$, all that we have to do is check that, for each $(m,n) \in \mathbb{Z}^+ \times \mathbb{N}$, $\ell \in \mathbb{N}$, and $(\boldsymbol{\xi}, \boldsymbol{\eta}) \in (\mathbb{R}^N)^2$,

$$\mathbb{E}^{\mathbb{P}}\big[(\boldsymbol{\xi}, X_{m,n+1})_{\mathbb{R}^N}(\boldsymbol{\eta}, B(\ell 2^{-n}))_{\mathbb{R}^N}\big] = 0.$$

But, since, for $s \le t$, $B(s)$ is independent of $B(t) - B(s)$,

$$\mathbb{E}^{\mathbb{P}}\big[(\boldsymbol{\xi}, B(s))_{\mathbb{R}^N}(\boldsymbol{\eta}, B(t))_{\mathbb{R}^N}\big] = \mathbb{E}^{\mathbb{P}}\big[(\boldsymbol{\xi}, B(s))_{\mathbb{R}^N}(\boldsymbol{\eta}, B(s))_{\mathbb{R}^N}\big] = s(\boldsymbol{\xi}, \boldsymbol{\eta})_{\mathbb{R}^N}$$

and therefore

$$\begin{aligned}
&2^{-\frac{n}{2}-1}\mathbb{E}^{\mathbb{P}}\big[(\boldsymbol{\xi}, X_{m,n+1})_{\mathbb{R}^N}(\boldsymbol{\eta}, B(\ell 2^{-n}))_{\mathbb{R}^N}\big] \\
&\quad = \mathbb{E}^{\mathbb{P}}\Big[\Big(\boldsymbol{\xi}, B\big((2m-1)2^{-n-1}\big)\Big)_{\mathbb{R}^N}\Big(\boldsymbol{\eta}, B(\ell 2^{-n})\Big)_{\mathbb{R}^N}\Big] \\
&\qquad - \frac{1}{2}\mathbb{E}^{\mathbb{P}}\Big[\Big(\boldsymbol{\xi}, B(m2^{-n}) + B\big((m-1)2^{-n}\big)\Big)_{\mathbb{R}^N}\Big(\boldsymbol{\eta}, B(\ell 2^{-n})\Big)_{\mathbb{R}^N}\Big] \\
&\quad = 2^{-n}(\boldsymbol{\xi}, \boldsymbol{\eta})_{\mathbb{R}^N}\left[\left(m - \tfrac{1}{2}\right) \wedge \ell - \frac{m \wedge \ell + (m-1) \wedge \ell}{2}\right] = 0.
\end{aligned}$$

We now know how to construct a Brownian motion. Namely, let $\{X_{m,n} : (m,n) \in \mathbb{Z}^+ \times \mathbb{N}\}$ be a family of mutually independent, $\mathbb{R}^N$-valued Gaussian random variables with mean $\mathbf{0}$ and covariance $\mathbf{I}$ on some probability space $(\Omega, \mathcal{F}, \mathbb{P})$, and, using induction on $n \ge 0$, define $\{B_n(t) : t \ge 0\}$ so that $B_n(0) = \mathbf{0}$, $t \rightsquigarrow B_n(t)$ is linear on each interval $I_{m,n}$, $B_0(m) = \sum_{1 \le \ell \le m} X_{\ell,0}$ for $m \in \mathbb{Z}^+$, $B_{n+1}(m2^{-n}) = B_n(m2^{-n})$ for $m \in \mathbb{N}$, and

$$B_{n+1}\big((2m-1)2^{-n-1}\big) = B_n\big((2m-1)2^{-n-1}\big) + 2^{-\frac{n}{2}-1}X_{m,n+1} \quad \text{for } m \in \mathbb{Z}^+.$$

If Brownian motion exists, then the distribution of $\{B_n(t) : t \ge 0\}$ is the distribution of the process obtained by linearizing it on each of the intervals $I_{m,n}$, and so the limit $\lim_{n\to\infty} B_n(t)$ should exist (a.s., $\mathbb{P}$) uniformly on compacts and should be a Brownian motion.

To see that this procedure works, one must first verify that the preceding definition of $\{B_n(t) : t \ge 0\}$ gives a process with the correct distribution. That is, we need to show that, for each $n \ge 0$, $\big\{B_n\big((m+1)2^{-n}\big) - B_n\big(m2^{-n}\big) : m \in \mathbb{N}\big\}$ is a sequence of mutually independent Gaussian random variables with mean $\mathbf{0}$ and covariance $2^{-n}\mathbf{I}$. But, since this sequence is contained in the centered Gaussian family $\mathfrak{G}$ spanned by $\{X_{m,n} : (m,n) \in \mathbb{Z}^+ \times \mathbb{N}\}$, Lemma 2.1.1 says that we need only show that

$$\begin{aligned}
&\mathbb{E}^{\mathbb{P}}\Big[\Big(\boldsymbol{\xi}, B_n\big((m+1)2^{-n}\big) - B_n\big(m2^{-n}\big)\Big)_{\mathbb{R}^N} \\
&\qquad \times \Big(\boldsymbol{\xi}', B_n\big((m'+1)2^{-n}\big) - B_n\big(m'2^{-n}\big)\Big)_{\mathbb{R}^N}\Big] = 2^{-n}(\boldsymbol{\xi}, \boldsymbol{\xi}')_{\mathbb{R}^N}\delta_{m,m'}
\end{aligned}$$

for $\boldsymbol{\xi}, \boldsymbol{\xi}' \in \mathbb{R}^N$ and $m, m' \in \mathbb{N}$. When $n = 0$, this is obvious. Now assume that it is true for n, and observe that

$$B_{n+1}(m2^{-n}) - B_{n+1}\big((2m-1)2^{-n-1}\big) = \frac{B_n(m2^{-n}) - B_n\big((m-1)2^{-n}\big)}{2} - 2^{-\frac{n}{2}-1}X_{m,n+1}$$

and

$$B_{n+1}\big((2m-1)2^{-n-1}\big) - B_{n+1}\big((m-1)2^{-n}\big) = \frac{B_n(m2^{-n}) - B_n\big((m-1)2^{-n}\big)}{2} + 2^{-\frac{n}{2}-1}X_{m,n+1}.$$

Using these expressions and the induction hypothesis, one sees that the required equality holds.

Second, and more challenging, we must show that, $\mathbb{P}$-almost surely, these processes are converging uniformly on compact time intervals. For this purpose, consider the difference $t \rightsquigarrow B_{n+1}(t) - B_n(t)$. Since this path is linear on each interval $[m2^{-n-1}, (m+1)2^{-n-1}]$,

$$\max_{t\in[0,2^L]} \big|B_{n+1}(t) - B_n(t)\big| = \max_{1\le m\le 2^{L+n+1}} \big|B_{n+1}(m2^{-n-1}) - B_n(m2^{-n-1})\big|$$

$$= 2^{-\frac{n}{2}-1} \max_{1\le m\le 2^{L+n}} |X_{m,n+1}| \le 2^{-\frac{n}{2}-1}\left(\sum_{m=1}^{2^{L+n}} |X_{m,n+1}|^4\right)^{\frac{1}{4}}.$$

Thus,

$$\mathbb{E}^{\mathbb{P}}\big[\|B_{n+1}(\,\cdot\,) - B_n(\,\cdot\,)\|_{[0,2^L]}^4\big]^{\frac{1}{4}} \le 2^{\frac{n}{2}-1}\left(\sum_{m=1}^{2^{L+n}} \mathbb{E}^{\mathbb{P}}\big[|X_{m,n+1}|^4\big]\right)^{\frac{1}{4}} = 2^{-\frac{n-L-4}{4}}C_N,$$

where $C_N^4 := \int_{\mathbb{R}^N} |\mathbf{y}|^4\, \gamma_{\mathbf{0},\mathbf{I}}(d\mathbf{y})$.

From the preceding we know that, for any $T > 0$,

$$\mathbb{E}^{\mathbb{P}}\left[\sup_{n>m} \|B_n(\,\cdot\,) - B_m(\,\cdot\,)\|_{[0,T]}^4\right]^{\frac{1}{4}} \le \sum_{n=m}^{\infty} \mathbb{E}^{\mathbb{P}}\left[\|B_{n+1}(\,\cdot\,) - B_n(\,\cdot\,)\|_{[0,T]}^4\right]^{\frac{1}{4}} \longrightarrow 0 \quad \text{as } m\to\infty,$$

and so there exists a measurable $B : [0,\infty)\times\Omega \longrightarrow \mathbb{R}^N$ such that $B(0) = \mathbf{0}$, $B(\,\cdot\,,\omega) \in C\big([0,\infty);\mathbb{R}^N\big)$ for each $\omega\in\Omega$, and $\|B_n - B\|_{[0,t]} \longrightarrow 0$ both $\mathbb{P}$-almost surely and in $L^4(\mathbb{P};\mathbb{R})$ for every $t\in[0,\infty)$. Furthermore, since $B(m2^{-n}) = B_n(m2^{-n})$ $\mathbb{P}$-almost surely for all $(m,n)\in\mathbb{N}^2$, it is clear that, for all $n\ge 0$, $\big\{B\big((m+1)2^{-n}\big) - B(m2^{-n}) : m \ge 0\big\}$ is a sequence of mutually independent, Gaussian random variables with mean $\mathbf{0}$ and covariance $2^{-n}\mathbf{I}$. Hence, by continuity, it follows that $\{B(t) : t \ge 0\}$ is a Brownian motion.

2.1.2 *Kolmogorov's continuity criterion*

There are elements of Lévy's construction that admit interesting generalizations, perhaps the most important of which is **Kolmogorov's continuity criterion**.

Remember that when dealing with an uncountable number of random variables, in general one can measure only events that depend on countably many of them. As a consequence, one can change individual members on sets of measure 0 without effecting the probability of events that can be measured. With this in mind, one says that two families of random variables are **versions** of one another if one family can be obtained from the other by changing individual random variables on sets of measure 0.

In the proof of the following theorem, we use the fact that if Q is a closed cube in $\mathbb{R}^N$ and, for each vertex $\mathbf{v}$ of Q, $a_{\mathbf{v}}$ is an element of a vector space E, then there is a unique function $f\colon Q \longrightarrow E$, known as the multilinear extention of the $a_{\mathbf{v}}$'s, such that $f(\mathbf{v}) = a_{\mathbf{v}}$ for each vertex $\mathbf{v}$ and f is an affine function of each coordinate. For example, if $Q = [0,1]^2$, then

$$f(x_1,x_2) = (1-x_1)(1-x_2)a_{(0,0)}+(1-x_1)x_2a_{(0,1)}+x_1(1-x_2)a_{(1,0)}+x_1x_2a_{(1,1)}.$$

The general case can be proved by translation, scaling, and induction on N.

Theorem 2.1.2. *Suppose that $\{X(\mathbf{x}) : \mathbf{x} \in [0,R]^N\}$ is a family of random variables taking values in a Banach space E, and assume that, for some $p \in [1,\infty)$, $C < \infty$, and $r \in (0,1]$,*

$$\mathbb{E}\big[\|X(\mathbf{y}) - X(\mathbf{x})\|_E^p\big]^{\frac{1}{p}} \le C|\mathbf{y}-\mathbf{x}|^{\frac{N}{p}+r} \quad \text{for all } \mathbf{x},\mathbf{y} \in [0,R]^N.$$

Then there exists a version $\{\widetilde{X}(\mathbf{x}) : \mathbf{x} \in [0,R]^N\}$ of $\{X(\mathbf{x}) : \mathbf{x} \in [0,R]^N\}$ such that $\mathbf{x} \in [0,R]^N \longmapsto \widetilde{X}(\mathbf{x})(\omega) \in E$ is continuous for all $\omega \in \Omega$. In fact, for each $\alpha \in [0,r)$, there is a $K < \infty$, depending only on N, p, r, and α, such that

$$\mathbb{E}\left[\sup_{\substack{\mathbf{x},\mathbf{y}\in[0,R]^N\\ \mathbf{x}\ne\mathbf{y}}}\left(\frac{\|\widetilde{X}(\mathbf{y}) - \widetilde{X}(\mathbf{x})\|_E}{|\mathbf{y}-\mathbf{x}|^\alpha}\right)^p\right]^{\frac{1}{p}} \le KCR^{\frac{n}{p}+r-\alpha}.$$

Proof. First note that, by an elementary rescaling argument, it suffices to treat the case when $R = 1$.

Given $n \geq 0$, set[1]

$$M_n = \max_{\substack{\mathbf{k},\mathbf{m}\in\mathbb{N}^N\cap[0,2^n]^N \\ \|\mathbf{m}-\mathbf{k}\|_\infty=1}} \big\|X(\mathbf{m}2^{-n}) - X(\mathbf{k}2^{-n})\big\|_E \leq \left(\sum_{\substack{\mathbf{k},\mathbf{m}\in\mathbb{N}^N\cap[0,2^n]^N \\ \|\mathbf{m}-\mathbf{k}\|_\infty=1}} \big\|X(\mathbf{m}2^{-n}) - X(\mathbf{k}2^{-n})\big\|_E^p \right)^{\frac{1}{p}},$$

and observe that

$$\mathbb{E}\big[M_n^p\big]^{\frac{1}{p}} \leq \left(\sum_{\substack{\mathbf{k},\mathbf{m}\in\mathbb{N}^N\cap[0,2^n]^N \\ \|\mathbf{m}-\mathbf{k}\|_\infty=1}} \mathbb{E}\Big[\big\|X(\mathbf{m}2^{-n}) - X(\mathbf{k}2^{-n})\big\|_E^p\Big] \right)^{\frac{1}{p}} \leq C(2^{N+1}N)^{\frac{1}{p}} 2^{-nr}.$$

Let $n \geq 0$ be given, and take $X_n(\,\cdot\,)$ to be the function that equals $X(\,\cdot\,)$ at the vertices of and is multilinear on each cube $\mathbf{m}2^{-n} + [0, 2^{-n}]^N$. Because $X_{n+1}(\mathbf{x}) - X_n(\mathbf{x})$ is a multilinear function on $\mathbf{m}2^{-n-1} + [0, 2^{-n-1}]^N$,

$$\sup_{\mathbf{x}\in[0,1]^N} \|X_{n+1}(\mathbf{x}) - X_n(\mathbf{x})\|_E = \max_{\mathbf{m}\in\mathbb{N}^N\cap[0,2^{n+1}]^N} \|X_{n+1}(\mathbf{m}2^{-n-1}) - X_n(\mathbf{m}2^{-n-1})\|_E.$$

Since $X_{n+1}(\mathbf{m}2^{-n-1}) = X(\mathbf{m}2^{-n-1})$ and either $X_n(\mathbf{m}2^{-n-1}) = X(\mathbf{m}2^{-n-1})$ or

$$X_n(\mathbf{m}2^{-n-1}) = \sum_{\substack{\mathbf{k}\in\mathbb{N}^N\cap[0,2^{n+1}] \\ \|\mathbf{k}-\mathbf{m}\|_\infty=1}} \theta_{\mathbf{m},\mathbf{k}} X(\mathbf{k}2^{-n-1}),$$

where the $\theta_{\mathbf{m},\mathbf{k}}$'s are non-negative and sum to 1, it follows that

$$\sup_{\mathbf{x}\in[0,1]^N} \|X_{n+1}(\mathbf{x}) - X_n(\mathbf{x})\|_E \leq M_{n+1}$$

and therefore that

$$\mathbb{E}\left[\sup_{\mathbf{x}\in[0,1]^N} \|X_{n+1}(\mathbf{x}) - X_n(\mathbf{x})\|_E^p\right]^{\frac{1}{p}} \leq C(2^{N+1}N)^{\frac{1}{p}} 2^{-nr}.$$

Hence, for $0 \leq n < n'$,

$$\mathbb{E}\left[\sup_{n'>n}\sup_{\mathbf{x}\in[0,1]^N} \|X_{n'}(\mathbf{x}) - X_n(\mathbf{x})\|_E^p\right]^{\frac{1}{p}} \leq \frac{C(2^{N+1}N)^{\frac{1}{p}}}{1-2^{-r}} 2^{-nr},$$

---

[1] Given $\mathbf{x} \in \mathbb{R}^N$, $\|\mathbf{x}\|_\infty = \max_{1\leq j\leq N} |x_j|$.

and so $\{X_n : n \geq 0\}$ converges in $C([0,1]^N; E)$ both $\mathbb{P}$-almost surely and in $L^p(\mathbb{P}; C([0,1]^N; E))$. Therefore there exists a measurable map $\widetilde{X}\colon [0,1]^N \times \Omega \longrightarrow E$ such that $\mathbf{x} \rightsquigarrow \widetilde{X}(\mathbf{x},\omega)$ is continuous for each $\omega \in \Omega$ and

$$\mathbb{E}\left[\sup_{\mathbf{x}\in[0,1]^N} \|\widetilde{X}(\mathbf{x}) - X_n(\mathbf{x})\|_E^p\right]^{\frac{1}{p}} \leq \frac{C(2^{N+1}N)^{\frac{1}{p}}}{1-2^{-r}} 2^{-nr}.$$

Furthermore, $\widetilde{X}(\mathbf{x}) = X(\mathbf{x})$ (a.s., $\mathbb{P}$) if $\mathbf{x} = \mathbf{m}2^{-n}$ for some $n \geq 0$ and $\mathbf{m} \in \mathbb{N}^N \cap [0, 2^n]^N$, and therefore, since $\mathbf{x} \rightsquigarrow \widetilde{X}(\mathbf{x})$ is continuous and

$$\mathbb{E}\big[\|X(\mathbf{m}2^{-n}) - X(\mathbf{x})\|_E^p\big]^{\frac{1}{p}} \leq C2^{-n(\frac{N}{p}+r)}$$
$$\text{if } m_j 2^{-n} \leq x_j < (m_j+1)2^{-n} \text{ for } 1 \leq j \leq N,$$

it follows that $X(\mathbf{x}) = \widetilde{X}(\mathbf{x})$ (a.s., $\mathbb{P}$) for each $\mathbf{x} \in [0,1]^N$.

To prove the final estimate, suppose that $2^{-n-1} < |\mathbf{y}-\mathbf{x}| \leq 2^{-n}$. Then

$$\|X_n(\mathbf{y}) - X_n(\mathbf{x})\|_E \leq N^{\frac{1}{2}} 2^n |\mathbf{x}-\mathbf{y}| M_n,$$

and so, $\mathbb{P}$-almost surely,

$$\|\widetilde{X}(\mathbf{y}) - \widetilde{X}(\mathbf{x})\|_E \leq 2 \sup_{\boldsymbol{\xi}\in[0,1]^N} \|\widetilde{X}(\boldsymbol{\xi}) - X_n(\boldsymbol{\xi})\|_E + N^{\frac{1}{2}} 2^n |\mathbf{x}-\mathbf{y}| M_n.$$

Hence, by the preceding,

$$\mathbb{E}\left[\sup_{\substack{\mathbf{x},\mathbf{y}\in[0,1]^N\\ 2^{-n-1}<|\mathbf{y}-\mathbf{x}|\leq 2^{-n}}} \left(\frac{\|\widetilde{X}(\mathbf{y}) - \widetilde{X}(\mathbf{x})\|_E}{|\mathbf{y}-\mathbf{x}|^\alpha}\right)^p\right]^{\frac{1}{p}}$$
$$\leq C(2^{N+1}N)^{\frac{1}{p}}\left(\frac{2}{1-2^{-r}} + N^{\frac{1}{2}}\right)2^{-n(r-\alpha)},$$

and therefore

$$\mathbb{E}\left[\left(\sup_{\substack{\mathbf{x},\mathbf{y}\in[0,1]^N\\ \mathbf{y}\neq\mathbf{x}}} \frac{\|\widetilde{X}(\mathbf{y}) - \widetilde{X}(\mathbf{x})\|_E}{|\mathbf{y}-\mathbf{x}|^\alpha}\right)^p\right]^{\frac{1}{p}} \leq KC,$$

where $K = (2^{N+1}N)^{\frac{1}{p}}\left(\frac{2}{1-2^{-r}} + N^{\frac{1}{2}}\right)(1-2^{-(r-\alpha)})^{-1}$ □

Corollary 2.1.3. *Assume that there is a* $p \in [1,\infty)$, $\beta > \frac{N}{p}$, *and* $C < \infty$ *such that*

$$\mathbb{E}\big[\|\widetilde{X}(\mathbf{y}) - \widetilde{X}(\mathbf{x})\|_E^p\big]^{\frac{1}{p}} \le C|\mathbf{y}-\mathbf{x}|^\beta \quad \textit{for all } \mathbf{x},\mathbf{y} \in [0,\infty)^N.$$

Then, for each $\gamma > \beta$,

$$\lim_{|\mathbf{x}|\to\infty} \frac{\widetilde{X}(\mathbf{x})}{|\mathbf{x}|^\gamma} = 0 \quad (\text{a.s.}, \mathbb{P}) \textit{ and in } L^p(\mathbb{P};E).$$

Proof. Take $\alpha = 0$ in Theorem 2.1.2. Then

$$\begin{aligned}
&\mathbb{E}\left[\left(\sup_{\mathbf{x}\in[2^{n-1},2^n]^N} \frac{\|\widetilde{X}(\mathbf{x}) - \widetilde{X}(\mathbf{0})\|_E}{|\mathbf{x}|^\gamma}\right)^p\right]^{\frac{1}{p}} \\
&\quad \le 2^{-\gamma(n-1)}\mathbb{E}\left[\sup_{\mathbf{x}\in[0,2^n]^N} \|\widetilde{X}(\mathbf{x}) - \widetilde{X}(\mathbf{0})\|_E^p\right]^{\frac{1}{p}} \le 2^\gamma K C 2^{(\beta-\gamma)n},
\end{aligned}$$

and so

$$\mathbb{E}\left[\left(\sup_{\mathbf{x}\in[2^{m-1},\infty)^N} \frac{\|\widetilde{X}(\mathbf{x}) - \widetilde{X}(\mathbf{0})\|_E}{|\mathbf{x}|^\gamma}\right)^p\right]^{\frac{1}{p}} \le \frac{2^\gamma KC}{1-2^{\beta-\gamma}} 2^{(\beta-\gamma)m}.$$

□

2.1.3 *Brownian motion and Wiener measure*

For various reasons, it has become common to give a more flexible description of what is meant by a Brownian motion. Given a probability space $(\Omega, \mathcal{F}, \mathbb{P})$, a **filtration** (i.e., a non-decreasing family) $\{\mathcal{F}_t : t \ge 0\}$ of sub-σ-algebras, and a family $\{B(t) : t \ge 0\}$ of $\mathbb{R}^N$-valued random variables, one says that the triple $\big(B(t), \mathcal{F}_t, \mathbb{P}\big)$ is an $\mathbb{R}^N$-valued **Brownian motion** if

(i) $\mathbb{P}$-almost surely, $B(0) = \mathbf{0}$ and $t \rightsquigarrow B(t)$ is continuous.
(ii) For each $s \ge 0$, $B(s)$ is $\mathcal{F}_s$-measurable, and, for $t > s$, $B(t) - B(s)$ is independent of $\mathcal{F}_s$ and has distribution $\gamma_{\mathbf{0},(t-s)\mathbf{I}}$.

It should be clear that $\{B(t) : t \ge 0\}$ is a Brownian motion in the sense that (2.1.1) holds if and only if $\big(B(t), \mathcal{F}_t, \mathbb{P}\big)$ is a Brownian motion in the preceding sense with $\mathcal{F}_t = \sigma\big(\{B(\tau) :\in [0,t]\}\big)$, the σ-algebra generated by the path up until time t. However, there are many times when the natural choice of $\mathcal{F}_t$'s are larger than this choice. For example, if $N \ge 2$ and $\big(B(t), \mathcal{F}_t, \mathbb{P}\big)$ is an $\mathbb{R}^N$-valued Brownian motion, then, for each $\mathbf{e} \in \mathbb{S}^{N-1}$, $\big((\mathbf{e}, B(t))_{\mathbb{R}^N}, \mathcal{F}_t, \mathbb{P}\big)$ will be an $\mathbb{R}$-valued Brownian motion even though

$$\mathcal{F}_t \ne \sigma\Big(\big\{\big(\mathbf{e}, B(\tau)\big)_{\mathbb{R}^N} : \tau \in [0,t]\big\}\Big).$$

Another way to think about Brownian motion is in terms of Gaussian families. Suppose that $\{X(t) : t \ge 0\}$ is a family of centered, Gaussian,

$\mathbb{R}^N$-valued random variables on $(\Omega, \mathcal{F}, \mathbb{P})$, and assume that the span of $\{(\boldsymbol{\xi}, X(t))_{\mathbb{R}^N} : t \geq 0 \ \&\ \boldsymbol{\xi} \in \mathbb{R}^N\}$ is a centered Gaussian family. If $\mathcal{F}_t$ is the $\mathbb{P}$-completion of the σ-algebra generated by $\{X(\tau) : \tau \in [0,t]\}$, then there is a Brownian motion $(B(t), \mathcal{F}_t, \mathbb{P})$ such that $X(t) = B(t)$ (a.s., $\mathbb{P}$) for each $t \geq 0$ if and only if

$$\mathbb{E}^{\mathbb{P}}\left[\big(\boldsymbol{\xi}, X(s)\big)_{\mathbb{R}^N}\big(\boldsymbol{\eta}, X(t)\big)_{\mathbb{R}^N}\right] = s \wedge t(\boldsymbol{\xi}, \boldsymbol{\eta})_{\mathbb{R}^N} \text{ for all } s,t \in [0,\infty) \text{ and } \boldsymbol{\xi}, \boldsymbol{\eta} \in \mathbb{R}^N.$$

The necessity requires no comment. To check the sufficiency, first note that if $s < t$ and $s' < t'$, then

$$\begin{aligned}\mathbb{E}^{\mathbb{P}}&\left[\big(\boldsymbol{\xi}, X(t) - X(s)\big)_{\mathbb{R}^N}\big(\boldsymbol{\xi}', X(t') - X(s')\big)_{\mathbb{R}^N}\right]\\ &= \begin{cases} |t-s|(\boldsymbol{\xi}, \boldsymbol{\xi}')_{\mathbb{R}^N} & \text{if } s = s' \text{ and } t = t'\\ 0 & \text{if } (s,t) \cap (s',t') = \emptyset.\end{cases}\end{aligned}$$

Now let $0 \leq t_0 < \cdots < t_n$ and $\boldsymbol{\xi}_1, \ldots, \boldsymbol{\xi}_n \in \mathbb{R}^N$ be given. Then

$$\begin{aligned}\mathbb{E}^{\mathbb{P}}&\left[\exp\left(i\sum_{m=1}^n \big(\boldsymbol{\xi}_m, X(t_m) - X(t_{m-1})\big)_{\mathbb{R}^N}\right)\right]\\ &= \exp\left(-\frac{1}{2}\mathbb{E}^{\mathbb{P}}\left[\left(\sum_{m=1}^n \big(\boldsymbol{\xi}_m, X(t_m) - X(t_{m-1})\big)_{\mathbb{R}^N}\right)^2\right]\right)\\ &= \exp\Bigg(-\frac{1}{2}\sum_{m,m'=1}^n \mathbb{E}^{\mathbb{P}}\Big[\big(\boldsymbol{\xi}_m, X(t_m) - X(t_{m-1})\big)_{\mathbb{R}^N}\\ &\qquad\qquad \times \big(\boldsymbol{\xi}_{m'}, X(t_{m'}) - X(t_{m'-1})\big)_{\mathbb{R}^N}\Big]\Bigg)\\ &= \exp\left(-\frac{1}{2}\sum_{m=1}^n (t_m - t_{m-1})|\boldsymbol{\xi}_m|^2\right) = \prod_{m=1}^n \widehat{\gamma_{\mathbf{0},(t_m - t_{m-1})\mathbf{I}}}(\boldsymbol{\xi}_m),\end{aligned}$$

and therefore $\{X(t) : t \geq 0\}$ satisfies (2.1.1). Further, since

$$\mathbb{E}^{\mathbb{P}}\left[|X(t) - X(s)|^p\right]^{\frac{1}{p}} = \left(\int_{\mathbb{R}^N} |\mathbf{y}|^p\, \gamma_{\mathbf{0},\mathbf{I}}(d\mathbf{y})\right)^{\frac{1}{p}} |t-s|^{\frac{1}{2}},$$

one can use Theorem 2.1.2 to construct a continuous version $\{B(t) : t \geq 0\}$ of $\{X(t) : t \geq 0\}$, in which case $(B(t), \mathcal{F}_t, \mathbb{P})$ will be a Brownian motion.

As a consequence of Theorem 2.1.2 it is easy to see that Brownian paths are locally Hölder continuous of any order $\alpha \in \left(0, \frac{1}{2}\right)$. Indeed, for any $p \in [1,\infty)$,

$$\mathbb{E}^{\mathbb{P}}\left[|B(t) - B(s)|^p\right]^{\frac{1}{p}} \leq C_p|t-s|^{\frac{1}{2}} \quad \text{where } C_p = \left(\int_{\mathbb{R}^N} |\mathbf{y}|^p\, \gamma_{\mathbf{0},\mathbf{I}}(d\mathbf{y})\right)^{\frac{1}{p}}.$$

Hence, if $\alpha \in \left(0, \frac{1}{2}\right)$ and $p > \frac{2}{1-2\alpha}$, then, by Theorem 2.1.2,

$$\mathbb{E}^{\mathbb{P}}\left[\sup_{0\le s<t\le T}\left(\frac{|B(t)-B(s)|}{(t-s)^\alpha}\right)^p\right]^{\frac{1}{p}} < \infty$$

for all $T > 0$. In addition, by Corollary 2.1.3, for any $\gamma > \frac{1}{2}$,

$$\lim_{t\to\infty} \frac{|B(t)|}{t^\gamma} = 0 \quad (\text{a.s.}, \mathbb{P}).$$

On the other hand, the paths of a Brownian motion are not of locally bounded variation. In fact,[2]

$$\begin{aligned}\sum_{0\le k<2^n t} \Big(B\big((k+1)2^{-n}\big) - B(k2^{-n})\Big) \otimes \Big(B\big((k+1)2^{-n}\big) - B(k2^{-n})\Big)& \\ \longrightarrow t\mathbf{I} \quad (\text{a.s.}, \mathbb{P})&\end{aligned} \tag{2.1.2}$$

uniformly for t in finite intervals. To see this, set

$$\Delta_{k,n} = B\big((k+1)2^{-n}\big) - B\big(k2^{-n}\big)$$

and $D_{k,n} = \Delta_{k,n}\otimes\Delta_{k,n} - 2^{-n}\mathbf{I}$. Then $\{D_{k,n} : k \ge 0\}$ is a sequence of mutually independent, $\mathrm{Hom}(\mathbb{R}^N;\mathbb{R}^N)$-valued random variable with mean value 0, and so, by Kolmogorov's inequality,

$$\begin{aligned}&\mathbb{P}\left(\sup_{t\in[0,T]}\left\|\sum_{0\le k<2^n t} D_{k,n}\right\|_{\text{H.S.}} \ge \epsilon\right) \le \epsilon^{-2} \sum_{0\le k<2^n T} \mathbb{E}^{\mathbb{P}}\left[\|D_{k,n}\|_{\text{H.S.}}^2\right] \\ &\le \epsilon^{-2} \sum_{0\le k<2^n T} \mathbb{E}^{\mathbb{P}}\left[\|\Delta_{k,n}\otimes\Delta_{k,n}\|_{\text{H.S.}}^2\right] \le 2^{-n}\epsilon^{-2}N^2 T \int |\mathbf{y}|^4\,\gamma_{0,\mathbf{I}}(dy).\end{aligned}$$

Since

$$\left\|\sum_{0\le k<2^n t} \Delta_{k,n}\otimes\Delta_{k,n} - t\mathbf{I} - \sum_{0\le k<2^n t} D_{k,n}\right\|_{\text{H.S.}} \le N2^{-n},$$

(2.1.2) follows. On the other hand, if $B(\,\cdot\,)$ were of locally bounded variation, then, since $B(\,\cdot\,)$ is continuous, we would have

$$\begin{aligned}&\sum_{0\le k<2^n} \left|B\big((k+1)2^{-n}\big) - B(k2^{-n})\right|^2 \\ &\le \left(\max_{0\le k<2^n t}\left|B\big((k+1)2^{-n}\big) - B(k2^{-n})\right|\right) \mathrm{var}_{[0,1]}\big(B(\,\cdot\,)\big),\end{aligned}$$

[2] If $\mathbf{v}$ and $\mathbf{w}$ are elements of $\mathbb{R}^N$, then their tensor product $\mathbf{v}\otimes\mathbf{w}$ is the $N\times N$-matrix whose (i,j)th entry is $v_i w_j$.

which would tend to 0 as $n \to \infty$.

There are lots of transformations of Brownian paths that leave their distribution unchanged. For example, the easily verified **Brownian scaling** property says that if $\big(B(t), \mathcal{F}_t, \mathbb{P}\big)$ is a Brownian motion and $\lambda > 0$, then $\big(\lambda^{-\frac{1}{2}} B(\lambda t), \mathcal{F}_{\lambda t}, \mathbb{P}\big)$ is again a Brownian motion. Less obvious is the following **Brownian time inversion** property. Given a Brownian motion $\big(B(t), \mathcal{F}_t, \mathbb{P}\big)$, set $\breve{B}(0) = \mathbf{0}$ and $\breve{B}(t) = tB\big(\frac{1}{t}\big)$ for $t > 0$. Because $\lim_{t\to\infty} \frac{B(t)}{t} = \mathbf{0}$ (a.s., $\mathbb{P}$), $t \rightsquigarrow \breve{B}(t)$ is $\mathbb{P}$-almost surely continuous. In addition, for $s, t > 0$ and $\boldsymbol{\xi}, \boldsymbol{\eta} \in \mathbb{R}^N$,

$$\mathbb{E}^{\mathbb{P}}\big[\big(\boldsymbol{\xi}, sB(s^{-1})\big)_{\mathbb{R}^N}\big(\boldsymbol{\eta}, tB(t^{-1})\big)_{\mathbb{R}^N}\big] = st\big(\frac{1}{s} \wedge \frac{1}{t}\big)(\boldsymbol{\xi}, \boldsymbol{\eta})_{\mathbb{R}^N} = s \wedge t(\boldsymbol{\xi}, \boldsymbol{\eta})_{\mathbb{R}^N}.$$

Thus, because $\{(\boldsymbol{\xi}, B(t))_{\mathbb{R}^N} : t \geq 0 \ \& \ \boldsymbol{\xi} \in \mathbb{R}^N\}$ spans a centered Gaussian family and has the same covariance as a Brownian motion, $\big(\breve{B}(t), \breve{\mathcal{F}}_t, \mathbb{P}\big)$ is a Brownian motion when $\breve{\mathcal{F}}_t = \sigma\big(\{\breve{B}(\tau) : \tau \in [0, t]\}\big)$.

We will now introduce the canonical setting for Brownian motion. Set $\mathcal{P}(\mathbb{R}^N) = C\big([0, \infty); \mathbb{R}^N\big)$, and endow it with the topology of uniform convergence on compacts. That is, $\{\psi_n : n \geq 1\} \subseteq \mathcal{P}(\mathbb{R}^N)$ converges to ψ in $\mathcal{P}(\mathbb{R}^N)$ if $\|\psi_n - \psi\|_{[0,t]} \longrightarrow 0$ for all $t > 0$. As an application (cf. Lemma 9.1.4 in [20]) of the Stone–Weierstrass approximation theorem, one sees that this topology is separable. In addition, if

$$\rho(\psi_1, \psi_2) = \sum_{m=1}^{\infty} 2^{-m} \frac{\|\psi_2 - \psi_1\|_{[0,m]}}{1 + \|\psi_2 - \psi_1\|_{[0,m]}} \quad \text{for } \psi_1, \psi_2 \in \mathcal{P}(\mathbb{R}^N), \tag{2.1.3}$$

then it is equally easy to check that ρ is a complete metric for the topology of uniform convergence on compacts. Next, consider the Borel σ-algebra $\mathcal{B}_{\mathcal{P}(\mathbb{R}^N)}$. Obviously, $\psi \rightsquigarrow \psi(t)$ is $\mathcal{B}_{\mathcal{P}(\mathbb{R}^N)}$-measurable for each $t \geq 0$, and so $\sigma\big(\{\psi(t) : t \geq 0\}\big) \subseteq \mathcal{B}_{\mathcal{P}(\mathbb{R}^N)}$. To prove the opposite inclusion, it suffices to show that, for given $\psi_0 \in \mathcal{P}(\mathbb{R}^N)$, $r > 0$, and $T > 0$,

$$\{\psi : \|\psi - \psi_0\|_{[0,T]} \leq r\} \in \sigma\big(\{\psi(t) : t \geq 0\}\big).$$

But if $\mathbb{Q}$ denotes the set of rational numbers, then

$$\begin{aligned}&\{\psi : \|\psi - \psi_0\|_{[0,T]} \leq r\} \\ &= \{\psi : |\psi(t) - \psi_0(t)| \leq r \text{ for } t \in \mathbb{Q} \cap [0, T]\} \in \sigma\big(\{\psi(t) : t \geq 0\}\big).\end{aligned}$$

Thus $\mathcal{B}_{\mathcal{P}(\mathbb{R}^N)} = \sigma\big(\{\psi(t) : t \geq 0\}\big)$, and so we will know that two Borel probability measures μ and ν on $\mathcal{P}(\mathbb{R}^N)$ are equal if they assign the same measure to sets of the form $\{\psi(t_m) \in \Gamma_m \text{ for } 1 \leq m \leq n\}$ for all

$$n \geq 1,\ 0 \leq t_1 < \cdots < t_n, \quad \text{and} \quad \Gamma_1, \ldots, \Gamma_n \in \mathcal{B}_{\mathbb{R}^N}.$$

Therefore, if $\{X(t) : t \geq 0\}$ is a family of $\mathbb{R}^N$-valued random variables and $t \rightsquigarrow X(t)$ is $\mathbb{P}$-almost surely continuous, then $\{X(t) : t \geq 0\}$ determines a unique Borel probability measure μ on $\mathcal{P}(\mathbb{R}^N)$, known as its **distribution**, such that

$$\mu\big(\psi(t_m) \in \Gamma_m \text{ for } 1 \leq m \leq n\big) = \mathbb{P}\big(X(t_m) \in \Gamma_m \text{ for } 1 \leq m \leq n\big).$$

Finally, in the future, $\mathcal{B}_t$ will be used to denote $\sigma\big(\{\psi(\tau) : \tau \in [0,t]\}\big)$, the σ-algebra over $\mathcal{P}(\mathbb{R}^N)$ generated by the path up to time t.

Let $\mathbb{W}(\mathbb{R}^N)$ denote the set of $w \in \mathcal{P}(\mathbb{R}^N)$ such that $w(0) = \lim_{t\to\infty} \frac{w(t)}{t} = 0$, and define the norm

$$\|w\|_{\mathbb{W}(\mathbb{R}^N)} = \sup_{t\geq 0} \frac{|w(t)|}{1+t}.$$

Then $\mathbb{W}(\mathbb{R}^N)$ is a separable Banach with this norm, although its topology is not the one that it inherits from $\mathcal{P}(\mathbb{R}^N)$. On the other hand, because, for each $(k,m) \in (\mathbb{Z}^+)^2$, $\{\psi : |\psi(t)| \leq \frac{t}{k} \text{ for } t \geq m\}$ is a closed subset of $\mathcal{P}(\mathbb{R}^N)$,

$$\mathbb{W}(\mathbb{R}^N) = \bigcap_{k=1}^{\infty} \bigcup_{m=1}^{\infty} \{\psi : \psi(0) = \mathbf{0} \ \&\ |\psi(t)| \leq \frac{t}{k} \text{ for } t \geq m\} \in \mathcal{B}_{\mathcal{P}(\mathbb{R}^N)},$$

and $\mathcal{B}_{\mathbb{W}(\mathbb{R}^N)} = \{\Gamma \cap \mathbb{W}(\mathbb{R}^N); \Gamma \in \mathcal{B}_{\mathcal{P}}(\mathbb{R}^N)\}$. Thus,

$$\mathcal{B}_{\mathbb{W}(\mathbb{R}^N)} = \sigma\big(\{w(t) : t \geq 0\}\big),$$

and if μ is a Borel probability measure on $\mathcal{P}(\mathbb{R}^N)$ and $\mu\big(\mathbb{W}(\mathbb{R}^N)\big) = 1$, then the restriction of μ to $\mathbb{W}(\mathbb{R}^N)$ determines a Borel probability measure there.

Arguably the most important probability measure on $\mathbb{W}(\mathbb{R}^N)$ is **Wiener measure** $\mathcal{W}$, the distribution of an $\mathbb{R}^N$-valued Brownian motion. Because it will be playing a central role in what follows, we will reserve special notation for it. Besides using $\mathcal{W}$ to denote Wiener measure, we will use W_t to denote the $\mathcal{W}$-completion of the σ-algebra generated by $\{w(\tau) : \tau \in [0,t]\}$ and will think of $\big(w(t), W_t, \mathcal{W}\big)$ as the canonical $\mathbb{R}^N$-valued Brownian motion.

2.2 Itô's coupling procedure

Let L be given by (1.2.5), and assume that $a = \sigma\sigma^\top$ for some $\sigma\colon \mathbb{R}^N \longrightarrow \mathrm{Hom}(\mathbb{R}^M; \mathbb{R}^N)$.

We are now ready to describe Itô's procedure, which is the pathspace implementation of the Euler approximation scheme used in §1.2.1, and the first step is to construct the random variables used in his coupling procedure. Given $\mathbf{x} \in \mathbb{R}^N$, $n \geq 0$, and $w \in \mathbb{W}(\mathbb{R}^M)$, define $X_n(0,\mathbf{x})(w) = \mathbf{x}$ and

$$
\begin{aligned}
X_n(t,\mathbf{x})(w) = X_n(m2^{-n},\mathbf{x})(w) \\
+\sigma\big(X_n(m2^{-n},\mathbf{x})(w)\big)\big(w(t)-w(m2^{-n})\big) \\
+b\big(X_n(m2^{-n},\mathbf{x})(w)\big)(t-m2^{-n})
\end{aligned}
\tag{2.2.1}
$$

for $m \geq 0$ and $t \in I_{m,n} = [m2^{-n},(m+1)2^{-n}]$. Clearly $X_n(t,\mathbf{x})$ is $W_{\lfloor t\rfloor_n}$-measurable. Hence, since $w(t)-w(\lfloor t\rfloor_n)$ is independent of $W_{\lfloor t\rfloor_n}$,

$$
\mathcal{W}\big(X_n(t)\in\Gamma \mid W_{\lfloor t\rfloor_n}\big) = Q\big(t-\lfloor t\rfloor_n, X_n(\lfloor t\rfloor_n),\Gamma\big)
$$
$$
\text{where } Q(\tau,\mathbf{y}) = \gamma_{\mathbf{y}+\tau b(\mathbf{y}),\tau a(\mathbf{y})},
$$

and so, using induction on $m \geq 0$, one can check that the distribution of $X_n(t,\mathbf{x})$ under $\mathcal{W}$ is the measure $\mu_{t,n}$ in (1.2.12) with $\nu = \delta_{\mathbf{x}}$. Now assume that σ and b are uniformly Lipschitz continuous, and set

$$
\|\sigma\|_{{}_\mathrm{u}Lip} = \sup_{\substack{\mathbf{y},\mathbf{y}'\in\mathbb{R}^N\\ \mathbf{y}'\neq\mathbf{y}}} \frac{\|\sigma(\mathbf{y}')-\sigma(\mathbf{y})\|_{\mathrm{H.S.}}}{|\mathbf{y}'-\mathbf{y}|}
$$
$$
\|b\|_{\mathrm{Lip}} = \sup_{\substack{\mathbf{y},\mathbf{y}'\in\mathbb{R}^N\\ \mathbf{y}'\neq\mathbf{y}}} \frac{|b(\mathbf{y}')-b(\mathbf{y})|}{|\mathbf{y}'-\mathbf{y}|}.
$$

Since $\|\sigma(\mathbf{y})\|_{\mathrm{H.S.}} \leq \|\sigma(\mathbf{0})\|_{{}_\mathrm{u}H.S.} + \|\sigma\|_{\mathrm{Lip}}|\mathbf{y}|$ and $|b(\mathbf{y})| \leq |b(\mathbf{0})| + \|b\|_{\mathrm{Lip}}|\mathbf{y}|$, one can use induction to see that $X_n(t,\mathbf{x}) \in L^p(\mathcal{W};\mathbb{R}^N)$ for all $t \geq 0$ and $p \in [1,\infty)$. Furthermore, $X_n(t,\mathbf{x})$ is W_t-measurable for each $t \geq 0$, and, if

$$
M_n(t,\mathbf{x}) = X_n(t,\mathbf{x}) - \mathbf{x} - \int_0^t b\big(X_n(\lfloor\tau\rfloor_n,\mathbf{x})\big)\,d\tau,
$$

then

$$
\mathbb{E}^{\mathcal{W}}\big[M_n(t,\mathbf{x}) \mid W_s\big] = M_n(s,\mathbf{x}) \quad (\text{a.s.}, \mathcal{W})
$$

for $s \leq t$. Indeed, if $s,t \in I_{m,n}$, then

$$
M_n(t,\mathbf{x}) - M_n(s,\mathbf{x}) = \sigma\big(X_n(m2^{-n},\mathbf{x})\big)\big(w(t)-w(s)\big),
$$

and $\mathbb{E}^{\mathcal{W}}\big[w(t)-w(s) \mid W_{m2^{-n}}\big] = \mathbf{0}$, and the general case follows easily from this. Thus we now know that $\big(M_n(t), W_t, \mathcal{W}\big)$ is a continuous, $\mathbb{R}^N$-valued martingale, and therefore, by Doob's inequality,[3]

$$
\mathbb{E}^{\mathcal{W}}\big[\|M_n(\,\cdot\,,\mathbf{x})\|^p_{[0,t]}\big]^{\frac{1}{p}} \leq \frac{p}{p-1}\mathbb{E}^{\mathcal{W}}\big[|M_n(t,\mathbf{x})|^p\big]^{\frac{1}{p}}
$$

for all $p \in (1,\infty)$. Note that, since

---

[3] This is actually a corollary of Doob's inequality, which states that, if $\big(M(t),\mathcal{F}_t,\mathbb{P}\big)$ is a continuous martingale, then $\mathbb{P}\big(\|M(\,\cdot\,)\|_{[0,t]} \geq R\big) \leq \mathbb{E}^{\mathbb{P}}\big[|M(t)|, |X(t)| \geq R\big]$, which implies that $\mathbb{E}^{\mathbb{P}}\big[\|M(\,\cdot\,)\|^p_{[0,t]}\big]^{\frac{1}{p}} \leq \frac{p}{p-1}\mathbb{E}^{\mathbb{P}}\big[|M(t)|^p\big]^{\frac{1}{p}}$ for $1 < p \leq \infty$. See Theorem 7.1.9 in [20].

$$\mathbb{E}^{\mathcal{W}}\big[\big(M_n(t_3,\mathbf{x})-M_n(t_2,\mathbf{x}),M_n(t_1,\mathbf{x})\big)_{\mathbb{R}^N}\big]=0\quad\text{for } 0\le t_1\le t_2\le t_3$$

and $\mathbb{E}^{\mathbb{P}}\big[\big(w(t)-w(s)\big)\otimes\big(w(t)-w(s)\big)\;\big|\;W_s\big]=(t-s)\mathbf{I}$ for $0\le s\le t$,

$$\begin{aligned}&\mathbb{E}^{\mathcal{W}}\big[|M_n(t,\mathbf{x})|^2\big]\\&\quad=\sum_{m<2^nt}\mathbb{E}^{\mathcal{W}}\Big[\Big|\sigma\big(X_n(m2^{-n},\mathbf{x})\big)\big(w(t\wedge(m+1)2^{-n})-w(m2^{-n})\big)\Big|^2\Big]\\&\quad=\int_0^t\mathbb{E}^{\mathcal{W}}\big[\|\sigma(X_n(\lfloor\tau\rfloor_n,\mathbf{x})\|^2_{\mathrm{u}H.S}\big]\,d\tau,\end{aligned}$$

and therefore that

$$\begin{aligned}\mathbb{E}^{\mathcal{W}}\big[\|X_n(\,\cdot\,,\mathbf{x})\|^2_{[0,t]}\big]\le 3|\mathbf{x}|^2+12\int_0^t\mathbb{E}^{\mathcal{W}}\big[\|\sigma(X_n(\lfloor\tau\rfloor_n,\mathbf{x})\|^2_{\mathrm{H.S.}}\big]\,d\tau\\+3t\int_0^t\mathbb{E}^{\mathcal{W}}\big[\big|b\big(X_n(\lfloor\tau\rfloor_n,\mathbf{x})\big)\big|^2\big]\,d\tau.\end{aligned}$$

Since

$$\|\sigma\big(X_n(\lfloor\tau\rfloor_n,\mathbf{x})\big)\|^2_{\mathrm{H.S.}}\vee\big|b\big(X_n(\lfloor\tau\rfloor_n,\mathbf{x})\big)\big|^2\le C\big(1+\|X_n(\,\cdot\,,\mathbf{x})\|^2_{[0,\tau]}\big)$$

for some $C<\infty$ depending only on the Lipschitz norms of σ and b, the preceding leads first to

$$\mathbb{E}^{\mathcal{W}}\big[\|X_n(\,\cdot\,,\mathbf{x})\|^2_{[0,t]}\big]\le 3|\mathbf{x}|^2+3(4+t)C\int_0^t\mathbb{E}^{\mathcal{W}}\big[1+\|X_n(\,\cdot\,,\mathbf{x})\|^2_{[0,\tau]}\big]\,d\tau,$$

and then, by Lemma 1.2.4, to the existence of a $C(t)<\infty$, depending only on $\|\sigma\|_{\mathrm{Lip}}\vee\|b\|_{\mathrm{Lip}}$, such that

$$\mathbb{E}^{\mathcal{W}}\big[\|X_n(\,\cdot\,,\mathbf{x})\|^2_{[0,t]}\big]\le C(t)(1+|\mathbf{x}|^2).\tag{2.2.2}$$

Next write the difference $\Delta_n(t,\mathbf{x})$ between $X_{n+1}(t,\mathbf{x})$ and $X_n(t,\mathbf{x})$ as

$$M_{n+1}(t,\mathbf{x})-M_n(t,\mathbf{x})+\int_0^t\Big(b\big(X_{n+1}(\lfloor\tau\rfloor_{n+1},\mathbf{x})\big)-b\big(X_n(\lfloor\tau\rfloor_n,\mathbf{x})\big)\Big)\,d\tau.$$

Since $\big(M_{n+1}(t,\mathbf{x})-M_n(t,\mathbf{x}),W_t,\mathcal{W}\big)$ is a martingale,

$$\begin{aligned}&\mathbb{E}^{\mathcal{W}}\big[\|\Delta_n(\,\cdot\,,\mathbf{x})\|^2_{[0,t]}\big]\\(*)\qquad&\le 8\mathbb{E}^{\mathcal{W}}\big[|M_{n+1}(t,\mathbf{x})-M_n(t,\mathbf{x})|^2\big]\\&\quad+2t\int_0^t\mathbb{E}^{\mathcal{W}}\big[\big|b\big(X_{n+1}(\lfloor\tau\rfloor_{n+1},\mathbf{x})\big)-b\big(X_n(\lfloor\tau\rfloor_n,\mathbf{x})\big)\big|^2\big]\,d\tau.\end{aligned}$$

Write $M_{n+1}(t,\mathbf{x})-M_n(t,\mathbf{x})$ as

$$\sum_{m<2^n t-\frac{1}{2}} \Big(\sigma\big(X_{n+1}((2m+1)2^{-n-1})\big)-\sigma\big(X_{n+1}(m2^{-n})\big)\Big)$$
$$\times\big(w(t\wedge(m+1)2^{-n})-w((2m+1)2^{-n-1})\big)$$
$$+\sum_{m<2^n t}\Big(\sigma\big(X_{n+1}(m2^{-n},\mathbf{x})\big)-\sigma\big(X_n(m2^{-n},\mathbf{x})\big)\Big)$$
$$\times\big(w(t\wedge(m+1)2^{-n})-w(m2^{-n})\big),$$

and conclude that

$$\begin{aligned}&\mathbb{E}^{\mathcal{W}}\big[|M_{n+1}(t,\mathbf{x})-M_n(t,\mathbf{x})|^2\big]\\&\le 2\int_0^t \mathbb{E}^{\mathcal{W}}\Big[\big\|\sigma\big(X_{n+1}(\lfloor\tau\rfloor_{n+1},\mathbf{x})\big)-\sigma\big(X_{n+1}(\lfloor\tau\rfloor_n,\mathbf{x})\big)\big\|_{\text{H.S.}}^2\Big]\,d\tau\\&\quad+2\int_0^t \mathbb{E}^{\mathcal{W}}\Big[\big\|\sigma\big(X_{n+1}(\lfloor\tau\rfloor_n,\mathbf{x})\big)-\sigma\big(X_n(\lfloor\tau\rfloor_n,\mathbf{x})\big)\big\|_{\text{H.S.}}^2\Big]\,d\tau.\end{aligned}$$

By (2.2.2), for $\tau\in[0,t]$,

$$\begin{aligned}&\mathbb{E}^{\mathcal{W}}\Big[\big\|\sigma\big(X_{n+1}(\lfloor\tau\rfloor_{n+1},\mathbf{x})\big)-\sigma\big(X_{n+1}(\lfloor\tau\rfloor_n,\mathbf{x})\big)\big\|_{\text{H.S.}}^2\Big]\\&\le\|\sigma\|_{\text{Lip}}^2\mathbb{E}^{\mathcal{W}}\Big[\big|X_{n+1}(\lfloor\tau\rfloor_{n+1},\mathbf{x})-X_{n+1}(\lfloor\tau\rfloor_n,\mathbf{x})\big|^2\Big]\\&\le 2\|\sigma\|_{\text{Lip}}^2\Big(\mathbb{E}^{\mathcal{W}}\Big[\big|\sigma\big(X_{n+1}(\lfloor\tau\rfloor_n,\mathbf{x})\big)\big(w(\lfloor\tau\rfloor_{n+1})-w(\lfloor\tau\rfloor_n)\big)\big|^2\Big]\\&\qquad\qquad+2^{-2n+1}\mathbb{E}^{\mathcal{W}}\Big[\big|b\big(X_{n+1}(\lfloor\tau\rfloor_n)\big)\big|^2\Big]\Big)\\&\le 2^{-n}\|\sigma\|_{\text{Lip}}^2\mathbb{E}^{\mathcal{W}}\Big[\big\|\sigma\big(X_{n+1}(\lfloor\tau\rfloor_n,\mathbf{x})\big)\big\|_{\text{H.S.}}^2+\big|b\big(X_{n+1}(\lfloor\tau\rfloor_n)\big)\big|^2\Big]\\&\le 2^{-n-1}C(t)(1+|\mathbf{x}|^2)\end{aligned}$$

for some $C(t)<\infty$, and therefore

$$\begin{aligned}&\mathbb{E}^{\mathcal{W}}\big[|M_{n+1}(t,\mathbf{x})-M_n(t,\mathbf{x})|^2\big]\\&\le 2^{-n}C(t)(1+|\mathbf{x}|^2)+2\|\sigma\|_{\text{Lip}}^2\int_0^t\mathbb{E}^{\mathcal{W}}\Big[\big|X_{n+1}(\lfloor\tau\rfloor_n,\mathbf{x})\big)-X_n(\lfloor\tau\rfloor_n,\mathbf{x})\big)\big|^2\Big]\,d\tau.\end{aligned}$$

By essentially the same, only easier, line of reasoning,

$$\begin{aligned}&\int_0^t\mathbb{E}^{\mathcal{W}}\Big[\big|b\big(X_{n+1}(\lfloor\tau\rfloor_{n+1},\mathbf{x})\big)-b\big(X_n(\lfloor\tau\rfloor_n,\mathbf{x})\big)\big|^2\Big]\,d\tau\\&\le 2^{-n}C(t)(1+|\mathbf{x}|^2)+2\int_0^t\mathbb{E}^{\mathcal{W}}\Big[\big|b\big(X_{n+1}(\lfloor\tau\rfloor_n,\mathbf{x})\big)-b\big(X_n(\lfloor\tau\rfloor_n,\mathbf{x})\big)\big|^2\Big]\,d\tau,\end{aligned}$$

and so, by $(*)$,

$$\mathbb{E}^{\mathcal{W}}\big[\|\Delta_n(\,\cdot\,,\mathbf{x})\|^2_{[0,t]}\big] \le 2^{-n+1}C(t)(1+|\mathbf{x}|^2) + 4\big(\|\sigma\|^2_{\mathrm{Lip}} + t\|b\|^2_{\mathrm{Lip}}\big)\int_0^t \mathbb{E}^{\mathcal{W}}\big[\|\Delta_n(\,\cdot\,,\mathbf{x})\|^2_{[0,\tau]}\big]\,d\tau,$$

which, after an application of Lemma 1.2.4, means that

$$\mathbb{E}^{\mathcal{W}}\big[\|\Delta_n(\,\cdot\,,\mathbf{x})\|^2_{[0,t]}\big] \le 2^{-n}C(t)(1+|\mathbf{x}|^2)$$

for some $C(t) < \infty$. Starting from this, we see that

$$\mathbb{E}^{\mathcal{W}}\left[\sup_{n>m}\|X_n(\,\cdot\,,\mathbf{x}) - X_m(\,\cdot\,,\mathbf{x})\|^2_{[0,t]}\right]^{\frac{1}{2}} \le C(t)^{\frac{1}{2}}(1+|\mathbf{x}|)\frac{2^{-\frac{m}{2}}}{1-2^{-\frac{1}{2}}}$$

for $n > m$, and there therefore exists a continuous $X(\,\cdot\,,\mathbf{x})$ such that, for all $t > 0$, $\|X_n(\,\cdot\,,\mathbf{x}) - X(\,\cdot\,,\mathbf{x})\|_{[0,t]} \longrightarrow 0$ (a.s., $\mathcal{W}$) and

$$\mathbb{E}^{W}\big[\|X_n(\,\cdot\,,\mathbf{x}) - X(\,\cdot\,,\mathbf{x})\|^2_{[0,t]}\big]^{\frac{1}{2}} \le K(t)(1+|\mathbf{x}|)2^{\frac{n}{2}} \tag{2.2.3}$$

for some $K(t) < \infty$ depending only on $\|\sigma\|_{\mathrm{Lip}}$ and $\|b\|_{\mathrm{Lip}}$.

2.2.1 *The Markov property*

For each $\mathbf{x} \in \mathbb{R}^N$, let $\mathbb{P}_{\mathbf{x}} \in \mathbf{M}_1\big(\mathcal{P}(\mathbb{R}^N)\big)$ be the distribution of $X(\,\cdot\,,\mathbf{x})$ under $\mathcal{W}$.

Lemma 2.2.1. *For each* $\Phi \in C_{\mathrm{b}}\big(\mathcal{P}(\mathbb{R}^N);\mathbb{R}\big)$,

$$\mathbb{E}^{\mathcal{W}}\big[\Phi\big(X_n(\,\cdot\,,\mathbf{x})\big)\big] \longrightarrow \mathbb{E}^{\mathbb{P}_{\mathbf{x}}}[\Phi]$$

uniformly for $\mathbf{x}$ *in compact subsets. In particular,* $\mathbf{x} \rightsquigarrow \mathbb{E}^{\mathbb{P}_{\mathbf{x}}}[\Phi]$ *is continuous. Further, if* $P_n(t,\mathbf{x})$ *is the distribution of* $X_n(t,\mathbf{x})$ *and* $P(t,\mathbf{x})$ *is the distribution of* $X(t,\mathbf{x})$ *under* $\mathcal{W}$*, then* $P_n \longrightarrow P$ *in* $C\big([0,\infty)\times\mathbb{R}^N;\mathbf{M}_1(\mathbb{R}^N)\big)$.

Proof. Suppose that $\{\mathbf{x}_n : n \ge 0\} \subseteq \mathbb{R}$ tends to $\mathbf{x}$, and let $\mathbb{P}_n$ be the distribution of $X_n(\,\cdot\,,\mathbf{x}_n)$ under $\mathcal{W}$. We need to show that $\mathbb{E}^{\mathbb{P}_n}[\Phi] \longrightarrow \mathbb{E}^{\mathbb{P}_{\mathbf{x}}}[\Phi]$ for $\Phi \in C_{\mathrm{b}}\big(\mathcal{P}(\mathbb{R}^N);\mathbb{R}\big)$, and, by Theorem 9.1.5 in [20], it suffices to do so when Φ is uniformly continuous with respect to the metric ρ in (2.1.3). To this end, note that, by (2.2.3),

$$\lim_{n\to\infty}\sup_{|\mathbf{y}|\le R}\mathcal{W}\Big(\rho\big(X_n(\,\cdot\,,\mathbf{y}),X(\,\cdot\,,\mathbf{y})\big) \ge \delta\Big) = 0$$

for all $\delta > 0$. Hence, if Φ is uniformly continuous,

$$\varlimsup_{n\to\infty} \sup_{|\mathbf{y}|\le R} \left|E^{\mathcal{W}}\left[\Phi\big(X_n(\,\cdot\,,\mathbf{y})\big)\right] - \mathbb{E}^{\mathbb{P}_{\mathbf{y}}}[\Phi]\right|$$
$$\le \sup_{n\ge 1}\sup_{|\mathbf{y}|\le R} \mathbb{E}^{\mathcal{W}}\left[\left|\Phi\big(X_n(\,\cdot\,,\mathbf{y})\big) - \Phi\big(X(\,\cdot\,,\mathbf{y})\big)\right|, \rho\big(X_n(\,\cdot\,,\mathbf{y}), X(\,\cdot\,,\mathbf{y})\big)\le\delta\right],$$

for all $\delta > 0$. Since the right hand side tends to 0 as $\delta \searrow 0$, this proves that $\mathbb{E}^{\mathbb{P}_n}[\Phi] \longrightarrow \mathbb{E}^{\mathbb{P}_{\mathbf{x}}}[\Phi]$ for all $\Phi \in C_{\mathrm{b}}\big(\mathcal{P}(\mathbb{R}^N);\mathbb{R}\big)$. In addition, because, for each $n \ge 0$, $\mathbf{x} \rightsquigarrow \mathbb{E}^{\mathcal{W}}\big[\Phi\big(X_n(\,\cdot\,,\mathbf{x})\big)\big]$ is continuous, we have also shown that $\mathbf{x} \rightsquigarrow \mathbb{E}^{\mathbb{P}_{\mathbf{x}}}[\Phi]$ is continuous.

Turning to the final assertion, apply the preceding to see that, for each $t \ge 0$ and $\varphi \in C_{\mathrm{b}}\big(\mathbb{R}^N;\mathbb{R}\big)$, $\langle\varphi, P_n(t,\mathbf{x})\rangle \longrightarrow \langle\varphi, P(t,\mathbf{x})\rangle$ uniformly for $\mathbf{x}$ in compact subsets. Further, using (2.2.2), one sees that, for each $T > 0$ and $R > 0$, there exists a $C(T,R) < \infty$ such that

$$\sup_{n\ge 0}\sup_{|\mathbf{x}|\le R} \mathbb{E}^{\mathcal{W}}\left[|X_n(t,\mathbf{x}) - X_n(s,\mathbf{x})|^2\right] \le C(T,R)(t-s) \quad \text{for } 0 \le s < t \le T.$$

Hence, if φ is uniformly continuous,

$$\lim_{\delta\searrow 0} \sup_{\substack{s,t\in[0,T]\\ |t-s|\le\delta}} \sup_{\substack{n\ge 0\\ |\mathbf{x}|\le R}} \left|\langle\varphi, P_n(t,\mathbf{x})\rangle - \langle\varphi, P_n(s,\mathbf{x})\rangle\right| = 0,$$

and so, if $(t_n,\mathbf{x}_n) \longrightarrow (t,\mathbf{x})$, then

$$\langle\varphi, P_n(t_n,\mathbf{x}_n)\rangle \longrightarrow \langle\varphi, P(t,\mathbf{x})\rangle$$

if φ is uniformly continuous. Thus, again by Theorem 9.1.5 in [20], it follows that $P_n \longrightarrow P$ in $C\big([0,\infty)\times\mathbb{R}^N; \mathbf{M}_1(\mathbb{R}^N)\big)$. □

As a consequence of the last part of Lemma 2.2.1 combined with the reasoning at the end of Chapter 1, we know that $(t,\mathbf{x}) \rightsquigarrow P(t,\mathbf{x})$ is a transition probability function that satisfies Kolmogorov's forward equation

$$\langle\varphi, P(t,\mathbf{x})\rangle - \varphi(\mathbf{x}) = \int_0^t \langle\varphi, P(\tau,\mathbf{x})\rangle\, d\tau$$

for all $\varphi \in C^2_{\mathrm{b}}(\mathbb{R}^N;\mathbb{R})$. My goal now is to prove that $\{\mathbb{P}_{\mathbf{x}} : \mathbf{x} \in \mathbb{R}^N\}$ has the Markov property, which is the pathspace analog of the Chapman–Kolmogorov equation. Namely, for $s \in [0,\infty)$, define the time shift map $\Sigma_s\colon \mathcal{P}(\mathbb{R}^N) \longrightarrow \mathcal{P}(\mathbb{R}^N)$ by $\Sigma_s\psi(t) = \psi(s+t)$. Then, the **Markov property** for the family $\{\mathbb{P}_{\mathbf{x}} : \mathbf{x} \in \mathbb{R}^N\}$ is the statement that $\mathbf{x} \rightsquigarrow \mathbb{E}^{\mathbb{P}_{\mathbf{x}}}[\Phi]$ is measurable and

$$\mathbb{E}^{\mathbb{P}_{\mathbf{x}}}\left[\Phi\circ\Sigma_s \,\middle|\, \mathcal{B}_s\right] = \mathbb{E}^{\mathbb{P}_{\psi(s)}}[\Phi] \quad (\text{a.s.}, \mathbb{P}_{\mathbf{x}}) \tag{2.2.4}$$

for Borel measurable $\Phi\colon \mathcal{P}(\mathbb{R}^N) \longrightarrow \mathbb{R}$ that are bounded below. To prove this, first observe that is suffices to do so in the case when Φ is bounded and continuous. The key to proving (2.2.4) in this case is the observation that

$$X_n(m2^{-n}+t,\mathbf{x})(w) = X_n\big(t, X_n(m2^{-n},\mathbf{x})(w)\big)(\delta_{m2^{-n}}w),$$

where $\delta_s\colon \mathbb{W}(\mathbb{R}^M) \longrightarrow \mathbb{W}(\mathbb{R}^M)$ is given by $\delta_s w(t) = w(t+s) - w(s)$. Notice that $\mathcal{W}$ is the distribution of $w \rightsquigarrow \delta_s w$ and that $\sigma(\{\delta_s w(t) : t \geq 0\})$ is independent of $\mathcal{W}_s$. Hence, if $\Psi \in C_b\big(\mathcal{P}(\mathbb{R}^N);\mathbb{C}\big)$ is $\mathcal{B}_{m2^{-n}}$-measurable and therefore $\Psi \circ X_n(\,\cdot\,,\mathbf{x})$ is $W_{m2^{-n}}$-measurable, then

$$\begin{aligned}
&\mathbb{E}^{\mathcal{W}}\big[\Psi(X_n(\,\cdot\,,\mathbf{x}))\Phi\circ\Sigma_{m2^{-n}}(X_n(\,\cdot\,,\mathbf{x}))\big]\\
&\quad= \int \Psi(X_n(\,\cdot\,,\mathbf{x})(w))\left(\int \Phi(X_n(\,\cdot\,,X_n(m2^{-n},\mathbf{x}))(w')\,\mathcal{W}(dw')\right)\mathcal{W}(dw)\\
&\quad= \mathbb{E}^{\mathcal{W}}\big[\Psi(X_n(\,\cdot\,,\mathbf{x}))\varphi_n(X_n(m2^{-n},\mathbf{x}))\big],
\end{aligned}$$

where $\varphi_n(\mathbf{y}) := \mathbb{E}^{\mathcal{W}}\big[\Phi(X_n(\,\cdot\,,\mathbf{y}))\big]$. Now let $s \in [0,\infty)$ be given, and set $m_n = \min\{m : m2^{-n} \geq s\}$ and $s_n = m_n 2^{-n}$ for $n \geq 0$. Then by (2.2.3),

$$\mathbb{E}^{\mathcal{W}}\big[\Psi(X_n(\,\cdot\,,\mathbf{x}))\Phi\circ\Sigma_{s_n}(X_n(\,\cdot\,,\mathbf{x}))\big] \longrightarrow \mathbb{E}^{\mathbb{P}_{\mathbf{x}}}\big[\Psi(\Phi\circ\Sigma_s)\big].$$

In addition, if $\varphi(\mathbf{y}) = \mathbb{E}^{\mathbb{P}_{\mathbf{y}}}[\Phi]$, then, by (2.2.3) and Lemma 2.2.1, $\varphi_n\big(X_n(s_n,\mathbf{x})\big) \longrightarrow \varphi\big(X(s,\mathbf{x})\big)$ (a.s., $\mathcal{W}$), and so

$$\mathbb{E}^{\mathcal{W}}\big[\Psi(X_n(\,\cdot\,,\mathbf{x}))\varphi_n(X_n(s_n,\mathbf{x}))\big] \longrightarrow \int \Psi(\psi)\varphi(\psi(s))\,\mathbb{P}_{\mathbf{x}}(d\psi).$$

Therefore we have shown that

$$\mathbb{E}^{\mathbb{P}_{\mathbf{x}}}\big[\Psi(\Phi\circ\Sigma_s)\big] = \int_{\mathcal{P}(\mathbb{R}^N)} \Psi(\psi)\mathbb{E}^{\mathbb{P}_{\psi(s)}}[\Phi]\,\mathbb{P}_{\mathbf{x}}(d\psi)$$

for all $\mathcal{B}_s$-measurable $\Psi \in C_b\big(\mathcal{P}(\mathbb{R}^N);\mathbb{R}\big)$ and $\Phi \in C_b\big(\mathcal{P}(\mathbb{R}^N);\mathbb{R}\big)$, and, starting from this, it is an elementary exercise to show that (2.2.4) holds for all Borel measurable Φ that are bounded below.

There is a useful extension of (2.2.4). Given a filtration $\{\mathcal{F}_t : t \geq 0\}$ of sub-σ-algebras over Ω, a function $\zeta : \Omega \longrightarrow [0,\infty]$ is called a **stopping time** relative to $\{\mathcal{F}_t : t \geq 0\}$ if $\{\zeta \leq t\} \in \mathcal{F}_t$ for all $t \geq 0$, the idea being that one can determine whether ζ has occured by time t from the information in the σ-algebra $\mathcal{F}_t$. Associated with ζ is the σ-algebra $\mathcal{F}_\zeta$ of $\Gamma \subseteq \Omega$ with the property that $\Gamma \cap \{\zeta \leq t\} \in \mathcal{F}_t$ for all $t \geq 0$. Now suppose that ζ is a stopping time on $\mathcal{P}(\mathbb{R}^N)$ relative to $\{\mathcal{B}_t : t \geq 0\}$, and set $\zeta_n = \inf\{m2^{-n} : m2^{-n} \geq \zeta\}$. Then ζ_n is again a stopping time and $\mathcal{B}_\zeta \subseteq \mathcal{B}_{\zeta_n}$. Given $\Gamma \in \mathcal{B}_\zeta$ and a bounded, continuous $\Phi\colon \mathcal{P}(\mathbb{R}^N) \longrightarrow \mathbb{C}$, (2.2.4) implies that

$$\begin{aligned}
\mathbb{E}^{\mathbb{P}_{\mathbf{x}}}\big[\Phi\circ\Sigma_{\zeta_n}, \Gamma\cap\{\zeta<\infty\}\big] &= \sum_{m=0}^{\infty} \mathbb{E}^{\mathbb{P}_{\mathbf{x}}}\big[\Phi\circ\Sigma_{m2^{-n}}, \Gamma\cap\{\zeta_n = m2^{-n}\}\big]\\
= \sum_{m=0}^{\infty}\int_{\Gamma\cap\{\zeta_n=m2^{-n}\}} \mathbb{E}^{\mathbb{P}_{\psi(m2^{-n})}}[\Phi]\,\mathbb{P}_{\mathbf{x}}(d\psi) &= \int_{\Gamma\cap\{\zeta<\infty\}} \mathbb{E}^{\mathbb{P}_{\psi(\zeta_n)}}[\Phi]\,\mathbb{P}_{\mathbf{x}}(d\psi).
\end{aligned}$$

Hence, after letting $n \to \infty$, one sees that

$$\mathbb{E}^{\mathbb{P}_{\mathbf{x}}}\big[\Phi \circ \Sigma_\zeta, \Gamma \cap \{\zeta < \infty\}\big] = \int_{\Gamma \cap \{\zeta < \infty\}} \mathbb{E}^{\mathbb{P}_{\psi(\zeta)}}[\Phi]\, \mathbb{P}_{\mathbf{x}}(d\psi),$$

which means that

$$\mathbb{E}^{\mathbb{P}_{\mathbf{x}}}\big[\Phi \circ \Sigma_\zeta, \{\zeta < \infty\} \,\big|\, \mathcal{B}_\zeta\big] = \mathbb{E}^{\mathbb{P}_{\psi(\zeta)}}[\Phi] \text{ (a.s., } \mathbb{P}_{\mathbf{x}}) \quad \text{on } \{\zeta < \infty\}, \tag{2.2.5}$$

first for bounded continuous functions Φ and then for all Borel measurable ones which are bounded below. The generalization of (2.2.4) in (2.2.5) is sometimes called the **strong Markov property**.

The Markov property provides an important pathspace interpretation of Kolmogorov's forward equation. We know that

$$\frac{d}{dt}\langle \varphi, P(t, \mathbf{x}, \cdot\,)\rangle = \langle L\varphi, P(t, \mathbf{x}, \cdot\,)\rangle$$

for $\varphi \in C^2_{\mathrm{b}}(\mathbb{R}^N; \mathbb{C})$, and from this it follows that

$$\frac{d}{dt}\langle \varphi(t, \cdot\,), P(t, \mathbf{x}, \cdot\,)\rangle = \langle (\partial_t + L)\varphi(t, \cdot\,), P(t, \mathbf{x}, \cdot\,)\rangle, \quad t \in (0, T)$$

for $\varphi \in C^{1,2}_{\mathrm{b}}\big((0,T) \times \mathbb{R}^N; \mathbb{C}\big)$. Hence, if $\varphi \in C^{1,2}_{\mathrm{b}}\big((0,T) \times \mathbb{R}^N; \mathbb{C}\big)$, then, for $0 < s < t < T$,

$$\begin{aligned}
&\mathbb{E}^{\mathbb{P}_{\mathbf{x}}}\big[\varphi(t, \psi(t)) - \varphi(s, \psi(s)) \,\big|\, \mathcal{B}_s\big] \\
&= \mathbb{E}^{\mathbb{P}_{\mathbf{x}}}\big[\langle \varphi(t, \cdot\,), P(t-s, \psi(s))\rangle - \varphi(s, \psi(s)) \,\big|\, \mathcal{B}_s\big] \\
&= \mathbb{E}^{\mathbb{P}_{\mathbf{x}}}\left[\int_0^{t-s} \big\langle (\partial_\tau + L)\varphi(\tau + s, \cdot\,), P(\tau, \psi(s))\big\rangle\, d\tau \,\middle|\, \mathcal{B}_s\right] \\
&= \mathbb{E}^{\mathbb{P}_{\mathbf{x}}}\left[\int_0^{t-s} (\partial_\tau + L)\varphi(\tau + s, \psi(\tau + s))\, d\tau \,\middle|\, \mathcal{B}_s\right] \\
&= \mathbb{E}^{\mathbb{P}_{\mathbf{x}}}\left[\int_s^{t} (\partial_\tau + L)\varphi(\tau, \psi(\tau))\, d\tau \,\middle|\, \mathcal{B}_s\right],
\end{aligned}$$

which means that

$$\begin{aligned}
&\mathbb{E}^{\mathbb{P}_{\mathbf{x}}}\left[\varphi(t, \psi(t)) - \int_0^t (\partial_\tau + L)\varphi(\tau, \psi(\tau))\, d\tau \,\middle|\, \mathcal{B}_s\right] \\
&\qquad = \varphi(s, \psi(s)) - \int_0^s (\partial_\tau + L)\varphi(\tau, \psi(\tau))\, d\tau
\end{aligned}$$

$\mathbb{P}$-almost surely. In other words,

$$\left(\varphi(t\wedge T,\psi(t\wedge T)) - \int_0^{t\wedge T} (\partial_\tau + L)\varphi(\tau,\psi(\tau))\,d\tau, \mathcal{B}_t, \mathbb{P}_{\mathbf{x}}\right) \tag{2.2.6}$$

is a martingale for $\varphi \in C^{1,2}_{\mathrm{b}}((0,T)\times\mathbb{R}^N;\mathbb{C})\cap C_{\mathrm{b}}([0,T]\times\mathbb{R}^N;\mathbb{C})$.

This important fact can be thought of as saying that, under $\mathbb{P}_{\mathbf{x}}$, paths behave like "integral curves of L." To understand this idea, remember that if V vector field V and $\mathcal{L}_V = \sum_{i=1}^N V_j\partial_{y_j}$ is the associated directional derivative operator, then $X(\,\cdot\,)$ is an integral curve of V if and only if

$$t \rightsquigarrow \varphi(t,X(t)) - \int_0^t (\partial_\tau + \mathcal{L}_V)\varphi(\tau,X(\tau))\,d\tau$$

is constant. Although (2.2.6) does not say that

$$t \rightsquigarrow \varphi(t\wedge T,\psi(t\wedge T)) - \int_0^{t\wedge T} (\partial_\tau + L)\varphi(\tau,\psi(\tau))\,d\tau$$

is constant, it does say that, under $\mathbb{P}_{\mathbf{x}}$, it is *conditionally constant.* (See Exercise 5.1.)

2.2.2 *A digression on square roots*

The preceding brings up an annoying technical question. Namely, knowing L means that one knows its coefficients a and b. On the other hand, a does not uniquely determine a σ for which $a = \sigma\sigma^\top$, and the choice of σ can be critical. In particular, one would like σ to be as smooth as possible. From that point of view, when L is **uniformly elliptic** (i.e., $a \geq \epsilon\mathbf{I}$ for some $\epsilon > 0$), the following lemma shows that one cannot do better than take σ to be the positive definite square root $a^{\frac{1}{2}}$ of a. In its statement and proof, we use the notation

$$\|\boldsymbol{\alpha}\| = \sum_{i=1}^N \alpha_i \quad\text{and}\quad \partial_{\mathbf{x}}^{\boldsymbol{\alpha}}\varphi = \partial_{x_1}^{\alpha_1}\cdots\partial_{x_N}^{\alpha_N}\varphi$$

for $\boldsymbol{\alpha} \in \mathbb{N}^N$.

Lemma 2.2.2. *Assume that $a \geq \epsilon\mathbf{I}$. If a is continuously differentiable in a neighborhood of $\mathbf{x}$, then so is $a^{\frac{1}{2}}$ and*

$$\max_{1\leq i\leq n}\left\|\partial_{x_i}a^{\frac{1}{2}}(\mathbf{x})\right\|_{\mathrm{op}} \leq \frac{\|\partial_{x_i}a(\mathbf{x})\|_{\mathrm{op}}}{2\epsilon^{\frac{1}{2}}}.$$

Moreover, for each $n \geq 2$, there is a $C_n < \infty$ such that

$$\max_{\|\boldsymbol{\alpha}\|=n} \left\|\partial_{\mathbf{x}}^{\boldsymbol{\alpha}} a^{\frac{1}{2}}(\mathbf{x})\right\|_{\mathrm{op}} \le C_n \epsilon^{\frac{1}{2}} \sum_{k=1}^{n} \left(\frac{\max_{\|\boldsymbol{\alpha}\|\le n} \left\|\partial^{\boldsymbol{\alpha}} a(\mathbf{x})\right\|_{\mathrm{op}}}{\epsilon} \right)^k$$

when a is n-times continuously differentiable in a neighborhood of $\mathbf{x}$. Hence, if $a \in C_{\mathrm{b}}^n(\mathbb{R}^N; \mathrm{Hom}(\mathbb{R}^N; \mathbb{R}^N))$, then so is $a^{\frac{1}{2}}$.

Proof. Without loss in generality, assume that $\mathbf{x} = \mathbf{0}$ and that there is a $\Lambda < \infty$ such that $a \le \Lambda \mathbf{I}$ on $\mathbb{R}^N$.

Set $d = \mathbf{I} - \frac{a}{\Lambda}$. Obviously d is symmetric, $0\mathbf{I} \le d \le (1 - \frac{\epsilon}{\Lambda})\mathbf{I}$, and $a = \Lambda(\mathbf{I} - d)$. Thus, if $\binom{\frac{1}{2}}{0} = 1$ and

$$\binom{\frac{1}{2}}{m} = \frac{\prod_{\ell=0}^{m-1} \left(\frac{1}{2} - \ell\right)}{m!} \quad \text{for } m \ge 1$$

are the coefficients in the Taylor expansion of $x \rightsquigarrow (1+x)^{\frac{1}{2}}$ around 0, then

$$\sum_{m=0}^{\infty} (-1)^m \binom{\frac{1}{2}}{m} d^m$$

converges in the operator norm uniformly on $\mathbb{R}^N$. In addition, if λ is an eigenvalue of $a(\mathbf{y})$ and $\boldsymbol{\xi}$ is an associated eigenvector, then $d(\mathbf{y})\boldsymbol{\xi} = (1 - \frac{\lambda}{\Lambda})\boldsymbol{\xi}$, and so

$$\left(\Lambda^{\frac{1}{2}} \sum_{m=0}^{\infty} (-1)^m \binom{\frac{1}{2}}{m} d^m(\mathbf{y}) \right) \boldsymbol{\xi} = \lambda^{\frac{1}{2}} \boldsymbol{\xi}.$$

Hence,

$$a^{\frac{1}{2}} = \Lambda^{\frac{1}{2}} \sum_{m=0}^{\infty} (-1)^m \binom{\frac{1}{2}}{m} d^m. \tag{2.2.7}$$

Now assume that a is continuously differentiable in a neighborhood of $\mathbf{0}$. Then from (2.2.7) one gets

$$\partial_{x_i} a^{\frac{1}{2}}(\mathbf{0}) = -\Lambda^{-\frac{1}{2}} \left(\sum_{m=1}^{\infty} m(-1)^m \binom{\frac{1}{2}}{m} d^{m-1}(\mathbf{0}) \right) \partial_{x_i} a(\mathbf{0}),$$

where again the series converges in the operator norm. Furthermore, because $(-1)^m \binom{\frac{1}{2}}{m} \ge 0$ for all $m \ge 0$ and $d(\mathbf{0}) \le (1 - \frac{\epsilon}{\Lambda})\mathbf{I}$,

$$\begin{aligned} &\left\| \sum_{m=1}^{\infty} m(-1)^m \binom{\frac{1}{2}}{m} d^{m-1}(\mathbf{0}) \right\|_{\mathrm{op}} \\ &\le \sum_{m=1}^{\infty} m(-1)^m \binom{\frac{1}{2}}{m} \|d(\mathbf{0})\|_{\mathrm{op}}^{m-1} = \tfrac{1}{2}\left(1 - \|d(\mathbf{0})\|_{\mathrm{op}}\right)^{-\frac{1}{2}} = \frac{\Lambda^{\frac{1}{2}}}{2\epsilon^{\frac{1}{2}}}, \end{aligned}$$

and so the first assertion is now proved.

Turning to the second assertion, one again uses (2.2.7) to see that if $\|\boldsymbol{\alpha}\| = n$ then

$$\partial_{\mathbf{x}}^{\boldsymbol{\alpha}} a^{\frac{1}{2}}(\mathbf{0}) = \sum_{k=1}^{n} (-1)^k \Lambda^{\frac{1}{2}-k} \left(\sum_{m=k}^{\infty} \frac{m!}{(m-k)!} (-1)^m \binom{\frac{1}{2}}{m} d^{m-k}(\mathbf{0}) \right) \times \left(\sum_{\boldsymbol{\alpha}_1+\cdots+\boldsymbol{\alpha}_k=\boldsymbol{\alpha}} \partial^{\boldsymbol{\alpha}_1} a(\mathbf{0}) \cdots \partial^{\boldsymbol{\alpha}_k} a(\mathbf{0}) \right).$$

Proceeding as above, one see that

$$\left\| \sum_{m=k}^{\infty} \frac{m!}{(m-k)!} (-1)^m \binom{\frac{1}{2}}{m} d^{m-k}(\mathbf{0}) \right\|_{\mathrm{op}} \leq k! \left| \binom{\frac{1}{2}}{k} \right| \left(\frac{\Lambda}{\epsilon} \right)^{k-\frac{1}{2}},$$

and the asserted estimate follows from this. □

In view of Lemma 2.2.2, what remains to examine are a's that can degenerate. In this case, $a^{\frac{1}{2}}$ will often not be the optimal choice of σ. For example, if $N = 1$ and $a(x) = x^2$, then $a^{\frac{1}{2}}(x) = |x|$, which is Lipschitz continuous but not continuously differentiable, and so $\sigma(x) = x$ is a preferable choice. Another example of the same sort is

$$a(\mathbf{x}) = \begin{pmatrix} 1 & 0 \\ 0 & |\mathbf{x}|^2 \end{pmatrix} \quad \text{for } \mathbf{x} \in \mathbb{R}^2.$$

Again $a^{\frac{1}{2}}$ is Lipschitz continuous but not differentiable. On the other hand, if

$$\sigma(\mathbf{x}) = \begin{pmatrix} 1 & 0 & 0 \\ 0 & x_1 & x_2 \end{pmatrix},$$

then $a = \sigma\sigma^\top$ and σ is smooth. However, it can be shown that in general there is no smooth choice of σ even when a is smooth. The reason why stems from a result of D. Hilbert in classical algebraic geometry. Specifically, he showed that there are non-negative polynomials that cannot be written as the finite sum of squares of polynomials. By applying his result to the Taylor series for a, one can show that it rules out the possibility of always being able to find a smooth σ. Nonetheless, the following lemma shows that if all that one wants is Lipschitz continuity, then it suffices to know that a has two continuous derivatives.

Lemma 2.2.3. *Assume that a has two continuous derivatives, and let $K < \infty$ be a bound on the operator norm of its second derivatives. Then*

$$\|a^{\frac{1}{2}}(\mathbf{y}) - a^{\frac{1}{2}}(\mathbf{x})\|_{\mathrm{H.S.}} \leq N^{\frac{3}{2}} \sqrt{K} |\mathbf{y} - \mathbf{x}|.$$

Proof. The proof turns on a simple fact about functions $f : \mathbb{R} \longrightarrow [0, \infty)$ that have two continuous derivatives and whose second derivatives are bounded.

Namely,

$$(*) \qquad |f'(0)| \le \sqrt{2\|f''\|_{\mathrm{u}} f(0)}.$$

To prove this, use Taylor's theorem to write

$$0 \le f(h) \le f(0) + hf'(0) + \frac{h^2}{2}\|f''\|_{\mathrm{u}}.$$

Hence $|f'(0)| \le h^{-1}f(0) + \frac{h}{2}\|f''\|_{\mathrm{u}}$ for all $h > 0$, and so $(*)$ follows when one minimizes the right hand side with respect to h.

Turning to the stated result, first observe that it suffices to prove it when a is uniformly positive definite, since, if that is not already the case, one can replace a by $a + \epsilon\mathbf{I}$ and then let $\epsilon \searrow 0$. Assuming uniform positivity, we know that $a^{\frac{1}{2}}$ has two continuous derivatives, and what we need to show is that $|\partial_{x_k} a^{\frac{1}{2}}_{ij}| \le \sqrt{K}$. For this purpose, let $\mathbf{x}$ be given, and, without loss in generality, assume that $a(\mathbf{x})$ is diagonal. Then, because $a = a^{\frac{1}{2}}a^{\frac{1}{2}}$,

$$\big|\partial_{x_k} a_{ij}(\mathbf{x})\big| = \big|\partial_{x_k} a^{\frac{1}{2}}_{ij}(\mathbf{x})\big|\Big(\sqrt{a_{ii}(\mathbf{x})} + \sqrt{a_{jj}(\mathbf{x})}\Big) \ge \big|\partial_{x_k} a^{\frac{1}{2}}_{ij}(\mathbf{x})\big|\sqrt{a_{ii}(\mathbf{x}) + a_{jj}(\mathbf{x})},$$

and so

$$\big|\partial_{x_k} a^{\frac{1}{2}}_{ij}(\mathbf{x})\big| \le \frac{\big|\partial_{x_k} a_{ij}(\mathbf{x})\big|}{\sqrt{a_{ii}(\mathbf{x}) + a_{jj}(\mathbf{x})}}.$$

When $i = j$, apply $(*)$ to $f(h) = a_{ii}(\mathbf{x} + h\mathbf{e}_k)$, and conclude that

$$\big|\partial_{x_k} a_{ii}(\mathbf{x})\big| \le \sqrt{2K a_{ii}(\mathbf{x})},$$

which means that $\big|\partial_{x_k} a^{\frac{1}{2}}_{ii}(\mathbf{x})\big| \le \sqrt{K}$. When $i \ne j$, set

$$f_\pm(h) = a_{ii}(\mathbf{x} + h\mathbf{e}_k) \pm 2a_{ij}(\mathbf{x} + h\mathbf{e}_k) + a_{jj}(\mathbf{x} + h\mathbf{e}_k).$$

Then $f_\pm \ge 0$, and so, by $(*)$,

$$\big|\partial_{x_k} a_{ij}(\mathbf{x})\big| \le \frac{|f'_+(0)| + |f'_-(0)|}{4} \le \sqrt{K\big(a_{ii}(\mathbf{x}) + a_{jj}(\mathbf{x})\big)}.$$

Hence, once again, $\big|\partial_{x_k} a^{\frac{1}{2}}_{ij}(\mathbf{x})\big| \le \sqrt{K}$. □

2.3 Exercises

Exercise 2.1. Let $(\Omega, \mathcal{F}, \mathbb{P})$ be a probability space. A linear subspace $\mathfrak{G}$ of $L^2(\mathbb{P}; \mathbb{R})$ is called a **Gaussian family** if all its elements are Gaussian random variables.

In this exercise, if L is a closed, linear subspace of $L^2(\mathbb{P};\mathbb{R})$, then Π_L denotes the orthogonal projection operator onto L.

(i) If $\mathfrak{G}$ is a Gaussian family, show that its closure in $L^2(\mathbb{P};\mathbb{R})$ is again a Gaussian family. In addition, show that $\mathbf{1} \oplus \mathfrak{G}$ is a Gaussian family. Finally, if $\widetilde{\mathfrak{G}} = \{Y - \mathbb{E}^{\mathbb{P}}[Y] : Y \in \mathfrak{G}\}$, show that $\widetilde{\mathfrak{G}}$ is a centered Gaussian family.

(ii) Suppose that L is a closed linear subspace of a Gaussian family $\mathfrak{G}$ and that $X \in \mathfrak{G}$, and set $\tilde{L} = \{Y - \mathbb{E}^{\mathbb{P}}[Y] : Y \in L\}$ and $\widetilde{X} = X - \mathbb{E}^{\mathbb{P}}[X]$. Show that $\sigma(L) = \sigma(\tilde{L})$ and that $X - \Pi_{1\oplus L}X = \widetilde{X} - \Pi_{\tilde{L}}\widetilde{X}$. Conclude that $X - \Pi_{1\oplus L}X$ is independent of $\sigma(L)$ and therefore that $\Pi_{1\oplus L}X = \mathbb{E}^{\mathbb{P}}[X \mid \sigma(L)]$.

Exercise 2.2. Let $\big(B(t), \mathcal{F}_t, \mathbb{P}\big)$ be an $\mathbb{R}^N$-valued Brownian motion. Given $T > 0$, set $h_T(t) = \frac{t\wedge T}{T}$ and $\theta_T(t) = B(t) - h_T(t)B(T)$. Show that $B(T)$ is independent of $\sigma\big(\{\theta_T(t) : t \geq 0\}\big)$, and use this to show that, for any Borel measurable $\Phi\colon \mathcal{P}(\mathbb{R}^N) \longrightarrow \mathbb{R}$ which is bounded below and any $\Gamma \in \mathcal{B}_{\mathbb{R}^N}$,

$$\mathbb{E}^{\mathbb{P}}\big[\Phi \circ B, B(T) \in \Gamma\big] = \int_{\mathbb{R}^N} \mathbb{E}^{\mathbb{P}}\big[\Phi \circ \theta_{T,\mathbf{y}}\big]\, \gamma_{0,T}(d\mathbf{y}),$$

where $\theta_{T,\mathbf{y}}(t) = h_T(t)\mathbf{y} + \theta_T(t)$. Conclude that, if

$$\varphi_T(y) = \mathbb{E}^{\mathbb{P}}\big[\Phi \circ \theta_{T,\mathbf{y}}\big],$$

then $\varphi\big(B(T)\big)$ is the conditional expectation value of $\Phi \circ B$ given $B(T)$. Hence, $\theta_{T,\mathbf{y}}$ can be thought of as *Brownian motion pinned to* $\mathbf{y}$ *at time* T.

Exercise 2.3. Let $\big(B(t), \mathcal{F}_t, \mathbb{P}\big)$ be an $\mathbb{R}$-valued Brownian motion, and set

$$E_\xi(t) = e^{\xi B(t) - \frac{t\xi^2}{2}} \quad \text{for } t \geq 0 \text{ and } \xi \in \mathbb{R}.$$

Show that $\big(E_\xi(t), \mathcal{F}_t, \mathbb{P}\big)$ is a martingale, and use Doob's inequality to show that

$$\begin{aligned}\mathbb{P}\big(\|B(\,\cdot\,)\|_{[0,t]} \geq R\big) &\leq 2\mathbb{P}\left(\max_{\tau\in[0,t]} B(\tau) \geq R\right)\\ &= 2\mathbb{P}\left(\max_{\tau\in[0,t]} E_\xi(\tau) \geq e^{\xi R - \frac{t\xi^2}{2}}\right) \leq 2e^{-\xi R + \frac{t\xi^2}{2}}\end{aligned}$$

for all $\xi \geq 0$, and conclude from this that

$$\mathbb{P}\big(\|B(\,\cdot\,)\|_{[0,t]} \geq R\big) \leq 2e^{-\frac{R^2}{2t}}.$$

Next, let $\big(B(t), \mathcal{F}_t, \mathbb{P}\big)$ be an $\mathbb{R}^N$-valued Brownian motion, and use the preceding to show that

$$\mathbb{P}\big(\|B(\,\cdot\,)\|_{[0,t]} \geq R\big) \leq 2Ne^{-\frac{R^2}{2Nt}}. \tag{2.3.1}$$

Exercise 2.4. Let $(B(t), \mathcal{F}_t, \mathbb{P})$ be an $\mathbb{R}^N$-valued Brownian motion, and use the Brownian time inversion property to give another proof that $\lim_{t\to\infty} \frac{B(t)}{t} = \mathbf{0}$ (a.s., $\mathbb{P}$). Next, given an $A \in \mathrm{Hom}(\mathbb{R}^N;\mathbb{R}^N)$, define $T_A\colon \mathbb{W}(\mathbb{R}^N) \longrightarrow \mathbb{W}(\mathbb{R}^N)$ so that $T_A w(t) = Aw(t)$. Show that $(T_A)_*\mathcal{W} = \mathcal{W}$ if A is an orthogonal transformation and that $(T_A)_*\mathcal{W} \perp \mathcal{W}$ otherwise. In particular, conclude that an orthogonal transformation of a Brownian motion is again a Brownian motion.

Exercise 2.5. Let $(\Omega, \mathcal{F}, \mathbb{P})$ be a probability space and $\{\mathcal{F}_t : t \geq 0\}$ a filtration of sub-σ-algebras.

(i) If $X\colon [0,\infty)\times\Omega \longrightarrow \mathbb{R}^N$ has the properties that $t \rightsquigarrow X(t,\omega)$ is continuous for $\mathbb{P}$-almost every ω and $X(t,\,\cdot\,)$ is $\mathcal{F}_t$-measurable for each $t \geq 0$, show that $(X(t), \mathcal{F}_t, \mathbb{P})$ is a Brownian motion if and only if $\left(e^{i(\boldsymbol{\xi}, X(t))_{\mathbb{R}^N} + \frac{|\boldsymbol{\xi}|^2}{2}t}, \mathcal{F}_t, \mathbb{P}\right)$ is a martingale for all $\boldsymbol{\xi} \in \mathbb{R}^N$.

(ii) Let $(B(t), \mathcal{F}_t, \mathbb{P})$ be an $\mathbb{R}^N$-valued Brownian motion and ζ a bounded stopping time. As an application of (i) and Hunt's stopping time theorem[4] show that $(B(t+\zeta)-B(\zeta), \mathcal{F}_{t+\zeta}, \mathbb{P})$ is a Brownian motion that is independent of $\mathcal{F}_\zeta$.

(iii) Let ζ be a stopping time, and set

$$\breve{B}(t) = B(t\wedge\zeta) - \big(B(t) - B(t\wedge\zeta)\big) = 2B(t\wedge\zeta) - B(t).$$

In other words, $\breve{B}(\,\cdot\,)$ is the path obtained by *reflecting* $B(\,\cdot\,)$ at time ζ. Show that $(\breve{B}(t), \mathcal{F}_t, \mathbb{P})$ is a Brownian motion. This is known as the **reflection principle** for Brownian motion.

(iv) Assume that $N = 1$, and, for $R > 0$, define

$$\zeta_R = \inf\{t \geq 0 : B(t) \geq R\} \quad \text{and} \quad \breve{B}_R(t) = 2B(t\wedge\zeta_R) - B(t),$$

and observe that $\zeta_R = \breve{\zeta}_R := \inf\{t \geq 0 : \breve{B}_R(t) \geq R\}$. Obviously,

$$\mathbb{P}\big(B(t) > a \;\&\; \zeta_R < t\big) = \mathbb{P}\big(B(t) > a\big)$$

if $a > R$. Show that if $a \leq R$, then

$$\mathbb{P}\big(B(t) < a \;\&\; \zeta_R < t\big) = \mathbb{P}\big(\breve{B}_R(t) < a \;\&\; \breve{\zeta}_R < t\big) = \mathbb{P}\big(B(t) > 2R - a\big),$$

and from these conclude that

$$\mathbb{P}(\zeta_R < t) = \mathbb{P}(\zeta_R \leq t) = 2\mathbb{P}\big(B(t) > R\big)$$

and

$$\mathbb{P}\big(B(t) \leq a \;\&\; \zeta_R > t\big) = \mathbb{P}\big(B(t) \leq a\big) - \mathbb{P}\big(B(t) > 2R - a\big) \quad \text{for } a < R.$$

---

[4] This is the statement that if $\zeta_1 \leq \zeta_2$ are bounded stopping times and $(M(t), \mathcal{F}_t, \mathbb{P})$ is a martingale, then $\mathbb{E}^{\mathbb{P}}[X(\zeta_2)|\mathcal{F}_{\zeta_1}] = X(\zeta_1)$ (a.s., $\mathbb{P}$). See Theorem 7.1.14 in [20].

(v) Continuing in the setting of (iv), show that $\mathbb{P}(\zeta_R < \infty) = 1$ and that

$$d\mathbb{P}(\zeta_R \le t) = \mathbf{1}_{(0,\infty)}(t)\frac{R}{\sqrt{2\pi t^3}}e^{-\frac{R^2}{2t}}\,dt.$$

Next, use Doob's stopping time theorem (cf. Theorem 7.1.15 in [20]) to show that

$$\mathbb{E}^{\mathbb{P}}\big[e^{-\frac{\lambda^2}{2}\zeta_R}\big] = e^{-\lambda R} \quad \text{for } \lambda > 0,$$

and use this to recover the result

$$\int_0^\infty t^{-\frac{3}{2}}e^{-\frac{\lambda^2 t}{2}-\frac{R^2}{2t}}\,dt = (2\pi)^{\frac{1}{2}}R^{-1}e^{-\lambda R}$$

obtained in Exercise 1.4.

(vi) Let $\big(B(t), \mathcal{F}_t, \mathbb{P}\big)$ be an $\mathbb{R}^{N+1}$-valued Brownian motion. Given $a > 0$, set

$$\zeta = \inf\{t \ge 0 : B(t)_{N+1}(t) \ge a\},$$

note that $\mathbb{P}(\zeta \le t) = 2\mathbb{P}\big(B(t)_{N+1}(t) \ge a\big)$, and define

$$X = \big(B(\zeta)_1, \dots, B(\zeta)_N\big) \text{ on } \{\zeta < \infty\} \quad \text{and} \quad X = \mathbf{0} \text{ on } \{\zeta = \infty\}.$$

Using (v) and Exercise 1.4, show that

$$\mathbb{P}\big(X \in \Gamma\big) = \frac{2a}{\omega_N}\int_\Gamma \big(a^2 + |\mathbf{y}|^2\big)^{-\frac{N+1}{2}}\,d\mathbf{y} \quad \text{for } \Gamma \in \mathcal{B}_{\mathbb{R}^N}.$$

In particular, conclude that, with probability 1, a Brownian motion will eventually escape any half space and therefore any bounded set.

(vii) Using Exercise 2.4 and part (vi), show that if

$$\zeta^{B(\mathbf{0},r)} = \inf\{t \ge 0 : |w(t)| \ge r\},$$

then $\mathcal{W}(\zeta^{B(\mathbf{0},r)} < \infty) = 1$ and

$$\mathcal{W}\big(w(\zeta^{B(\mathbf{0},r)}) \in \Gamma\big) = \frac{1}{\omega_{N-1}}\int_{\mathbb{S}^{N-1}} \mathbf{1}_\Gamma(r\omega)\,\lambda_{\mathbb{S}^{N-1}}(d\omega) \quad \text{for } \Gamma \in \mathcal{B}_{\partial B(\mathbf{0},r)},$$

where $\lambda_{\mathbb{S}^{N-1}}$ is surface measure on the unit sphere

$$\mathbb{S}^{N-1} = \{\mathbf{y} \in \mathbb{R}^N : |\mathbf{y}| = 1\}$$

and $\omega_{N-1} = \lambda_{\mathbb{S}^{N-1}}(\mathbb{S}^{N-1})$.

Exercise 2.6. Let $\big(B(t), \mathcal{F}_t, \mathbb{P}\big)$ be an $\mathbb{R}$-valued Brownian motion. Using (2.1.2), show that, $\mathbb{P}$-almost surely,

$$\lim_{n\to\infty}\sum_{m=0}^{2^n-1} B(m2^{-n})\big(B((m+1)2^{-n})-B(m2^{-n})\big)=\frac{B(1)^2-1}{2}$$

$$\lim_{n\to\infty}\sum_{m=0}^{2^n-1} B\big((2m+1)2^{-n-1}\big)\big(B((m+1)2^{-n})-B(m2^{-n})\big)=\frac{B(1)^2}{2}$$

$$\lim_{n\to\infty}\sum_{m=0}^{2^n-1} B\big((m+1)2^{-n}\big)\big(B((m+1)2^{-n})-B(m2^{-n})\big)=\frac{B(1)^2+1}{2}.$$

These provide further evidence that Brownian paths have unbounded variation.

Chapter 3
Brownian Stochastic Integration

In §2.2 we used Itô's idea to construct a coupling of the measures $\mu_{t,n}$ in (1.2.12). As most readers will have realized, our treatment would have been less cumbersome if we had a notion of integration that allowed me to replace the prescription in (2.2.1) by

$$X_n(t,\mathbf{x}) = \mathbf{x} + \int_0^t \sigma\big(X_n(\lfloor\tau\rfloor_n,\mathbf{x})\big)\,dw(\tau) + \int_0^t b\big(X_n(\lfloor\tau\rfloor_n,\mathbf{x})\big)\,d\tau \qquad (3.0.1)$$

and allowed me to show that such an expression converges to an expression like

$$X(t,\mathbf{x}) = \mathbf{x} + \int_0^t \sigma\big(X(\tau,\mathbf{x})\big)\,dw(\tau) + \int_0^t b\big(X(\tau,\mathbf{x})\big)\,d\tau. \qquad (3.0.2)$$

However, because, as we showed at the end of §2.1.2 and in Exercise 2.6, Brownian paths do not have locally bounded variation, standard notions of integration are not going to suffice.

The first authors to perform integrals with respect to Brownian paths were R.C. Paley and Wiener, but they restricted their attention to non-random integrands. It was Itô who first figured out how to extend their theory to cover integrands that are random.

3.1 The Paley–Wiener integral

Recall (cf. §1.2 in [19]) that if $f\colon [0,\infty) \longrightarrow \mathbb{C}$ is continuous and $g\colon [0,\infty)\longrightarrow\mathbb{C}$ has locally bounded variation, then f and g are Riemann-Stieltjes integrable with respect to each other and

$$\int_s^t g(\tau)\,df(\tau) = f(t)g(t) - f(s)g(s) - \int_s^t f(\tau)dg(\tau),$$

D. W. Stroock, *Elements of Stochastic Calculus and Analysis*,
CRM Short Courses, https://doi.org/10.1007/978-3-319-77038-3_3

from which it follows that $\int_s^t g(\tau)\,df(\tau)$ is a continuous function of $t > s$.

Assume that $\big(B(t), \mathcal{F}_t, \mathbb{P}\big)$ is an $\mathbb{R}^M$-valued Brownian motion on a probability space $(\Omega, \mathcal{F}, \mathbb{P})$, and, without loss in generality, assume that the $\mathcal{F}_t$'s are complete and that $t \rightsquigarrow B(t)(\omega)$ is continuous for *all* $\omega \in \Omega$. Given a function $\boldsymbol{\eta}\colon [0,\infty) \longrightarrow \mathbb{R}^M$ of locally bounded variation, set

$$I_{\boldsymbol{\eta}}(t) = \int_0^t \big(\boldsymbol{\eta}(\tau), dB(\tau)\big)_{\mathbb{R}^M},$$

where the integral is taken in the sense of Riemann–Stieltjes. In particular, if $\boldsymbol{\eta}_n(t) = \boldsymbol{\eta}(\lfloor t \rfloor_n)$, then $I_{\boldsymbol{\eta}_n}(t) \longrightarrow I_{\boldsymbol{\eta}}(t)$. In addition, because

$$I_{\boldsymbol{\eta}_n}(t) = \sum_{m<2^n t} \Big(\boldsymbol{\eta}(m2^{-n}), B\big((m+1)2^{-n} \wedge t\big) - B(m2^{-n})\Big)_{\mathbb{R}^M},$$

for all $0 \le s < t$, the increment $I_{\boldsymbol{\eta}_n}(t) - I_{\boldsymbol{\eta}_n}(s)$ is an $\mathcal{F}_t$-measurable, centered Gaussian random variable that is independent of $\mathcal{F}_s$ and has variance $\int_s^t |\boldsymbol{\eta}(\lfloor \tau \rfloor_n)|^2\,d\tau$. Hence $I_{\boldsymbol{\eta}}(t) - I_{\boldsymbol{\eta}}(s)$ is an $\mathcal{F}_t$-measurable, centered Gaussian random variable that is independent of $\mathcal{F}_s$ and, because η has at most countably many discontinuities, has variance $\int_s^t |\boldsymbol{\eta}(\tau)|^2\,d\tau$. In particular,

$$\big(I_{\boldsymbol{\eta}}(t), \mathcal{F}_t, \mathbb{P}\big), \quad \left(I_{\boldsymbol{\eta}}(t)^2 - \int_0^t |\boldsymbol{\eta}(\tau)|^2\,d\tau, \mathcal{F}_t, \mathbb{P}\right), \tag{3.1.1}$$

and, for each $\zeta \in \mathbb{C}$,

$$\left(\exp\left(\zeta I_{\boldsymbol{\eta}}(t) - \frac{\zeta^2}{2}\int_0^t |\boldsymbol{\eta}(\tau)|^2\,d\tau\right), \mathcal{F}_t, \mathbb{P}\right) \tag{3.1.2}$$

are all martingales. Thus, by Doob's inequality,

$$\mathbb{E}^{\mathbb{P}}\left[\|I_{\boldsymbol{\eta}}(\,\cdot\,)\|_{[0,t]}^2\right] \le 4 \sup_{t\ge 0} \mathbb{E}^{\mathbb{P}}\big[I_{\boldsymbol{\eta}}(t)^2\big] = 4\int_0^\infty |\boldsymbol{\eta}(\tau)|^2\,d\tau. \tag{3.1.3}$$

Before taking the next step, here is an example that demonstrates the potential applicability of these considerations. Take $N = 1$, and consider the integral equation

$$(*) \qquad X(t,x) = x + B(t) - \int_0^t X(s,x)\,ds.$$

Since the difference $D(t)$ between two solutions of this equation will satisfy

$$D(t) = -\int_0^t D(\tau)\,d\tau,$$

Lemma 1.2.4 shows that there is at most one solution. Next set $X(t,x) = e^{-t}x + \int_0^t e^{s-t}\,dB(s)$. Then

$$\begin{aligned}\int_0^t X(\tau,x)\,d\tau &= (1-e^{-t})x + \int_0^t \left(\int_s^t e^{s-\tau}\,d\tau\right) dB(s)\\ &= (1-e^{-t})x + \int_0^t \left(1-e^{s-t}\right) dB(s) = x - X(t,x) + B(t).\end{aligned}$$

Hence this $X(\,\cdot\,,x)$ is the one and only solution to $(*)$. In particular, since

$$\int_0^t e^{2(\tau-t)}\,d\tau = \frac{1-e^{-2t}}{2},$$

the solution to $(*)$ at time t is a Gaussian random variable with mean $e^{-t}x$ and variance $\frac{1-e^{-2t}}{2}$. Further, observe that

$$X(s+t,x) = e^{-t}X(s,x) + \int_s^{s+t} e^{\tau-s-t}\,dB(\tau)$$

and that the second term on the right is independent of the first and has the same distribution as $X(t,0)$. Therefore,

$$\mathbb{E}^{\mathbb{P}}\big[\varphi(X(s+t)) \mid \mathcal{F}_s\big] = \int_{\mathbb{R}} \varphi(y)\,P\big(t,X(s,x),dy\big),$$

where

$$P(t,x,dy) = g\left(\frac{1-e^{-2t}}{2}, y - e^{-t}x\right) dy$$

and $g(t,y)$ is the heat kernel $(2\pi t)^{-\frac{1}{2}}e^{-\frac{y^2}{2}}$. Hence, $X(\,\cdot\,,x)$ is a Markov process whose transition function is the solution to the Ornstein–Uhlenbeck equation in Exercise 1.1. For this reason, $\{X(t,x) : t \geq 0\}$ is known as the **Ornstein–Uhlenbeck process**.

Given (3.1.3), it is easy to define $I_{\boldsymbol{\eta}}$ for all square-integrable $\boldsymbol{\eta}\colon [0,\infty) \longrightarrow \mathbb{R}^M$. Indeed, if $\boldsymbol{\eta}$ is such an function, choose $\{\boldsymbol{\eta}_k : k \geq 1\} \subseteq C^\infty([0,\infty);\mathbb{R}^M)$ so that

$$\int_0^\infty |\boldsymbol{\eta}_k(\tau) - \boldsymbol{\eta}(\tau)|^2\,d\tau \longrightarrow 0.$$

Then, by (3.1.3),

$$\mathbb{E}^{\mathbb{P}}\left[\|I_{\boldsymbol{\eta}_\ell}(\,\cdot\,) - I_{\boldsymbol{\eta}_k}(\,\cdot\,)\|^2_{[0,t]}\right] = \mathbb{E}^{\mathbb{P}}\left[\|I_{\boldsymbol{\eta}_\ell-\boldsymbol{\eta}_k}(\,\cdot\,)\|^2_{[0,t]}\right] \leq 4\int_0^\infty |\boldsymbol{\eta}_\ell(\tau)-\boldsymbol{\eta}_k(\tau)|^2\,d\tau,$$

and so there exists a measurable function $I_{\boldsymbol{\eta}}\colon \Omega \longrightarrow C([0,\infty);\mathbb{R})$, known as the **Paley–Wiener integral** of $\boldsymbol{\eta}$, such that $I_{\boldsymbol{\eta}}(t)$ is $\mathcal{F}_t$ measurable for each $t \geq 0$ and

$$\mathbb{E}^{\mathbb{P}}\left[\sup_{t\geq 0}|I_\eta(t) - I_{\eta_k}(t)|^2\right] \leq 4\int_0^\infty |\eta(\tau) - \eta_k(\tau)|^2\,d\tau \longrightarrow 0.$$

Clearly, apart from a $\mathbb{P}$-null set, I_η does not depend on the choice of the approximating sequence and is therefore a linear function of η. Moreover, for each $t \geq 0$, $I_\eta(t)$ is a centered Gaussian random variable with variance $\int_0^t |\eta(\tau)|^2\,d\tau$, the expressions in (3.1.1) and (3.1.2) are martingales, and (3.1.3) continues to hold. What is no longer true is that $I_\eta(t)(\omega)$ is given by a Riemann–Stieltjes integral or that it is even defined ω by ω. Nevertheless, we will use

$$\int_0^t \big(\eta(\tau), dB(\tau)\big)_{\mathbb{R}^M}$$

to denote $I_\eta(t)$ even though in general it is not a Riemann–Stieltjes integral.

3.1.1 *The Cameron–Martin–Segal theorem*

One of the most important applications of the ideas discussed in the preceding has its origins in the early work of R. Cameron and W. Martin and reached its final form in the work of I. Segal.

To fully appreciate what they did, one should know about the following theorem of V. Sudakov.

Theorem 3.1.1. *Let E be an infinite dimensional, separable Banach space over $\mathbb{R}$, and, for $a \in E$, define the translation map $T_a\colon E \longrightarrow E$ by $T_a x = x + a$. Given a Borel probability measure μ on E, there is a dense subset D such that $(T_a)_*\mu \perp \mu$ for all $a \in D$.*

To prove this theorem, we will need an elementary fact about compact subsets of infinite dimensional Banach spaces.

Lemma 3.1.2. *If K_1 and K_2 are compact subsets of E, then* $\operatorname{int}(K_2 - K_1) = \varnothing$.

Proof. Since $K_2 - K_1$, as the image of the compact set $K_1 \times K_2$ under the continuous map $(x_1, x_2) \rightsquigarrow x_1 - x_2$, is compact if K_1 and K_2 are compact, it suffices for us to prove that $\operatorname{int}(K) = \varnothing$ for every compact K.

Given K and $r > 0$, we will show that there is an $a \in E$ such that $\|a\| = r$ and $a \notin K - K$. For that purpose, choose $x_1, \dots, x_n \in K$ so that $K \subseteq \bigcup_{m=1}^n B_E\big(x_m, \frac{r}{4}\big)$. Because E is infinite dimensional, the Hahn–Banach theorem guarantees that there is an $a^* \in E^*$ such that $\|a^*\|_{E^*} = 1$ and[1] $\langle x_m, a^*\rangle = 0$ for $1 \leq m \leq n$. Now choose $a \in E$ so that $\|a\|_E = r$ and $\langle a, a^*\rangle \geq \frac{3r}{4}$. If x were in $K \cap (a + K)$, then there would exist $1 \leq m, m' \leq n$

[1] Here $\langle x, x^*\rangle$ denotes the action of $x^* \in E^*$ on $x \in E$.

such that $\|x - x_m\|_E < \frac{r}{4}$ and $\|x - a - x_{m'}\|_E < \frac{r}{4}$, which would lead to the contradiction that

$$\langle x, a^* \rangle = \langle x - x_m, a^* \rangle < \frac{r}{4} \text{ and } \langle x, a^* \rangle = \langle a, a^* \rangle + \langle x - a - x_{m'}, a^* \rangle \geq \frac{r}{2}.$$

Thus $K \cap (a + K) = \emptyset$, and so $a \notin K - K$, which means that $K - K$ contains no non-empty ball centered at the origin. Finally, if $K \supseteq B_E(b, r)$ for some $b \in E$ and $r > 0$, then $B_E(0, r) \subseteq K - K$, which cannot be. □

Proof of Theorem 3.1.1. Let μ be a Borel probability measure on E. By Ulam's lemma (cf. Lemma 9.1.7 in [20]), for each $m \geq 1$ there exists a compact set K_m such that $\nu(K_m) \geq 1 - \frac{1}{m}$. Set $A = \bigcup_{m=1}^{\infty} K_m$ and

$$B = A - A = \bigcup_{m,n=1}^{\infty} (K_n - K_m).$$

Obviously, $\mu(A) = 1$. At the same time, Lemma 3.1.2 says that

$$\operatorname{int}(K_n - K_m) = \emptyset \quad \text{for all } m, n \geq 1,$$

and so, by the Baire category theorem, $\operatorname{int}(B) = \emptyset$ and therefore $B\complement$ is dense. Now suppose that $a \notin B$. Then $A \cap (a + A) = \emptyset$, and so $(T_{-a}\mu)_*(A) = \mu(a + A) = 0$. Hence, since $a \notin B \iff -a \notin B$, we have shown that $(T_a)_*\mu \perp \mu$ for all $a \in D = B\complement$. □

As Theorem 3.1.1 makes clear, translation of a Borel probability measure on an infinite dimensional Banach space in most directions will produce a measure that is singular to the original one, and it is not obvious that there are any measures on an infinite dimensional Banach that are quasi-invariant under translation in any, much less a dense set, of directions. Thus it was a significant discovery when Cameron and Martin showed that Wiener measure is quasi-invariant under translation by paths h in the Hilbert space $H^1(\mathbb{R}^M) \subseteq \mathbb{W}(\mathbb{R}^M)$ with norm

$$\|h\|_{H^1(\mathbb{R}^M)} = \|\dot{h}\|_{L^2([0,\infty);\mathbb{R}^M)}$$

of absolutely continuous paths whose derivatives are square integrable. Somewhat later, Segal showed that translates of Wiener measure by any other paths are singular.

Theorem 3.1.3. *Given $h \in H^1(\mathbb{R}^M)$, set*

$$I(\dot{h}) = \int_0^{\infty} \big(\dot{h}(\tau), dw(\tau)\big)_{\mathbb{R}^M} \quad \text{and} \quad R_h = \exp\big(I(\dot{h}) - \tfrac{1}{2}\|h\|^2_{H^1(\mathbb{R}^M)}\big).$$

Then $(T_h)_*\mathcal{W} \ll \mathcal{W}$ and $\frac{d(T_h)_*\mathcal{W}}{d\mathcal{W}} = R_h$ for all $h \in H^1(\mathbb{R}^M)$. On the other hand, if $f \in \mathbb{W}(\mathbb{R}^M) \setminus H^1(\mathbb{R}^M)$, then $(T_f)_*\mathcal{W} \perp \mathcal{W}$.

Proof. To prove the first part, let $h \in H^1(\mathbb{R}^M)$, $n \geq 1$, $0 \leq t_1 < \cdots < t_n$, and $\boldsymbol{\xi}_1, \ldots, \boldsymbol{\xi}_n \in \mathbb{R}^M$ be given. Clearly

$$\mathbb{E}^{(T_h)_*\mathcal{W}}\left[e^{i\sum_{m=1}^n (\boldsymbol{\xi}_m, w(t_m))_{\mathbb{R}^M}}\right] = \exp\left(i\sum_{m=1}^n (\boldsymbol{\xi}_m, h(t_m))_{\mathbb{R}^M} - \frac{1}{2}\sum_{m,m'=1}^n (t_m \wedge t_{m'})(\boldsymbol{\xi}_m, \boldsymbol{\xi}_{m'})_{R^M}\right).$$

To compute the same integral with respect to $R_h d\mathcal{W}$, set $h_m(t) = (t \wedge t_m)\boldsymbol{\xi}_m$ and $f(t) = \dot{h}(t) + i\sum_{m=1}^n \dot{h}_m(t)$. Then

$$R_h(w)e^{i\sum_{m=1}^n (\boldsymbol{\xi}_m, w(t_m))_{\mathbb{R}^M}} = \exp\left(\int_0^\infty (f(\tau), dw(\tau))_{\mathbb{R}^M} - \frac{1}{2}\|h\|^2_{H^1(\mathbb{R}^M)}\right),$$

and so (cf. Exercise 3.2)

$$\begin{aligned}
&\mathbb{E}^{\mathcal{W}}\left[R_h(w)e^{i\sum_{m=1}^n (\boldsymbol{\xi}_m, w(t_m))_{\mathbb{R}^M}}\right] \\
&= \exp\left(\frac{1}{2}\sum_{j=1}^M \int_0^\infty f(\tau)_j^2\, d\tau - \frac{1}{2}\|h\|^2_{H^1(\mathbb{R}^M)}\right) \\
&= \exp\left(i\sum_{m=1}^n (h_m, h)_{H^1(\mathbb{R}^M)} - \frac{1}{2}\sum_{m,m'=1}^n (h_m, h_{m'})_{H^1(\mathbb{R}^M)}\right) \\
&= \exp\left(i\sum_{m=1}^n (\boldsymbol{\xi}_m, h(t_m))_{\mathbb{R}^M} - \frac{1}{2}\sum_{m,m'=1}^n (t_m \wedge t_{m'})(\boldsymbol{\xi}_m, \boldsymbol{\xi}_{m'})_{R^M}\right).
\end{aligned}$$

Hence $d(T_h)_*\mathcal{W} = R_h d\mathcal{W}$.

The proof of the second assertion requires some preparations. Let L denote the subspace of $H^1(\mathbb{R}^M)$ consisting of twice continuously differentiable functions whose first derivatives have compact support. Clearly L is dense in $H^1(\mathbb{R}^M)$, and so one can find an orthonormal basis $\{h_m : m \geq 1\} \subseteq L$ for $H^1(\mathbb{R}^M)$. Now define the linear functional Λ on L by

$$\Lambda(h) = -\int_0^\infty (f(\tau), \ddot{h}(\tau))_{\mathbb{R}^M}\, d\tau.$$

We need to show that $\sum_{m=1}^\infty \Lambda(h_m)^2 = \infty$. To this end, suppose that $C = \sqrt{\sum_{m=1}^\infty \Lambda(h_m)^2} < \infty$. Then, if h is in the span of $\{h_m : m \geq 1\}$,

$$|\Lambda(h)| \leq \sum_{m=1}^\infty |\Lambda(h_m)|\left|(h, h_m)_{H^1(\mathbb{R}^M)}\right| \leq C\|h\|_{H^1(\mathbb{R}^M)},$$

and so Λ would admit a unique extension as a continuous linear functional on $H^1(\mathbb{R}^M)$. Thus, by the Riesz representation theorem for Hilbert space, there would exist an $h_0 \in H^1(\mathbb{R}^M)$ such that $\Lambda(h) = \big(h, h_0\big)_{H^1(\mathbb{R}^M)}$, and so we would have that

$$\int_0^\infty \big(f(\tau), \ddot{h}(\tau)\big)_{\mathbb{R}^M}\, d\tau = \int_0^\infty \big(h_0(\tau), \ddot{h}(\tau)\big)_{\mathbb{R}^M}\, d\tau \quad \text{for all } h \in L.$$

Now choose $\rho \in C^\infty(\mathbb{R};\mathbb{R})$ so that $\rho = 0$ off of $(0,1)$ and $\int \rho(\tau)\, d\tau = 1$. Given $t > 0$, set $\rho_\epsilon(\tau) = \epsilon^{-1}\rho(\epsilon^{-1}\tau)$ and

$$\psi_\epsilon(\tau) = \int_0^\tau \left(\int_0^{\tau_1} \rho_\epsilon(t - \tau_2)\, d\tau_2\right) d\tau_1 - \tau$$

for $0 < \epsilon < t$. Then $\dot{\psi}_\epsilon = 0$ off $[0,t)$ and $\ddot{\psi}_\epsilon(\tau) = \rho_\epsilon(t-\tau)$. Hence, for $\boldsymbol{\xi} \in \mathbb{R}^N$, $\psi_\epsilon\boldsymbol{\xi} \in L$ and

$$\begin{aligned}
\big(\boldsymbol{\xi}, f(t)\big)_{\mathbb{R}^M} &= \lim_{\epsilon \searrow 0} \int_0^\infty \big(f(\tau), \rho_\epsilon(t-\tau)\boldsymbol{\xi}\big)_{\mathbb{R}^M}\, d\tau = -\lim_{\epsilon \searrow 0} \Lambda(\psi_\epsilon\boldsymbol{\xi}) \\
&= -\lim_{\epsilon \searrow 0} \big(\psi_\epsilon\boldsymbol{\xi}, h_0\big)_{H^1(\mathbb{R}^M)} = \lim_{\epsilon \searrow 0} \int_0^\infty \big(h_0(\tau), \rho_\epsilon(t-\tau)\boldsymbol{\xi}\big)_{\mathbb{R}^M}\, d\tau = \big(\boldsymbol{\xi}, h_0(t)\big)_{\mathbb{R}^M},
\end{aligned}$$

which leads to the contradiction $f = h_0 \in H^1(\mathbb{R}^M)$.

With this information, we can complete the proof as follows. Define $F\colon \mathbb{W}(\mathbb{R}^M) \longrightarrow \mathbb{R}^{\mathbb{Z}^+}$ so that

$$F(w)_m = \int_0^\infty \big(\dot{h}_m(\tau), dw(\tau)\big)_{\mathbb{R}^M} = -\int_0^\infty \big(w(\tau), \ddot{h}_m(\tau)\big)_{\mathbb{R}^M}\, d\tau.$$

Then $F_*\mathcal{W} = \gamma_{0,1}^{\mathbb{Z}^+}$, and, because $F(w+f)_m = F(w)_m + \Lambda(h_m)$,

$$F_*\big((T_f)_*\mathcal{W}\big) = \prod_{m=1}^\infty \gamma_{\Lambda(h_m),1}.$$

Since (cf. Exercise 5.2.42 in [20]) $\gamma_{0,1}^{\mathbb{Z}^+} \perp \prod_{m=1}^\infty \gamma_{a_m,1}$ if $\sum_{m=1}^\infty a_m^2 = \infty$, it follows that $F_*\big((T_f)_*\mathcal{W}\big) \perp F_*\mathcal{W}$, which means that $(T_f)_*\mathcal{W} \perp \mathcal{W}$. □

Among other things, Theorem 3.1.3 allows us to show that $\mathcal{W}$ gives positive measure to every open subset of $\mathbb{W}(\mathbb{R}^M)$. To see this, first observe that, because $H^1(\mathbb{R}^M)$ is dense in $\mathbb{W}(\mathbb{R}^M)$, it suffices to show that $\mathcal{W}\big(B_{\mathbb{W}(\mathbb{R}^N)}(h,r)\big) > 0$ for every $h \in H^1(\mathbb{R}^M)$ and $r > 0$. Second, note that

$$\begin{aligned}
\mathcal{W}\big(B_{\mathbb{W}(\mathbb{R}^N)}(0,r)\big) &= \mathbb{E}^{\mathcal{W}}\big[R_{-h}^{-\frac12} R_{-h}^{\frac12}, B_{\mathbb{W}(\mathbb{R}^N)}(0,r)\big] \\
&\le \mathbb{E}^{\mathcal{W}}\big[R_{-h}^{-1}\big]^{\frac12} (T_{-h})_*\mathcal{W}\big(B_{\mathbb{W}(\mathbb{R}^N)}(0,r)\big)^{\frac12} = e^{\frac{\|h\|^2_{H^1(\mathbb{R}^M)}}{2}} \mathcal{W}\big(B_{\mathbb{W}(\mathbb{R}^N)}(h,r)\big)^{\frac12},
\end{aligned}$$

and therefore it suffices to show that $\mathcal{W}\big(B_{\mathbb{W}(\mathbb{R}^N)}(0,r)\big) > 0$ for all $r > 0$. To this end, consider the function

$$u(t,x) = e^{\frac{\pi^2 t}{8r^2}} \sin \frac{\pi(x+r)}{2r},$$

and observe that $\partial_t u + \frac{1}{2}\partial_x^2 u = 0$. Thus, by (2.2.6) applied with $M = 1$, $\sigma = 1$, and $b = 0$, $\big(u(t, w(t)), \mathcal{W}_t, \mathcal{W}\big)$ is a martingale, and therefore, by Doob's stopping time theorem, if $\zeta_r = \inf\{t \geq 0 : |w(t)| \geq r\}$, then, because $u(t, \pm r) = 0$,

$$1 = \mathbb{E}^{\mathcal{W}}\big[u(t\wedge\zeta_r, w(t\wedge\zeta_r))\big] = e^{\frac{\pi^2 t}{8r^2}}\mathbb{E}^{\mathcal{W}}\left[\sin\frac{\pi(w(t)+r)}{2r}, \zeta_r > t\right],$$

and so

$$\mathcal{W}(\zeta_r > t) \geq e^{-\frac{\pi^2 t}{8r^2}}.$$

For general M, the preceding implies that

$$\begin{aligned}\mathcal{W}(\|w(\,\cdot\,)\|_{[0,T]} < r) &\geq \mathcal{W}\left(\max_{1\leq j\leq M}\|w(\,\cdot\,)_j\|_{[0,T]} < \frac{r}{M^{\frac{1}{2}}}\right)\\ &= \mathcal{W}\big(\zeta_{M^{-\frac{1}{2}}r} > T\big)^M \geq e^{-\frac{M^2\pi^2 T}{8r^2}},\end{aligned}$$

and therefore that, and any $T > 0$,

$$\begin{aligned}\mathcal{W}\big(B_{\mathbb{W}(\mathbb{R}^N)}(0,r)\big) &\geq \mathcal{W}\left(\|w(\,\cdot\,)\|_{[0,T]} < \frac{r}{2} \ \&\ \sup_{t\geq T}\frac{|w(t)-w(T)|}{1+t} \leq \frac{r}{2}\right)\\ &\geq e^{-\frac{M^2\pi^2 T}{2r^2}}\mathcal{W}\left(\sup_{t\geq 0}\frac{|w(t)|}{1+t+T} \leq \frac{r}{2}\right).\end{aligned}$$

Finally, because $\|w\|_{\mathbb{W}(\mathbb{R}^M)} < \infty$ and $\frac{|w(t)|}{1+t} \longrightarrow 0$, we can choose $T > 0$ so that

$$\mathcal{W}\left(\sup_{t\in[0,\sqrt{T}]}\frac{|w(t)|}{1+t+T} \geq \frac{r}{2}\right) \vee \mathcal{W}\left(\sup_{t\geq\sqrt{T}}\frac{|w(t)|}{1+t} \geq \frac{r}{2}\right) \leq \frac{1}{4},$$

in which case

$$\begin{aligned}&\mathcal{W}\left(\sup_{t\geq 0}\frac{|w(t)|}{1+t+T} \geq \frac{r}{2}\right)\\ &\quad\leq \mathcal{W}\left(\sup_{t\in[0,\sqrt{T}]}\frac{|w(t)|}{1+t+T} \geq \frac{r}{2}\right) + \mathcal{W}\left(\sup_{t\geq\sqrt{T}}\frac{|w(t)|}{1+t} \geq \frac{r}{2}\right) \leq \frac{1}{2}.\end{aligned}$$

We have therefore shown that the **support of** $\mathcal{W}$ is the whole of $\mathbb{W}(\mathbb{R}^N)$.

Besides what it says about Wiener measure, the preceding result has the following interesting application to partial differential equations. Let $\mathfrak{G} \ni (0, \mathbf{0})$ be an open subset of $\mathbb{R}\times\mathbb{R}^M$ which is "forward pathwise connected to $(0,\mathbf{0})$" in the sense that for each $(s,\mathbf{x}) \in \mathfrak{G}$ with $s > 0$ there is a continuous path $p\colon [0,s] \longrightarrow \mathbb{R}^M$ such that $p(0) = \mathbf{0}$, $p(s) = \mathbf{x}$, and $\big(t, p(t)\big) \in \mathfrak{G}$ for all $t \in [0,s]$. The **parabolic strong minimum principle**[2] says that if $u \in C^{1,2}(\mathfrak{G};\mathbb{R})$ is a solution to the heat equation $\partial_t u + \frac{1}{2}\Delta u = 0$ in $\mathfrak{G}$ which achieves its minimum value at $(0,\mathbf{0})$, then $u(s,\mathbf{x}) = u(0,\mathbf{0})$ for all $(s,\mathbf{x}) \in \mathfrak{G}$ with $s > 0$. The following is one way to prove this result. Suppose that $(s,\mathbf{x})$ is a point in $\mathfrak{G}$ for which $s > 0$ and $u(s,\mathbf{x}) > u(0,\mathbf{0})$, and set $\delta = \frac{u(s,\mathbf{x})-u(0,\mathbf{0})}{2}$. Choose a path p accordingly for $(s,\mathbf{x})$. Then there is an $r \in (0,s)$ such that $[0,s] \times \bar{H} \subseteq \mathfrak{G}$ when

$$H = \{\mathbf{y} : |\mathbf{y} - p(t)| < 2r \text{ for some } t \in [0,s]\}$$

and $u(t,\mathbf{y}) \geq u(0,\mathbf{0}) + \delta$ for $(t,\mathbf{y}) \in [s-r,s] \times \overline{B(\mathbf{x},r)}$. Set

$$\zeta^H = \inf\{t \geq 0 : w(t) \notin H\} \quad \text{and} \quad \zeta = \inf\{t \geq s-r : |w(t) - \mathbf{x}| \leq r\}.$$

Then (cf. Exercise 3.3) $\big(u(t\wedge s\wedge\zeta^H, w(t\wedge s\wedge\zeta^H)), W_t, \mathcal{W}\big)$ is a martingale, and therefore

$$\begin{aligned} u(0,\mathbf{0}) &= \mathbb{E}^{\mathcal{W}}\big[u(\zeta\wedge s\wedge\zeta^H, w(\zeta\wedge s\wedge\zeta^H))\big] \\ &\geq u(0,\mathbf{0})\mathcal{W}(\zeta > s\wedge\zeta^H) + \big(u(0,\mathbf{0})+\delta\big)\mathcal{W}\big(\zeta \leq s\wedge\zeta^H\big). \end{aligned}$$

Finally, observe that $\|w(\,\cdot\,) - p(\,\cdot\,)\|_{[0,s]} < r \implies \zeta(w) \leq s\wedge\zeta^H(w)$ and therefore that $\mathcal{W}\big(\zeta \leq s\wedge\zeta^H\big) > 0$, which leads to the contradiction $u(0,\mathbf{0}) > u(0,\mathbf{0})$.

A second application of Theorem 3.1.3 is to the development of a Sobolev calculus for functions on $\mathbb{W}(\mathbb{R}^M)$, one in which Wiener measure plays the role that Lebesgue measure plays in finite dimensions. First observe that

$$\mathbb{E}^{\mathcal{W}}\big[R_h^p\big] = e^{\frac{p(p-1)\|h\|^2_{H^1(\mathbb{R}^M)}}{2}} \quad \text{for all } p \in [1,\infty).$$

Hence, by Theorem 3.1.3 and Hölder's inequality,

$$\begin{gathered} \|\Phi\circ T_h\|_{L^q(\mathcal{W};\mathbb{R})} \leq e^{\frac{\|h\|^2_{H^1(\mathbb{R}^M)}}{2(p-q)}} \|\Phi\|_{L^p(\mathcal{W};\mathbb{R})} \\ \text{for } 1 \leq q < p < \infty \text{ and } \Phi \in L^p(\mathcal{W};\mathbb{R}). \end{gathered} \tag{3.1.4}$$

Now suppose that $\Phi \in L^p(\mathcal{W};\mathbb{R})$ for some $p \in (1,\infty)$ and that there exists a function $D_h\Phi \in L^p(\mathcal{W};\mathbb{R})$ such that

$$\frac{\Phi\circ T_{\xi h} - \Phi}{\xi} \longrightarrow D_h\Phi \text{ in } L^1(\mathcal{W};\mathbb{R}) \quad \text{as } \xi \longrightarrow 0.$$

Then, because

[2] Apart from a minus sign, this is equivalent to the corresponding strong maximum principle.

$$\mathbb{E}^{\mathcal{W}}\big[\Phi\circ T_{\xi h}-\Phi\big]=\mathbb{E}^{\mathcal{W}}\big[(R_{\xi h}-1)\Phi\big]$$

and $\frac{R_{\xi h}-1}{\xi}\longrightarrow I(\dot h)$ in $L^{p'}(\mathcal{W};\mathbb{R})$,[3] it follows that

$$\mathbb{E}^{\mathcal{W}}\big[D_h\Phi\big]=\mathbb{E}^{\mathcal{W}}\big[I(\dot h)\Phi\big]. \tag{3.1.5}$$

Next suppose that $\Phi_1,\Phi_2\in L^p(\mathcal{W};\mathbb{R})$ for some $p\in(2,\infty)$ and that

$$\lim_{\xi\to 0}\frac{\Phi_i\circ T_{\xi h}-\Phi_i}{\xi}\longrightarrow D_h\Phi_i \text{ in } L^p(\mathcal{W};\mathbb{R})\quad\text{for } i\in\{1,2\}.$$

Then another application of (3.1.4) and Hölder's inequality shows that

$$\frac{(\Phi_1\circ T_{\xi h})(\Phi_2\circ T_{\xi h})-\Phi_1\Phi_2}{\xi}\longrightarrow \Phi_1D_h\Phi_2+\Phi_2D_h\Phi_1 \text{ in } L^1(\mathcal{W};\mathbb{R}),$$

and therefore

$$\mathbb{E}^{\mathcal{W}}\big[\Phi_1D_h\Phi_2\big]=-\mathbb{E}^{\mathcal{W}}\big[\Phi_2D_h\Phi_1\big]+\mathbb{E}^{\mathcal{W}}\big[I(\dot h)\Phi_1\Phi_2\big]. \tag{3.1.6}$$

This formula is the starting point for the Sobolev type calculus on which P. Malliavin based his analysis of functions on Wiener space. See §5.5 for some examples that illustrate his ideas.

3.2 Itô's integral

Again let $\big(B(t),\mathcal{F}_t,\mathbb{P}\big)$ be an $\mathbb{R}^M$-valued Brownian motion on a probability space $(\Omega,\mathcal{F},\mathbb{P})$. It will be convenient to assume that $\mathcal{F}$ and, for all $t\ge 0$, $\mathcal{F}_t$ are $\mathbb{P}$-complete. Given any $\mathcal{B}_{[0,\infty)}\times\mathcal{F}$-measurable function $\boldsymbol{\eta}\colon[0,\infty)\times\Omega\longrightarrow\mathbb{R}^M$ for which $\boldsymbol{\eta}(\,\cdot\,,\omega)$ has locally bounded variation, there is no problem defining

$$I_{\boldsymbol{\eta}}(t)=\int_0^t\big(\boldsymbol{\eta}(\tau),dB(\tau)\big)_{\mathbb{R}^M}$$

as a Riemann–Stieltjes integral. However, $I_{\boldsymbol{\eta}}(t)$ need no longer be a Gaussian random variable. Worse, in general, we have to use the variation norm of $\boldsymbol{\eta}$ in order to control the size of $I_{\boldsymbol{\eta}}(t)$, and so the sort of extension that we made in the non-random case will not be possible. Indeed, if we want to make such an extension, we have to restrict ourselves to integrands for which some variant of (3.1.3) holds. What Itô realized is that if $\boldsymbol{\eta}$ is **adapted** to the filtration $\{\mathcal{F}_t:t\ge 0\}$ in the sense that $\boldsymbol{\eta}(t)$ is $\mathcal{F}_t$-measurable for all $t\ge 0$, then such a variant of (3.1.3) would exist. Namely, because

[3] Here and elsewhere, $p'=\frac{p}{p-1}$ denotes the Hölder conjugate of p.

$$I_{\boldsymbol{\eta}}(t) = \lim_{n\to\infty} \sum_{0\le m<2^n t} \Big(\boldsymbol{\eta}(m2^{-n}), B\big(t\wedge(m+1)2^{-n}\big) - B(m2^{-n})\Big)_{\mathbb{R}^M}$$

and, for each m, $B\big(t\wedge(m+1)2^{-n}\big) - B(m2^{-n})$ is independent of $\mathcal{F}_{m2^{-n}}$ and has mean value $\mathbf{0}$ and covariance $\big(t\wedge(m+1)2^{-n} - m2^{-n}\big)\mathbf{I}$, $\mathbb{E}^{\mathbb{P}}\big[I_{\boldsymbol{\eta}}(t)^2\big]$ equals

$$\lim_{n\to\infty} \sum_{0\le m<2^n t} \mathbb{E}^{\mathbb{P}}\Big[\Big(\boldsymbol{\eta}(m2^{-n}), B\big(t\wedge(m+1)2^{-n}\big) - B(m2^{-n})\Big)^2_{\mathbb{R}^M}\Big]$$
$$= \lim_{n\to\infty} \int_0^t \mathbb{E}^{\mathbb{P}}\big[|\boldsymbol{\eta}(\lfloor\tau\rfloor_n)|^2\big]\,d\tau.$$

Thus, if $\boldsymbol{\eta}$ is a bounded, adapted function of locally bounded variation,

$$\mathbb{E}^{\mathbb{P}}\big[I_{\boldsymbol{\eta}}(t)^2\big] = \int_0^t \mathbb{E}^{\mathbb{P}}\big[|\boldsymbol{\eta}(\tau)|^2\big]\,d\tau.$$

Furthermore, if

$$M_n(t) = \sum_{0\le m<2^n t} \Big(\boldsymbol{\eta}(m2^{-n}), B\big(t\wedge(m+1)2^{-n}\big) - B(m2^{-n})\Big)_{\mathbb{R}^M},$$

then, because $B(t) - B(s)$ is independent of $\mathcal{F}_s$ when $s \le t$,

$$\mathbb{E}^{\mathbb{P}}\big[M_n(t) - M_n(s) \mid \mathcal{F}_s\big] = \mathbb{E}^{\mathbb{P}}\big[(\boldsymbol{\eta}(m2^{-n}), B(t) - B(s))_{\mathbb{R}^M} \mid \mathcal{B}_s\big] = 0$$

and

$$\mathbb{E}^{\mathbb{P}}\big[M_n(t)^2 - M_n(s)^2 \mid \mathcal{F}_s\big] = \mathbb{E}^{\mathbb{P}}\big[\big(M_n(t) - M_n(s)\big)^2 \mid \mathcal{F}_s\big]$$
$$= \mathbb{E}^{\mathbb{P}}\big[(\boldsymbol{\eta}(m2^{-n}), B(t) - B(s))^2_{\mathbb{R}^M} \mid \mathcal{B}_s\big] = |\boldsymbol{\eta}(m2^{-n})|^2(t-s)$$

if $m2^{-n} \le s < t \le (m+1)2^{-n}$. Hence,

$$\big(M_n(t), \mathcal{F}_t, \mathbb{P}\big) \quad\text{and}\quad \left(M_n(t)^2 - \int_0^t |\boldsymbol{\eta}(\lfloor\tau\rfloor_n)|^2\,d\tau, \mathcal{F}_t, \mathbb{P}\right)$$

are continuous martingales, and so, after letting $n\to\infty$ and applying Doob's inequality, we have that

$$\mathbb{E}^{\mathbb{P}}\big[\|I_{\boldsymbol{\eta}}(\,\cdot\,)\|^2_{[0,\infty)}\big] \le 4\sup_{t\ge0}\mathbb{E}^{\mathbb{P}}\big[I_{\boldsymbol{\eta}}(t)^2\big] = 4\int_0^\infty \mathbb{E}^{\mathbb{P}}\big[|\boldsymbol{\eta}(\tau)|^2\big]\,d\tau \tag{3.2.1}$$

and

$$\big(I_{\boldsymbol{\eta}}(t), \mathcal{F}_t, \mathbb{P}\big) \quad\text{and}\quad \left(I_{\boldsymbol{\eta}}(t)^2 - \int_0^t |\boldsymbol{\eta}(\tau)|^2\,d\tau, \mathcal{F}_t, \mathbb{P}\right) \tag{3.2.2}$$

are continuous martingales.

Given (3.2.1), it is easy to extend these conclusions to any adapted $\boldsymbol{\eta}$ of locally bounded variation with the property that

$$\int_0^\infty \mathbb{E}^{\mathbb{P}}\left[|\boldsymbol{\eta}(\tau)|^2\right] d\tau < \infty. \tag{3.2.3}$$

All that one has to do is replace $\boldsymbol{\eta}$ by $t \rightsquigarrow \mathbf{1}_{[0,k]}(|\boldsymbol{\eta}(t)|)\boldsymbol{\eta}(t)$ and then use (3.2.1) to justify passing to the limit as $k \to \infty$.

Itô's next step[4] was to use (3.2.1) and (3.2.2) to extend his definition to adapted integrands satisfying (3.2.3), and for this purpose he chose a $\rho \in C^\infty(\mathbb{R}; [0, \infty))$ that vanishes off $(0, 1)$ and has total integral 1, and set

$$\boldsymbol{\eta}_k(t) = \int_0^\infty \rho_{\frac{1}{k}}(t - \tau)\boldsymbol{\eta}(\tau)\, d\tau,$$

where $\rho_\epsilon(t) = \epsilon^{-1}\rho\left(\frac{t}{\epsilon}\right)$. Clearly $\boldsymbol{\eta}_k$ has locally bounded variation, and, as is well known,

$$\int_0^\infty |\boldsymbol{\eta}_k(\tau)|^2\, d\tau \le \int_0^\infty |\boldsymbol{\eta}(\tau)|^2\, d\tau \quad \text{and} \quad \lim_{k\to\infty} \int_0^\infty |\boldsymbol{\eta}_k(\tau) - \boldsymbol{\eta}(\tau)|^2\, d\tau = 0.$$

Hence,

$$\lim_{k\to\infty} \mathbb{E}^{\mathbb{P}}\left[\int_0^\infty |\boldsymbol{\eta}_k(\tau) - \boldsymbol{\eta}(\tau)|^2\, d\tau\right] = 0.$$

Furthermore, because the construction of $\boldsymbol{\eta}_k(t)$ involves $\boldsymbol{\eta}(\tau)$ only for $\tau \le t$, Itô, without further comment, claimed that $\boldsymbol{\eta}_k$ must be adapted. However, as Doob realized, a rigorous proof of that requires an intricate argument, one that Doob gave when he explained Ito's ideas in his renowned book [3]. Fortunately, thanks to P.-A. Meyer, there is a way to circumvent this technical hurdle by replacing *adapted*' with the slightly more restrictive notion of **progressively measurable**. A function on $[0, \infty) \times \Omega$ with values in a measurable space is said to be progressively measurable if its restriction to $[0, t] \times \Omega$ is $\mathcal{B}_{[0,t]} \times \mathcal{F}_t$-measurable for each $t \ge 0$. The great advantage of this notion is that a function is progressively measurable if and only if it is measurable with respect to the σ-algebra PM of progressively measurable sets: those subsets of $[0, \infty) \times \Omega$ whose indicator functions are progressively measurable. Thus a function will be progressively measurable if it is the pointwise limit of progressively measurable functions. In particular, and for us most important, it is elementary to check that $\boldsymbol{\eta}_k$ is progressively measurable if $\boldsymbol{\eta}$ is. Finally, although every progressively measurable function is adapted, not all adapted functions are progressively measurable. Nonetheless, if E is a topological space and $\eta\colon [0, \infty) \times \Omega \longrightarrow E$ is an adapted function that is either right- or left-continuous with respect t, then η is progressively measurable.

[4] To be more accurate, Itô did not know martingale theory at the time when he wrote [7]. It was Doob who used martingales to simplify his presentation of Itô's theory in [3].

To see this, suppose that η is right-continuous with respect to t, and, for each $n \geq 0$ define $\eta_n\colon [0,\infty) \times \Omega \longrightarrow E$ so that $\eta_n(\tau,\omega) = \eta(m2^{-n},\omega)$ if $(m-1)2^{-n} \leq \tau < m2^{-n}$. Then, for each $t > 0$, $(\tau,\omega) \in [0,t] \times \Omega \longmapsto \eta_n(\tau \wedge t,\omega) \in E$ is $\mathcal{B}_{[0,t]} \times \mathcal{F}_t$-measurable and $\eta(\tau,\omega) = \lim_{n\to\infty} \eta_n(\tau \wedge t,\omega)$ for $\tau \in [0,t]$. Hence $\eta \restriction [0,t] \times \Omega$ is $\mathcal{B}_{[0,t]} \times \mathcal{F}_t$-measurable. When $t \rightsquigarrow \eta(t,\omega)$ is left continuous, define $\eta_n(\tau,\omega) = \eta(m2^{-n},\omega)$ if $m2^{-n} \leq \tau < (m+1)2^{-n}$, note that η_n is progressively measurable and $\eta_n \longrightarrow \eta$, and conclude that η is progressively measurable.

For the reason explained in the preceding paragraph, we will now restrict our attention to the class $PM^2(\mathbb{R}^M)$ of integrands $\boldsymbol{\eta}\colon [0,\infty) \times \Omega \longrightarrow \mathbb{R}^M$ which are progressively measurable and satisfy

$$\|\boldsymbol{\eta}\|_{PM^2(\mathbb{R}^M)} := \left(\int_0^\infty \mathbb{E}^{\mathbb{P}}\big[|\boldsymbol{\eta}(\tau)|^2\big]\,d\tau\right)^{\frac{1}{2}} < \infty.$$

Let $M^2(\mathbb{P},\mathbb{R})$ be the space of continuous, $\mathbb{R}$-valued martingales $\big(M(t),\mathcal{F}_t,\mathbb{P}\big)$ for which

$$\|M\|_{M^2(\mathbb{P};\mathbb{R})} := \sup_{t\geq 0} \mathbb{E}^{\mathbb{P}}\big[|M(t)|^2\big]^{\frac{1}{2}} < \infty.$$

Then, (3.2.1) and (3.2.2) can be used to construct an isometric linear mapping from $PM^2(\mathbb{R}^M)$ into $M^2(\mathbb{P};\mathbb{R})$. Indeed, given $\boldsymbol{\eta} \in PM^2(\mathbb{R}^M)$, construct $\{\boldsymbol{\eta}_k : k \geq 1\}$ as in the preceding paragraph. Then, by (3.2.1),

$$\sup_{\ell > k} \mathbb{E}^{\mathbb{P}}\big[\|I_{\boldsymbol{\eta}_\ell} - I_{\boldsymbol{\eta}_k}\|^2_{[0,\infty)}\big] \leq 4 \sup_{\ell > k} \int_0^\infty \mathbb{E}^{\mathbb{P}}\big[|\boldsymbol{\eta}_\ell(\tau) - \boldsymbol{\eta}_k(\tau)|^2\big]\,d\tau \longrightarrow 0$$

as $k \to \infty$. Hence there exists a continuous, progressively measurable[5] $I_{\boldsymbol{\eta}}$ such that

$$\mathbb{E}^{\mathbb{P}}\big[\|I_{\boldsymbol{\eta}} - I_{\boldsymbol{\eta}_k}\|^2_{[0,\infty)}\big] \leq 4 \int_0^\infty \mathbb{E}^{\mathbb{P}}\big[|\boldsymbol{\eta}(\tau) - \boldsymbol{\eta}_k(\tau)|^2\big]\,d\tau \longrightarrow 0$$

as $k \to \infty$, and clearly $I_{\boldsymbol{\eta}}$ inherits the properties in (3.2.1) and (3.2.2) from the $I_{\boldsymbol{\eta}_k}$'s.

Because it shares many properties with standard integrals, the quantity $I_{\boldsymbol{\eta}}(t)$ is usually denoted by

$$\int_0^t \big(\boldsymbol{\eta}(\tau), dB(\tau)\big)_{\mathbb{R}^M}$$

and is called the **Itô stochastic integral**, or just the **stochastic integral**, of $\boldsymbol{\eta}$ with respect to B. Of course, like the Paley–Wiener integral, Itô's is not in general a Riemann–Stieltjes integral and is defined only up to a set of measure 0. Nonetheless, we know that $I_{\boldsymbol{\eta}}(t)$ is a Riemann–Stieltjes integral when $\boldsymbol{\eta}$ is

[5] In the following and elsewhere, we will say that a progressively measurable function on $[0,\infty) \times \Omega$ is continuous if it is continuous as a function of time.

a bounded, progressively measurable function of locally finite variation, and the following simple fact about Riemann–Stieltjes integration allows us to see that the same is true even when the boundedness assumption is dropped.

Lemma 3.2.1. *Let* $\varphi \in C([0,t];\mathbb{R})$ *and a sequence* $\{\psi_n : n \geq 0\}$ *of functions on* $[0,t]$ *with* $|\psi_n(0)| \vee \mathrm{var}_{n\,[0,t]}(\psi_n) \leq C < \infty$ *be given. If* $\psi_n \longrightarrow \psi$ *pointwise, then* $\mathrm{var}_{[0,t]}(\psi) \leq C$ *and*

$$\lim_{n\to\infty} \int_0^t \psi_n(\tau)\, d\varphi(\tau) = \int_0^t \psi(\tau)\, d\varphi(\tau).$$

Proof. Clearly $\mathrm{var}_{[0,t]}(\psi) \leq C$. Next, choose $\{\varphi_k : k \geq 1\} \subseteq C^1(\mathbb{R};\mathbb{R})$ so that $\varphi_k(0) = \varphi(0)$, $\|\varphi_k\|_{\mathrm{u}} \leq \|\varphi\|_{[0,t]}$, and $\|\varphi - \varphi_k\|_{[0,t]} \leq \frac{1}{k}$. Then

$$\left|\int_0^t \psi_n(\tau)\, d\varphi(\tau) - \int_0^t \psi_n(\tau)\, d\varphi_k(\tau)\right|$$
$$\leq |\varphi(t) - \varphi_k(t)||\psi_n(t)| + \left|\int_0^t (\varphi(\tau) - \varphi_k(\tau))\, d\psi_n(\tau)\right| \leq 2C\|\varphi - \varphi_k\|_{[0,t]} \leq \frac{2C}{k},$$

and similarly

$$\left|\int_0^t \psi(\tau)\, d\varphi(\tau) - \int_0^t \psi(\tau)\, d\varphi_k(\tau)\right| \leq \frac{2C}{k}.$$

Hence, it suffices to show that

$$\lim_{n\to\infty} \int_0^t \psi_n(\tau)\, d\varphi_k(\tau) = \int_0^t \psi(\tau)\, d\varphi_k(\tau)$$

for each $k \geq 1$. But, by Lebesgue's dominated convergence theorem,

$$\begin{aligned}\lim_{n\to\infty} \int_0^t \psi_n(\tau)\, d\varphi_k(\tau) &= \lim_{n\to\infty} \int_0^t \psi_n(\tau)\dot{\varphi}_k(\tau)\, d\tau \\ &= \int_0^t \psi(\tau)\dot{\varphi}_k(\tau)\, d\tau = \int_0^t \psi(\tau)\, d\varphi_k(\tau).\end{aligned}$$

□

3.2.1 *Some properties and extentions*

Given $\eta_1, \eta_2 \in PM^2(\mathbb{R}^M)$, (3.2.2) plus a simple polarization argument[6] shows that

$$\left(I_{\eta_1}(t)I_{\eta_3}(t) - \int_0^t \big(\eta_1(\tau), \eta_2(\tau)\big)_{\mathbb{R}^M}\, d\tau, \mathcal{F}_t, \mathbb{P}\right)$$

[6] A polarization argument is one based on the identity $4ab = (a+b)^2 - (a-b)^2$.

is a martingale.

Now suppose that $\boldsymbol{\eta} \in PM^2(\mathbb{R}^M)$ and that ζ is a stopping time relative to $\{\mathcal{F}_t : t \geq 0\}$. Because $(t,\omega) \rightsquigarrow \mathbf{1}_{[0,\zeta(\omega))}(t)$ is adapted and left continuous, it, and therefore $(t,\omega) \rightsquigarrow \mathbf{1}_{[0,\zeta(\omega))}(t)\eta(t,\omega)$, are progressively measurable. Further, by Hunt's stopping time theorem,

$$\begin{aligned}
&\mathbb{E}^{\mathbb{P}}\left[I_{\boldsymbol{\eta}}(t\wedge\zeta)\int_0^t \mathbf{1}_{[0,\zeta)}(\tau)\big(\boldsymbol{\eta}(\tau),dB(\tau)\big)_{\mathbb{R}^M}\right]\\
&\quad= \mathbb{E}^{\mathbb{P}}\left[I_{\boldsymbol{\eta}}(t\wedge\zeta)\int_0^{t\wedge\zeta} \mathbf{1}_{[0,\zeta)}(\tau)\big(\boldsymbol{\eta}(\tau),dB(\tau)\big)_{\mathbb{R}^M}\right] = \mathbb{E}^{\mathbb{P}}\left[\int_0^{t\wedge\zeta}|\boldsymbol{\eta}(\tau)|^2\,d\tau\right],
\end{aligned}$$

and so

$$\mathbb{E}^{\mathbb{P}}\left[\left|I_{\boldsymbol{\eta}}(t\wedge\zeta) - \int_0^t \mathbf{1}_{[0,\zeta)}(\tau)\big(\boldsymbol{\eta}(\tau),dB(\tau)\big)_{\mathbb{R}^M}\right|^2\right] = 0.$$

Hence

$$I_{\boldsymbol{\eta}}(t\wedge\zeta) = \int_0^t \mathbf{1}_{[0,\zeta)}(\tau)\big(\boldsymbol{\eta}(\tau),dB(\tau)\big)_{\mathbb{R}^M}.$$

In particular, if ζ_1 and ζ_2 are a pair of stopping times and $\zeta_1 \leq \zeta_2$, then

$$\int_0^{t\wedge\zeta_1} \mathbf{1}_{[0,\zeta_2)}(\tau)\big(\boldsymbol{\eta}(\tau),dB(\tau)\big)_{\mathbb{R}^M} = \int_0^t \mathbf{1}_{[0,\zeta_1)}(\tau)\big(\boldsymbol{\eta}(\tau),dB(\tau)\big)_{\mathbb{R}^M}.$$

Similarly, if $\zeta_1 \leq \zeta_2 < \infty$ are stopping times, then

$$\begin{aligned}
\int_{\zeta_1}^{\zeta_2}\big(\boldsymbol{\eta}(\tau),dB(\tau)\big)_{\mathbb{R}^M} &:= I_{\boldsymbol{\eta}}(\zeta_2) - I_{\boldsymbol{\eta}}(\zeta_1)\\
&= \int_0^\infty \mathbf{1}_{[\zeta_1,\zeta_2)}(\tau)\big(\boldsymbol{\eta}(\tau),dB(\tau)\big)_{\mathbb{R}^M}.
\end{aligned} \tag{3.2.4}$$

The preceding considerations afford us the opportunity to integrate $\boldsymbol{\eta}$'s that are not in $PM^2(\mathbb{R}^M)$. Namely, let $PM^2_{\text{loc}}(\mathbb{R}^M)$ be the set of progressively measurable, $\mathbb{R}^M$-valued functions $\boldsymbol{\eta}$ with the property that $\int_0^t |\boldsymbol{\eta}(\tau)|^2\,d\tau < \infty$ for all $t \in [0,\infty)$. Then, if $\boldsymbol{\eta} \in PM^2_{\text{loc}}(\mathbb{R}^M)$ and $\boldsymbol{\eta}_k = \mathbf{1}_{[0,\zeta_k)}\boldsymbol{\eta}$ where

$$\zeta_k = \inf\left\{t \geq 0 : \int_0^t |\boldsymbol{\eta}(\tau)|^2\,d\tau \geq k\right\},$$

$\boldsymbol{\eta}_k \in PM^2(\mathbb{R}^M)$, $I_{\boldsymbol{\eta}_{k+1}}(t\wedge\zeta_k) = I_{\boldsymbol{\eta}_k}(t)$, and so not only does

$$I_{\boldsymbol{\eta}}(t) := \lim_{k\to\infty} I_{\boldsymbol{\eta}_k}(t)$$

exist, but also $I_{\boldsymbol{\eta}}(t\wedge\zeta_k) = I_{\boldsymbol{\eta}_k}(t)$ for all $k \geq 1$. Of course, in general, $\big(I_{\boldsymbol{\eta}}(t),\mathcal{F}_t,\mathbb{P}\big)$ will not be a martingale since $I_{\boldsymbol{\eta}}(t)$ need not be even $\mathbb{P}$-integrable. On the other hand, for each $k \geq 1$,

$$\big(I_{\boldsymbol{\eta}}(t\wedge\zeta_k),\mathcal{F}_t,\mathbb{P}\big) \quad \text{and} \quad \left(I_{\boldsymbol{\eta}}(t\wedge\zeta_k)^2 - \int_0^{t\wedge\zeta_k}|\boldsymbol{\eta}(\tau)|^2\,d\tau,\mathcal{F}_t,\mathbb{P}\right)$$

will be martingales. Such considerations motivate the introduction of continuous **local martingales**: progressively measurable maps M that are $\mathbb{P}$-almost surely continuous with respect to time and for which there exists a non-decreasing sequence $\zeta_k \nearrow \infty$ of stopping times with the property that $\big(M(t\wedge\zeta_k),\mathcal{F}_t,\mathbb{P}\big)$ is a martingale for each $k\geq 1$. Observe that if $\boldsymbol{\eta}\in PM^2_{\text{loc}}(\mathbb{R}^M)$ and ζ is a stopping time, then, since

$$I_{\boldsymbol{\eta}}(t\wedge\zeta) = \lim_{k\to\infty} I_{\boldsymbol{\eta}}(t\wedge\zeta_k\wedge\zeta) = \lim_{k\to\infty}\int_0^{t\wedge\zeta_k}\mathbf{1}_{[0,\zeta)}(\tau)\big(\boldsymbol{\eta}(\tau),dB(\tau)\big)_{\mathbb{R}^M},$$

(3.2.4) continues to hold. Moreover, if $\mathbb{E}^{\mathbb{P}}\left[\int_0^\zeta|\boldsymbol{\eta}(\tau)|^2\,d\tau\right] < \infty$, then, by Doob's inequality,

$$\mathbb{E}^{\mathbb{P}}\big[\|I_{\boldsymbol{\eta}}(\,\cdot\,)\|^2_{[0,\zeta)}\big] = \lim_{k\to\infty}\mathbb{E}^{\mathbb{P}}\big[\|I_{\boldsymbol{\eta}}(\,\cdot\,)\|^2_{[0,\zeta\wedge\zeta_k)}\big] \leq 4\mathbb{E}^{\mathbb{P}}\left[\int_0^\zeta|\boldsymbol{\eta}(\tau)|^2\,d\tau\right] < \infty,$$

and so

$$\big(I_{\boldsymbol{\eta}}(t\wedge\zeta),\mathcal{F}_t,\mathbb{P}\big) \quad \text{and} \quad \left(I_{\boldsymbol{\eta}}(t\wedge\zeta)^2 - \int_0^{t\wedge\zeta}|\boldsymbol{\eta}(\tau)|^2\,d\tau,\mathcal{F}_t,\mathbb{P}\right)$$

are martingales. In particular, if $\mathbb{E}^{\mathbb{P}}\left[\int_0^t|\boldsymbol{\eta}(\tau)|^2\,d\tau\right] < \infty$ for all $t\geq 0$, then

$$\big(I_{\boldsymbol{\eta}}(t),\mathcal{F}_t,\mathbb{P}\big) \quad \text{and} \quad \left(I_{\boldsymbol{\eta}}(t)^2 - \int_0^t|\boldsymbol{\eta}(\tau)|^2\,d\tau,\mathcal{F}_t,\mathbb{P}\right)$$

are martingales. Finally, if $\boldsymbol{\eta}$ is an $\mathbb{R}^M$-valued, adapted function and $\boldsymbol{\eta}(\,\cdot\,,\omega)$ is continuous for all $\omega\in\Omega$, then $\boldsymbol{\eta}$ is progressively measurable and, by taking

$$\zeta_k = \inf\left\{t\geq 0 : \int_0^t|\boldsymbol{\eta}(\tau)|^2\,d\tau \geq k\right\},$$

one sees that $\boldsymbol{\eta}\in PM^2_{\text{loc}}(\mathbb{R}^M)$.

Let $PM^2\big(\text{Hom}(\mathbb{R}^M;\mathbb{R}^N)\big)$ be the space of $\text{Hom}(\mathbb{R}^M;\mathbb{R}^N)$-valued progressively measurable functions σ with the property that

$$\mathbb{E}^{\mathbb{P}}\left[\int_0^\infty\|\sigma(\tau)\|^2_{\text{H.S.}}\,d\tau\right]^{\frac{1}{2}} < \infty,$$

and define the $\mathbb{R}^N$-valued random variable

$$I_\sigma(t) = \int_0^t \sigma(\tau)\, dB(\tau)$$

so that

$$\big(\boldsymbol{\xi}, I_\sigma(t)\big)_{\mathbb{R}^N} = \int_0^t \big(\sigma(\tau)^\top \boldsymbol{\xi}, dB(\tau)\big)_{\mathbb{R}^M} \quad \text{for each } \boldsymbol{\xi} \in \mathbb{R}^N.$$

It is then an elementary exercise to check that

$$\big(I_\sigma(t), \mathcal{F}_t, \mathbb{P}\big) \quad \text{and} \quad \left(I_\sigma(t) \otimes I_\sigma(t) - \int_0^t \sigma(\tau)\sigma(\tau)^\top \, d\tau, \mathcal{F}_t, \mathbb{P}\right) \tag{3.2.5}$$

are, respectively, $\mathbb{R}^N$-valued and $\mathrm{Hom}(\mathbb{R}^N;\mathbb{R}^N)$-valued martingales and that

$$\mathbb{E}^{\mathbb{P}}\left[\sup_{t\geq 0} |I_\sigma(t)|^2\right]^{\frac{1}{2}} \leq 4 \sup_{t\geq 0} \mathbb{E}^{\mathbb{P}}\big[|I_\sigma(t)|^2\big]^{\frac{1}{2}} = 4\mathbb{E}^{\mathbb{P}}\left[\int_0^\infty \|\sigma(\tau)\|_{\mathrm{H.S.}}^2 \, d\tau\right]^{\frac{1}{2}}.$$

Further, starting from (3.2.5) and using polarization, one sees that if $\widetilde{\sigma}$ is a second element of $PM^2\big(\mathrm{Hom}(\mathbb{R}^M;\mathbb{R}^N)\big)$, then

$$\left(I_\sigma(t) \otimes I_{\widetilde{\sigma}}(t) - \int_0^t \sigma(\tau)\widetilde{\sigma}(\tau)^\top \, d\tau, \mathcal{F}_t, \mathbb{P}\right)$$

is a martingale. Finally, define $PM^2_{\mathrm{loc}}\big(\mathrm{Hom}(\mathbb{R}^M;\mathbb{R}^N)\big)$ by analogy with $PM^2_{\mathrm{loc}}(\mathbb{R}^M)$, and define $I_\sigma(t)$ for $\sigma \in PM^2_{\mathrm{loc}}\big(\mathrm{Hom}(\mathbb{R}^M;\mathbb{R}^N)\big)$ accordingly.

3.2.2 *Stochastic integral equations*

Let $\sigma\colon \mathbb{R}^N \longrightarrow \mathrm{Hom}(\mathbb{R}^M;\mathbb{R}^N)$ and $b\colon \mathbb{R}^N \longrightarrow \mathbb{R}^N$ be uniformly Lipschitz continuous functions. We can now interpret the construction in §2.2 in terms of stochastic integrals. For each $n \geq 0$, $X_n(\,\cdot\,,\mathbf{x})$ is given by (3.0.1). Thus

$$\begin{aligned}\mathbb{E}^{\mathcal{W}}\big[\|X_n(\,\cdot\,,\mathbf{x})\|^2_{[0,t]}\big] \leq 3|\mathbf{x}|^2 + 12 \int_0^t \mathbb{E}^{\mathcal{W}}\big[\big\|\sigma\big(X_n(\lfloor\tau\rfloor_n,\mathbf{x})\big)\big\|^2_{\mathrm{H.S.}}\big]\, d\tau \\ + 3t \int_0^t \mathbb{E}^{\mathcal{W}}\big[\big|b\big(X_n(\lfloor\tau\rfloor_n,\mathbf{x})\big)\big|^2\big]\, d\tau,\end{aligned}$$

$$\begin{aligned}\mathbb{E}^{\mathcal{W}}\big[|X_n(t,\mathbf{x}) - X_n(s,\mathbf{x})|^2\big] \leq 2 \int_s^t \mathbb{E}^{\mathcal{W}}\big[\big\|\sigma\big(X_n(\lfloor\tau\rfloor_n,\mathbf{x})\big)\big\|^2_{\mathrm{H.S.}}\big]\, d\tau \\ + 2t \int_s^t \mathbb{E}^{\mathcal{W}}\big[\big|b\big(X_n(\lfloor\tau\rfloor_n,\mathbf{x})\big)\big|^2\big],\end{aligned}$$

and

$$\begin{aligned}&\mathbb{E}^{\mathcal{W}}\big[\|X_{n+1}(\,\cdot\,,\mathbf{x}) - X_n(\,\cdot\,,\mathbf{x})\|_{[0,t]}^2\big]\\&\quad\le 8\int_0^t \mathbb{E}^{\mathcal{W}}\Big[\big\|\sigma\big(X_{n+1}(\lfloor\tau\rfloor_{n+1},\mathbf{x})\big) - \sigma\big(X_n(\lfloor\tau\rfloor_n,\mathbf{x})\big)\big\|_{\mathrm{H.S.}}^2\Big]\,d\tau\\&\qquad + 2t\int_0^t \mathbb{E}^{\mathcal{W}}\Big[\big|b\big(X_{n+1}(\lfloor\tau\rfloor_{n+1},\mathbf{x})\big) - b\big(X_n(\lfloor\tau\rfloor_n,\mathbf{x})\big)\big|^2\Big]\,d\tau.\end{aligned}$$

Given these, one can proceed as in §2.2.1 and thereby recover (2.2.2) and (2.2.3). In addition, knowing (2.2.3), we see that $X(\,\cdot\,,\mathbf{x})$ is progressively measurable with respect to the filtration $\{W_t : t \ge 0\}$ and

$$\lim_{n\to\infty} \mathbb{E}^{\mathcal{W}}\left[\left|\int_0^t \sigma\big(X_n(\lfloor\tau\rfloor_n,\mathbf{x})\big)\,dw(\tau) - \int_0^t \sigma\big(X(\tau,\mathbf{x})\big)\,dw(\tau)\right|^2\right] = 0.$$

Therefore $X(\,\cdot\,,\mathbf{x})$ solves the **stochastic integral equation** in (3.0.2). In fact, it is the only solution, since if $\widetilde{X}(\,\cdot\,,\mathbf{x})$ were a second solution, then

$$\begin{aligned}&\mathbb{E}^{\mathcal{W}}\big[\big|\widetilde{X}(t,\mathbf{x}) - X(t,\mathbf{x})\big|^2\big]\\&\quad\le 2\int_0^t \mathbb{E}^{\mathcal{W}}\Big[\big\|\sigma\big(\widetilde{X}(\tau,\mathbf{x})\big) - \sigma\big(X(\tau,\mathbf{x})\big)\big\|_{\mathrm{H.S.}}^2\Big]\,d\tau\\&\qquad + 2t\int_0^t \mathbb{E}^{\mathcal{W}}\Big[\big|b\big(\widetilde{X}(\tau,\mathbf{x})\big) - b\big(X(\tau,\mathbf{x})\big)\big|^2\Big]\,d\tau\\&\quad\le 2\big(\|\sigma\|_{\mathrm{Lip}}^2 + t\|b\|_{\mathrm{Lip}}^2\big)\int_0^t \mathbb{E}^{\mathcal{W}}\big[|\widetilde{X}(\tau,\mathbf{x}) - X(\tau,\mathbf{x})|^2\big]\,d\tau,\end{aligned}$$

which, by Lemma 1.2.4, means that $\widetilde{X}(t,\mathbf{x}) = X(t,\mathbf{x})$ (a.s., $\mathcal{W}$).

Having described $X(\,\cdot\,,\mathbf{x})$ as the solution to (3.0.2), it is time for me to admit that the method that we used to construct the solution is not the one chosen by Itô. Instead of using Euler's approximation scheme, Itô chose to use a Picard iteration scheme. That is, he set $\widetilde{X}_0(t,\mathbf{x}) = \mathbf{x}$ and

$$\widetilde{X}_{n+1}(t,\mathbf{x}) = \mathbf{x} + \int_0^t \sigma\big(\widetilde{X}_n(\tau,\mathbf{x})\big)\,dw(\tau) + \int_0^t b\big(\widetilde{X}_n(\tau,\mathbf{x})\big)\,d\tau$$

for $n \ge 0$. If $\Delta_n(t) = \|\widetilde{X}_{n+1}(\,\cdot\,,\mathbf{x}) - \widetilde{X}_n(\,\cdot\,,\mathbf{x})\|_{[0,t]}$, then

$$\begin{aligned}\mathbb{E}^{\mathcal{W}}\big[\Delta_n(t)^2\big] &\le 8\int_0^t \mathbb{E}^{\mathcal{W}}\big[\|\sigma(\widetilde{X}_n(\tau,\mathbf{x})) - \sigma(\widetilde{X}_{n-1}(\tau,\mathbf{x}))\|_{\mathrm{H.S.}}^2\big]\,d\tau\\ &\quad + 2t\int_0^t \mathbb{E}^{\mathcal{W}}\big[\big|b(\widetilde{X}_n(\tau,\mathbf{x})) - b(\widetilde{X}_{n-1}(\tau,\mathbf{x}))\big|^2\big]\,d\tau\\ &\le C(1+t)\int_0^t \mathbb{E}^{\mathcal{W}}\big[\Delta_{n-1}(\tau)^2\big]\,d\tau\end{aligned}$$

for some $C < \infty$ and all $n \ge 1$. Working by induction, one concludes that, for $n \ge 1$,

$$\mathbb{E}^{\mathcal{W}}\big[\Delta_n(t)^2\big] \le \frac{\big(C(1+t)t\big)^{n-1}}{(n-1)!}\mathbb{E}^{\mathcal{W}}\big[\Delta_0(t)^2\big],$$

and from this one sees that there is an $\widetilde{X}(\,\cdot\,,\mathbf{x})$ such that

$$\lim_{n\to\infty} \mathbb{E}^{\mathcal{W}}\big[\|\widetilde{X}(\,\cdot\,,\mathbf{x}) - \widetilde{X}_n(\,\cdot\,,\mathbf{x})\|_{[0,t]}^2\big] = 0 \quad \text{for all } t \ge 0,$$

and clearly this $\widetilde{X}(\,\cdot\,,\mathbf{x})$ will solve (3.0.2). Although the Picard iteration scheme is more elegant than the Euler one, Euler approximations are more intuitive and have virtues we will be exploiting later.

3.3 The crown jewel: Itô's formula

Unless the integrand has locally bounded variation, stochastic integrals are somewhat inscrutable quantities. They exist, but, like Fourier series, they are not robust and converge only because of intricate cancellations. As a consequence, it is desirable to find more tractable quantities to which stochastic integrals are related, and (3.2.2) provides prime examples of the sort of relationship for which one should be looking.

The key to finding relationships like those in (3.2.2) was discovered by Itô. To describe his result, let $\big(B(t),\mathcal{F}_t,\mathbb{P}\big)$ be an $\mathbb{R}^M$-valued Brownian motion, $V\colon [0,\infty)\times\Omega \longrightarrow \mathbb{R}^{N_1}$ a continuous, progressively measurable function of locally bounded variation, and $\sigma \in PM^2_{\mathrm{loc}}\big(\mathrm{Hom}(\mathbb{R}^M;\mathbb{R}^{N_2}\big)$. Then **Itô's formula** says that any $\varphi \in C^{1,2}\big(R^{N_1}\times\mathbb{R}^{N_2};\mathbb{C}\big)$,

$$\begin{aligned}&\varphi\big(V(t),I_\sigma(t)\big) - \varphi(V(0),\mathbf{0})\\ &\quad = \int_0^t \big(\nabla_{(1)}\varphi(V(\tau),I_\sigma(\tau)),dV(\tau)\big)_{\mathbb{R}^{N_1}}\\ &\qquad + \int_0^t \big(\sigma(\tau)^\top\nabla_{(2)}\varphi(I_\sigma(\tau)),dB(\tau)\big)_{\mathbb{R}^M}\\ &\qquad + \frac{1}{2}\int_0^t \mathrm{Trace}\big(\sigma(\tau)\sigma(\tau)^\top\nabla^2_{(2)}\varphi(I_\sigma(\tau)\big)\,d\tau,\end{aligned} \tag{3.3.1}$$

where the subscripts on $\nabla_{(1)}$ and $\nabla_{(2)}$ are used to distinguish between differentiation with respect to variables in $\mathbb{R}^{N_1}$ and those in $\mathbb{R}^{N_2}$. Obviously, (3.3.1) is a version of the fundamental theorem of calculus in which second derivatives appear because, as (2.1.2) makes clear, $dB(t)$ is of order $\sqrt{dt}$, not dt, and one therefore has to go out two terms in Taylor's expansion before getting terms that are truly infinitesimal. For this reason, it is useful to think of the result in (2.1.2) as saying that $dB(t) \otimes dB(t) = \mathbf{I}dt$ and write (3.3.1) in differential form:

$$
\begin{aligned}
&d\varphi\big(V(t), I_\sigma(t)\big) \\
&\quad = \Big(\nabla_{(1)}\varphi\big(V(t), I_\sigma(t)\big), dV(t)\Big)_{\mathbb{R}^{N_1}} + \Big(\sigma(t)^\top \nabla_{(2)}\varphi\big(V(t), I_\sigma(t)\big), dB(t)\Big)_{\mathbb{R}^{N_2}} \\
&\qquad + \tfrac{1}{2}\operatorname{Trace}\Big(\nabla^2_{(2)}\varphi\big(V(t), I_\sigma(t)\big)\big(\sigma(\tau)dB(t)\big) \otimes \big(\sigma(\tau)dB(t)\big)\Big).
\end{aligned}
$$

To prove (3.3.1), first observe that, by using standard approximation methods and stopping times, one can easily show that it suffices to prove it in the case when $\varphi \in C^\infty_{\mathrm{c}}(\mathbb{R}^{N_1} \times \mathbb{R}^{N_2}; \mathbb{R})$ and $\int_0^t \|\sigma(\tau)\|^2_{\mathrm{H.S}}\, d\tau$ and $V(\mathbf{0}) + \mathrm{var}_{[0,t]}(V)$ are bounded. Further, under these conditions, one can reduce to the case when $\tau \rightsquigarrow \sigma(\tau)$ is bounded and continuous. Thus we will proceed under these assumptions.

Define $\sigma_n(t) = \sigma(\lfloor t \rfloor_n)$. Then

$$
\varphi\big(V(t), I_\sigma(t)\big) - \varphi\big(V(0), \mathbf{0}\big) = \lim_{n\to\infty} \Big(\varphi\big(V(t), I_{\sigma_n}(t)\big) - \varphi\big(V(0), \mathbf{0}\big)\Big)
$$

and

$$
\begin{aligned}
&\varphi\big(V(t), I_{\sigma_n}(t)\big) - \varphi\big(V(0), \mathbf{0}\big) \\
&\qquad = \sum_{m<2^n t} \Big(\varphi\big(V(t_{m+1,n}), I_{m+1,n}\big) - \varphi\big(V(t_{m,n}), I_{m,n}\big)\Big),
\end{aligned}
$$

where $t_{m,n} = m2^{-n} \wedge t$ and $I_{m,n} = I_\sigma(t_{m,n})$. Since

$$
\begin{aligned}
&\varphi\big(V(t_{m+1,n}), I_{m+1,n}\big) - \varphi\big(V(t_{m,n}, I_{m+1,n})\big) \\
&\qquad = \int_{t_{m,n}}^{t_{m+1,n}} \big(\nabla_{(1)}\varphi(V(\tau), I_{m+1,n}), dV(\tau)\big)_{\mathbb{R}^{N_1}},
\end{aligned}
$$

$\varphi\big(V(t), I_\sigma(t)\big) - \varphi\big(V(0), \mathbf{0}\big)$ equals

$$
\begin{aligned}
&\int_0^t \big(\nabla_{(1)}\varphi(V(\tau), I_{\sigma_n}(\lfloor \tau \rfloor_n + 2^{-n})), dV(\tau)\big)_{\mathbb{R}^{N_1}} \\
&\qquad + \sum_{m<2^n t} \Big(\varphi\big(V(t_{m,n}), I_{m+1,n}\big) - \varphi\big(V(t_{m,n}, I_{m,n})\big)\Big),
\end{aligned}
$$

and

$$\int_0^t \big(\nabla_{(1)}\varphi(V(\tau), I_\sigma(\lfloor\tau\rfloor_n + 2^{-n})), dV(\tau)\big)_{\mathbb{R}^{N_1}} \longrightarrow \int_0^t \big(\nabla_{(1)}\varphi(V(\tau), I_\sigma(\tau)), dV(\tau)\big)_{\mathbb{R}^{N_1}} \text{ in } L^2(\mathbb{P};\mathbb{R}).$$

Next, by Taylor's theorem,

$$\begin{aligned}&\varphi\big(V(t_{m,n}), I_{m+1,n}\big) - \varphi\big(V(t_{m,n}), I_{m,n}\big)\\&= \big(\nabla_{(2)}\varphi\big(V(t_{m,n}), I_{m,n}\big), \Delta_{m,n}\big)_{\mathbb{R}^{N_2}}\\&\quad + \tfrac{1}{2}\operatorname{Trace}\big(\nabla^2_{(2)}\varphi\big(V(t_{m,n}), I_{m,n}\big)\Delta_{m,n}\otimes\Delta_{m,n}\big) + E_{m,n},\end{aligned}$$

where $\Delta_{m,n} = I_{m+1,n} - I_{m,n} = \sigma(t_{m,n})\big(B(t_{m+1,n}) - B(t_{m,n})\big)$ and there is a $C<\infty$ such that $|E_{m,n}| \le C|\Delta_{m,n}|^3$. Because

$$\mathbb{E}^{\mathbb{P}}\big[|B(t_{m+1,n}) - B(t_{m,n})|^3\big] \le E^{\mathbb{P}}\big[|B(2^{-n})|^3\big] = 2^{-\frac{3n}{2}}\mathbb{E}^{\mathbb{P}}\big[|B(1)|^3\big],$$

$\sum_{0\le m<2^n} E_{m,n}$ tends to 0 (a.s, $\mathbb{P}$). Obviously,

$$\begin{aligned}&\sum_{m<2^n}\Big(\nabla_{(2)}\varphi\big(V(t_{m,n}), I_{m,n}\big), \Delta_{m,n}\Big)_{\mathbb{R}^{N_2}}\\&= \int_0^t \Big(\sigma(\lfloor\tau\rfloor_n)^\top\nabla_{(2)}\varphi\big(V(\lfloor\tau\rfloor_n), I_{\sigma_n}(\lfloor\tau\rfloor_n)\big)), dB(\tau)\Big)_{\mathbb{R}^M}\\&\longrightarrow \int_0^t \Big(\sigma(\tau)^\top\nabla_{(2)}\varphi\big(V(\tau), I_\sigma(\tau)\big)), dB(\tau)\Big)_{\mathbb{R}^M}\end{aligned}$$

in $L^2(\mathbb{P};\mathbb{R})$. Finally, write

$$\begin{aligned}&\operatorname{Trace}\Big(\nabla^2_{(2)}\varphi\big(V(t_{m,n}), I_{m,n}\big)\Delta_{m,n}\otimes\Delta_{m,n}\Big)\\&= \operatorname{Trace}\Big(\nabla^2_{(2)}\varphi\big(V(t_{m,n}), I_{m,n}\big)A_{m,n}\Big)\\&\quad + \operatorname{Trace}\Big(\nabla^2_{(2)}\varphi\big(V(t_{m,n}), I_{m,n}\big)\big(\Delta_{m,n}\otimes\Delta_{m,n} - A_{m,n}\big)\Big),\end{aligned}$$

where $A_{m,n} = \int_{t_{m,n}}^{t_{m+1,n}} \sigma_n(\tau)\sigma_n(\tau)^\top\,d\tau = (t_{m+1,n} - t_{m,n})\sigma(t_{m,n})\sigma(t_{m,n})^\top$. Clearly

$$\begin{aligned}&\sum_{m<2^n t}\operatorname{Trace}\Big(\nabla^2_{(2)}\varphi\big(V(t_{m,n}), I_{m,n}\big)A_{m,n}\Big)\\&= \int_0^t \operatorname{Trace}\Big(\nabla^2_{(2)}\varphi(V(\lfloor\tau\rfloor_n), I_{\sigma_n}(\lfloor\tau\rfloor_n))\sigma(\lfloor\tau\rfloor_n)\sigma(\lfloor\tau\rfloor_n)^\top\Big)\,d\tau\\&\longrightarrow \int_0^t \operatorname{Trace}\Big(\sigma(\tau)\sigma(\tau)^\top\nabla^2_{(2)}\varphi(V(\tau), I_\sigma(\tau))\Big)\,d\tau.\end{aligned}$$

At the same time, because $\mathbb{E}^{\mathbb{P}}\big[\Delta_{m,n} \mid \mathcal{F}_{t_{m,n}}\big] = \mathbf{0}$ and therefore

$$\begin{aligned}
&\mathbb{E}^{\mathbb{P}}\big[\Delta_{m,n} \otimes \Delta_{m,n} \mid \mathcal{F}_{t_{m,n}}\big] \\
&\quad = \mathbb{E}^{\mathbb{P}}\big[I_{\sigma_n}(t_{m+1,n}) \otimes I_{\sigma_n}(t_{m+1,n}) - \Delta_{m,n} \otimes I_{\sigma_n}(t_{m,n}) \\
&\qquad\qquad - I_{\sigma_n}(t_{m,n}) \otimes \Delta_{m,n} - I_{\sigma_n}(t_{m,n}) \otimes I_{\sigma_n}(t_{m,n}) \mid \mathcal{F}_{t_{m,n}}\big] \\
&\quad = \mathbb{E}^{\mathbb{P}}\big[I_{\sigma_n}(t_{m+1,n}) \otimes I_{\sigma_n}(t_{m+1,n}) - I_{\sigma_n}(t_{m,n}) \otimes I_{\sigma_n}(t_{m,n}) \mid \mathcal{F}_{t_{m,n}}\big] \\
&\quad = \mathbb{E}^{\mathbb{P}}\big[A_{m,n} \mid \mathcal{F}_{t_{m,n}}\big],
\end{aligned}$$

the terms

$$\mathrm{Trace}\Big(\nabla^2_{(2)}\varphi(V(t_{m,n}), I_{m,n})\big(\Delta_{m,n} \otimes \Delta_{m,n} - A_{m,n}\big)\Big)$$

are orthogonal in $L^2(\mathbb{P};\mathbb{R})$, and so

$$\begin{aligned}
&\mathbb{E}^{\mathbb{P}}\left[\left(\sum_{m<2^n t} \mathrm{Trace}\Big(\nabla^2_{(2)}\varphi(V(t_{m,n}), I_{m,n})\big(\Delta_{m,n} \otimes \Delta_{m,n} - A_{m,n}\big)\Big)\right)^2\right] \\
&\quad = \sum_{m<2^n t} \mathbb{E}^{\mathbb{P}}\left[\left(\mathrm{Trace}\Big(\nabla^2_{(2)}\varphi(V(t_{m,n}), I_{m,n})\big(\Delta_{m,n} \otimes \Delta_{m,n} - A_{m,n}\big)\Big)\right)^2\right].
\end{aligned}$$

Obviously,

$$\|A_{m,n}\|_{\mathrm{H.S.}} \le C2^{-n} \quad \text{and} \quad \|\Delta_{m,n} \otimes \Delta_{m,n}\|_{\mathrm{H.S.}} \le C|B(t_{m+1,n} - B(t_{m,n})|^2$$

for some $C < \infty$. Hence, since

$$\mathbb{E}^{\mathbb{P}}\big[|B(t_{m+1,n}) - B(t_{m,n})|^4\big] = 2^{-2n}\mathbb{E}^{\mathbb{P}}\big[|B(1)|^4\big],$$

it follows that

$$\sum_{m<2^n t} \mathrm{Trace}\Big(\nabla^2_{(2)}\varphi(V(t_{m,n}), I_{m,n})\big(\Delta_{m,n} \otimes \Delta_{m,n} - A_{m,n}\big)\Big) \longrightarrow 0$$

in $L^2(\mathbb{P};\mathbb{R})$, and so the proof of (3.3.1) is complete.

It should be clear that when $X(\,\cdot\,, \mathbf{x})$ is the solution to (3.0.2), the martingale in (2.2.6) is the stochastic integral

$$\int_0^t \big(\sigma(X(\tau,\mathbf{x}))^\top \nabla\varphi(\tau, X(\tau,\mathbf{x})), dw(\tau)\big)_{\mathbb{R}^M}.$$

3.3.1 *Burkholder's inequality*

Let $\sigma \in PM^2_{\mathrm{loc}}(\mathrm{Hom}(\mathbb{R}^M;\mathbb{R}^N))$, and set $A(t) = \int_0^t \sigma(\tau)\sigma(\tau)^\top\,d\tau$. When σ is deterministic, $I_\sigma(t)$ is a centered Gaussian random variable with covariance $A(t)$, and therefore, for each $p \in [1,\infty)$

$$\mathbb{E}^{\mathbb{P}}\big[\big|(\boldsymbol{\xi}, I_\sigma(t))_{\mathbb{R}^N}\big|^p\big] = C_p\big(\boldsymbol{\xi}, A(t)\boldsymbol{\xi}\big)_{\mathbb{R}^N}^{\frac{p}{2}} \quad \text{for all } \boldsymbol{\xi} \in \mathbb{R}^N,$$

where $C_p = \int |y|^p\,\gamma_{0,1}(dy)$. Our first application of (3.3.1) shows that moments of $I_\sigma(\,\cdot\,)$ can be estimated in terms of $A(\,\cdot\,)$ even when σ is random. Namely, given $p \in [2,\infty)$, observe that

$$\nabla^2|\mathbf{y}|^p = p(p-2)|\mathbf{y}|^{p-2}\frac{\mathbf{y}\otimes\mathbf{y}}{|\mathbf{y}|^2} + p|\mathbf{y}|^{p-2}\mathbf{I}.$$

Therefore, by (3.3.1),

$$|I_\sigma(t)|^p - \frac{1}{2}\int_0^t \left(p(p-2)|I_\sigma(\tau)|^{p-2}\frac{|\sigma^\top(\tau)I_\sigma(\tau)|^2}{|I_\sigma(\tau)|^2} + p\|\sigma(\tau)\|_{\mathrm{H.S.}}^2|I_\sigma(\tau)|^{p-2}\right)d\tau$$

is a local $\mathbb{P}$-martingale relative to $\{\mathcal{F}_t : t \ge 0\}$. Thus, if

$$\zeta_R = \inf\{t \ge 0 : |I_\sigma(\tau)| \vee \mathrm{Trace}(A(\tau)) \ge R\},$$

then, by Hölder's inequality, one sees that

$$\begin{aligned}
\mathbb{E}^{\mathbb{P}}\big[|I_\sigma(t\wedge\zeta_R)|^p\big] &\le \frac{p(p-1)}{2}\mathbb{E}^{\mathbb{P}}\left[\int_0^{t\wedge\zeta_R} \|\sigma(\tau)\|_{\mathrm{H.S.}}^2|I_\sigma(\tau)|^{p-2}\,d\tau\right]\\
&\le \frac{p(p-1)}{2}\mathbb{E}^{\mathbb{P}}\left[\mathrm{Trace}\big(A(t\wedge\zeta_R)\big)\|I_\sigma(\,\cdot\,)\|_{[0,t\wedge\zeta_R]}^{p-2}\right]\\
&\le \frac{p(p-1)}{2}\mathbb{E}^{\mathbb{P}}\big[\mathrm{Trace}\big(A(t)\big)^{\frac{p}{2}}\big]^{\frac{2}{p}}\mathbb{E}^{\mathbb{P}}\big[\|I_\sigma(\,\cdot\,)\|_{[0,t\wedge\zeta_R]}^p\big]^{1-\frac{2}{p}},
\end{aligned}$$

and then, using Doob's inequality, one concludes that

$$\mathbb{E}^{\mathbb{P}}\big[\|I_\sigma(\,\cdot\,)\|_{[0,t\wedge\zeta_R]}^p\big]^{\frac{2}{p}} \le \left(\frac{p^{p+1}}{2(p-1)^{p-1}}\right)\mathbb{E}^{\mathbb{P}}\big[\mathrm{Trace}\big(A(t)\big)^{\frac{p}{2}}\big]^{\frac{2}{p}}.$$

Therefore, after letting $R \to \infty$, we have proved that, for each $p \in [2,\infty)$,

$$\mathbb{E}^{\mathbb{P}}\big[|I_\sigma(t)|^p\big] \le \mathbb{E}^{\mathbb{P}}\big[\|I_\sigma(\,\cdot\,)\|_{[0,t]}^p\big] \le K_p\mathbb{E}^{\mathbb{P}}\big[\mathrm{Trace}\big(A(t)\big)^{\frac{p}{2}}\big] \tag{3.3.2}$$

where $K_p = \left(\frac{p^{p+1}}{2(p-1)^{p-1}}\right)^{\frac{p}{2}}$. This estimate is a very special case of a much more general inequality known as **Burkholder's inequality**. See (4.3.1) for a generalization and Exercise 3.6 for an entirely different approach.

3.3.2 *Tanaka's formula*

The preceding application of (3.3.1) is mundane by comparison to the one made by H. Tanaka to represent what Lévy called *local time* for an $\mathbb{R}$-valued Brownian motion $\big(B(t), \mathcal{F}_t, \mathbb{P}\big)$. To describe local time, define the *occupation time* measures $L(t, \cdot\,)$ for $t \geq 0$ by

$$L(t, \Gamma) = \int_0^t \mathbf{1}_\Gamma\big(B(\tau)\big)\, d\tau, \quad \Gamma \in \mathcal{B}_\mathbb{R}.$$

One of Lévy's many remarkable discoveries is that there is a map $\omega \in \Omega \longmapsto \ell(\,\cdot\,, \cdot\cdot\,)(\omega) \in C\big([0,\infty) \times \mathbb{R}; [0,\infty)\big)$ such that $\ell(\,\cdot\,, y)$ a is progressively measurable, non-decreasing function with the property that

$$\mathbb{P}\left(L(t, \Gamma) = \int_\Gamma \ell(t, y)\, dy \text{ for all } t \geq 0 \;\&\; \Gamma \in \mathcal{B}_\mathbb{R}\right) = 1. \tag{3.3.3}$$

In other words, with probability 1, $L(t, \cdot\,)$ is absolutely continuous with respect to Lebesgue measure $\lambda_\mathbb{R}$ for all $t \geq 0$, and

$$\frac{L(t, dy)}{\lambda_\mathbb{R}(dy)} = \ell(t, y).$$

Notice that this result is another manifestation of the non-differentiability of Brownian paths. Indeed, suppose that $p\colon [0,\infty) \longrightarrow \mathbb{R}$ is a continuously differentiable path, and let $t \rightsquigarrow \mu(t, \cdot\,)$ be its occupation time measures. If $\dot{p} = 0$ on an interval $[a, b]$, then it is clear that, for $t > a$, $\mu(t, \{p(a)\}) \geq (t \wedge b - a)$, and therefore μ_t can't be absolutely continuous with respect to $\lambda_\mathbb{R}$. On the other hand, if $\dot{p} > 0$ on $[a, b]$, then, for $t \in (a, b]$,

$$\frac{\mu(t, dy) - \mu(a, dy)}{\lambda_\mathbb{R}(dy)} = \frac{\mathbf{1}_{[a,t]}(y)}{\dot{p} \circ (p \restriction [p(a), p(t)])^{-1}(y)},$$

and so $\mu(t, \cdot\,) - \mu(a, \cdot\,)$ is absolutely continuous but its Radon–Nikodym derivative cannot be continuous. It is only because a Brownian path *dithers* as it leaves points that its occupation time measure can admit a continuous density.

Tanaka's idea to prove the existence of local time is based on the heuristic representation of $\ell(t, y)$ as

$$\int_0^t \delta_y\big(B(\tau)\big)\, d\tau,$$

where δ_y is the Dirac delta function at y. Because $h_y' = \mathbf{1}_{[y,\infty)}$ and $h_y'' = \delta_y$ when $h_y(x) = x \vee y$, a somewhat cavalier application of (3.3.1) to h_y led Tanaka to guess that

$$\tfrac{1}{2}\ell(t,y) = B(t)\vee y - 0\vee y - \int_0^t \mathbf{1}_{[y,\infty)}\big(B(\tau)\big)\,dB(\tau).$$

The key step in the justification of his idea is the proof that there is a continuous map $(t,y) \rightsquigarrow I(t,y)$ such that

$$\mathbb{P}\left(I(t,y) = \int_0^t \mathbf{1}_{[y,\infty)}\big(B(\tau)\big)\,dB(\tau) \text{ for all } t\ge 0\right) = 1 \quad \text{for each } y\in\mathbb{R}.$$

To prove this, use (3.3.2) to see that there is a $C<\infty$ such that

$$\begin{aligned}\mathbb{E}^{\mathbb{P}}&\left[\sup_{t\in[0,T]}\left|\int_0^t \mathbf{1}_{[y,\infty)}\big(B(\tau)\big)\,dB(\tau) - \int_0^t \mathbf{1}_{[x,\infty)}\big(B(\tau)\big)\,dB(\tau)\right|^4\right]\\ &\le C\mathbb{E}^{\mathbb{P}}\left[\left(\int_0^T \mathbf{1}_{[x,y]}\big(B(\tau)\big)\,d\tau\right)^2\right]\end{aligned}$$

for $T>0$ and $x<y$. Next use the Markov property for Brownian motion to write

$$\begin{aligned}\mathbb{E}^{\mathbb{P}}&\left[\left(\int_0^T \mathbf{1}_{[x,y]}\big(B(\tau)\big)\,d\tau\right)^2\right]\\ &= 2\mathbb{E}^{\mathbb{P}}\left[\int_{0\le\tau_1<\tau_2\le T} \mathbf{1}_{[x,y]}\big(B(\tau_1)\big)\mathbf{1}_{[x,y]}\big(B(\tau_2)\big)\,d\tau_1 d\tau_2\right]\\ &= \frac{1}{\pi}\int_{0\le\tau_1<\tau_2\le T} \tau_1^{-\frac12}(\tau_2-\tau_1)^{-\frac12}\left(\int_{[x,y]^2} e^{-\frac{\xi_1^2}{2\tau_1}} e^{-\frac{(\xi_2-\xi_1)^2}{2(\tau_2-\tau_1)}}\,d\xi_1 d\xi_2\right) d\tau_1 d\tau_2\\ &\le T(y-x)^2.\end{aligned}$$

Hence the existence of $(t,y)\rightsquigarrow I(t,y)$ follows from Kolmogorov's continuity criterion in Theorem 2.1.2.

We now define

$$\ell(t,y) = 2\big(B(t)\vee y - B(t)\vee 0 - I(t,y)\big).$$

Given $\varphi\in C_c(\mathbb{R};\mathbb{R})$, set $f(x) = \int_{\mathbb{R}}(x\vee y)\varphi(y)\,dy$. Then

$$f'(x) = \int_{-\infty}^x \varphi(y)\,dy \quad \text{and} \quad f'' = \varphi,$$

and so, by (3.3.1),

$$f\big(B(t)\big) = f(0) + \int_0^t f'\big(B(\tau)\big)\,dB(\tau) + \frac12\int_0^t \varphi\big(B(\tau)\big)\,d\tau.$$

At the same time,

$$\frac{1}{2}\int \varphi(y)\ell(t,y)\,dy = f\big(B(t)\big) - f(0) - \int \varphi(y)I(t,y)\,dy,$$

and because φ and $I(t,\,\cdot\,)$ are continuous and φ has compact support,

$$\begin{aligned}\int \varphi(y)I(t,y)\,dy &= \lim_{n\to\infty}\frac{1}{n}\sum_{m\in\mathbb{N}}\varphi\Big(\frac{m}{n}\Big)I\Big(t,\frac{m}{n}\Big)\\ &= \lim_{n\to\infty}\int_0^t\bigg(\frac{1}{n}\sum_{m\in\mathbb{N}}\varphi\Big(\frac{m}{n}\Big)\mathbf{1}_{(-\infty,B(\tau)]}\Big(\frac{m}{n}\Big)\bigg)\,dB(\tau) = \int_0^t f'\big(B(\tau)\big)\,dB(\tau)\end{aligned}$$

in $L^2(\mathbb{P};\mathbb{R})$. Hence, we have shown that

$$\int \varphi(y)\ell(t,y)\,dy = \int_0^t \varphi\big(B(\tau)\big)\,d\tau \quad (\text{a.s.}, \mathbb{P})$$

for each $\varphi \in C_c(\mathbb{R}^N;\mathbb{R})$. Starting from this, an elementary argument shows that (3.3.3) holds, and once one has (3.3.3), there is no reason not to replace $\ell(t,y)$ by $\|\ell(\,\cdot\,,y)\|_{[0,t]}$ so that it becomes non-decreasing.

The function $\ell(\,\cdot\,,y)$ has many strange and intriguing properties. Since it is non-decreasing, it determines a Borel measure $\ell(dt,y)$ on $[0,\infty)$ for which

$$\ell\big((s,t],y\big) = \ell(t,y) - \ell(s,y) \quad \text{for } 0 \le s < t,$$

and we will now show

$$\mathbb{P}\Big(\ell\big(\{t : B(t) \ne y\},y\big) = 0\Big) = 1, \tag{3.3.4}$$

which is the reason why $\ell(\,\cdot\,,y)$ is called the **local time at** y. To this end, first observe that if f is a bounded $\mathcal{B}_{[0,\infty)} \times \mathcal{B}_{\mathbb{R}} \times \mathcal{F}$-measurable $\mathbb{R}$-valued function and $(t,x) \rightsquigarrow f(t,x,\omega)$ is continuous for each $\omega \in \Omega$, then, by using Riemann sum approximations, one can check that

$$(*) \qquad \int\bigg(\int_0^t f(\tau,x,\omega)\,\ell(d\tau,x)(\omega)\bigg)\,dx = \int_0^t f\big(\tau,B(\tau)(\omega),\omega\big)\,d\tau$$

for $\mathbb{P}$-almost every ω. Now choose continuous functions $\rho : \mathbb{R} \longrightarrow [0,\infty)$ and $\eta\colon \mathbb{R} \longrightarrow [0,1]$ so that $\rho = 0$ off $(-1,1)$ and has total integral 1, and $\eta = 0$ on $[-1,1]$ and 1 off $(-2,2)$. Next, for $\epsilon, R > 0$, define $\rho_\epsilon(x) = \epsilon^{-1}\rho(\epsilon^{-1}x)$ and $\eta_R(x) = \eta(Rx)$, and set

$$f_{\epsilon,R}(t,x,\omega) = \rho_\epsilon(x-y)\eta_R\big(B(t)(\omega) - y\big).$$

If $0 < \epsilon < R^{-1}$, then $f_{\epsilon,R}\big(\tau,B(\tau)(\omega),\omega\big) = 0$, and therefore, by $(*)$,

$$\int \rho_\epsilon(x-y)\left(\int_0^t \eta_R\big(B(\tau)-y\big)\,\ell(d\tau,x)\right)dx = 0 \quad (\text{a.s.}, \mathbb{P}).$$

Thus (3.3.4) follows after one first lets $\epsilon \searrow 0$ and then $R \nearrow \infty$.

As a consequence of (3.3.4), we know that, with probability 1, $t \rightsquigarrow \ell(t,y)$ is a singular, continuous, non-decreasing function. Indeed, if

$$S(\omega) = \{\tau \in [0,\infty) : B(\tau)(\omega) = y\},$$

then the expected value of the Lebesgue measure of S equals

$$\mathbb{E}^{\mathbb{P}}\left[\int_0^\infty \mathbf{1}_{\{y\}}\big(B(\tau)\big)\,d\tau\right] = \int_0^\infty \gamma_{0,\tau}(\{y\})\,d\tau = 0.$$

Here is an interesting application of local time. Because $x = x \vee 0 + x \wedge 0$ and $|x| = x \vee 0 - x \wedge 0$,

$$\begin{aligned} B(t) \wedge 0 &= B(t) - \int_0^t \mathbf{1}_{[0,\infty)}\big(B(\tau)\big)\,dB(\tau) - \tfrac{1}{2}\ell(t,0) \\ &= \int_0^t \mathbf{1}_{(-\infty,0)}\big(B(\tau)\big)\,dB(\tau) - \tfrac{1}{2}\ell(t,0), \end{aligned}$$

and so

$$|B(t)| = \widetilde{B}(t) + \ell(t,0) \quad \text{where } \widetilde{B}(t) := \int_0^t \operatorname{sgn}\big(B(\tau)\big)\,dB(\tau). \qquad (3.3.5)$$

Now observe that, for all $\xi \in \mathbb{R}$,

$$\left(e^{i\xi\widetilde{B}(t)+\frac{\xi^2}{2}}, \mathcal{F}_t, \mathbb{P}\right)$$

is a martingale, and (cf. Exercise 2.5) conclude that $\big(\widetilde{B}(t), \mathcal{F}_t, \mathbb{P}\big)$ is a Brownian motion. Hence, since $\ell(\,\cdot\,,0)$ is $\mathbb{P}$-almost surely constant on time intervals during which B stays away from 0, (3.3.5) says that $t \rightsquigarrow |B(t)|$ behaves like a Brownian motion on such intervals and is prevented from becoming negative because $\ell(\,\cdot\,,0)$ gives it a kick to the right whenever it visits 0. It should also be noted that

$$\widetilde{\mathcal{F}}_t := \overline{\sigma\big(\{\widetilde{B}(\tau) : \tau \in [0,t]\}\big)}^{\mathbb{P}} \subseteq \overline{\sigma\big(\{|B(\tau)| : \tau \in [0,t]\}\big)}^{\mathbb{P}}.$$

To check this, remember that, $\mathbb{P}$-almost surely,

$$\ell(t,0) = \lim_{\delta \searrow 0} \frac{1}{2\delta}\int_0^t \mathbf{1}_{[-\delta,\delta]}\big(B(\tau)\big)\,d\tau = \lim_{\delta \searrow 0} \frac{1}{2\delta}\int_0^t \mathbf{1}_{[0,\delta]}\big(|B(\tau)|\big)\,d\tau,$$

and therefore both $\ell(t,0)$ and $\widetilde{B}(t)=|B(t)|-\ell(t,0)$ are $\overline{\sigma\big(\{|B(\tau)|:\tau\in[0,t]\}\big)}^{\mathbb{P}}$-measurable. In particular, this means that $\operatorname{sgn}\big(B(t)\big)$ is not $\widetilde{\mathcal{F}}_t$-measurable, and so $B(t)$ isn't either. Even so, $B(\,\cdot\,)$ is a solution to the stochastic integral equation

$$(*)\qquad X(t)=\int_0^t \operatorname{sgn}\big(X(\tau)\big)\,d\widetilde{B}(\tau).$$

To see this, all that one has to show is that

$$\int_0^t \xi(\tau)\,d\widetilde{B}(\tau)=\int_0^t \xi(\tau)\operatorname{sgn}\big(B(\tau)\big)\,dB(\tau)$$

for bounded $\xi\in PM^2_{\text{loc}}(\mathbb{R})$ and then take $\xi(\tau)=\operatorname{sgn}\big(B(\tau)\big)$. When $\xi(\,\cdot\,)$ is continuous (cf. Exercise 3.1)

$$\begin{aligned}&\int_0^t \xi(\tau)\operatorname{sgn}\big(B(\tau)\big)\,dB(\tau)\\&\quad=\lim_{n\to\infty}\sum_{m<2^nt}\xi(m2^{-n})\big(\widetilde{B}(t\wedge(m+2)2^{-n})-\widetilde{B}(m2^{-n})\big)=\int_0^t \xi(\tau)\,d\widetilde{B}(\tau),\end{aligned}$$

and so the general case follows when one uses the same approximation procedure that allowed us to define stochastic integrals. The interest of this observation is that it shows that solutions to stochastic integral equations need not be measurable with respect to the driving Brownian motion. In addition, it shows that there are stochastic integral equations all of whose solutions have the same distribution even though there is more than one solution. Indeed, every solution to $(*)$ must be a Brownian motion, but both $B(\,\cdot\,)$ and $-B(\,\cdot\,)$ are solutions.

We will now examine some further properties of $\ell(t,y)$.

Lemma 3.3.1. *For $s,t>0$ and $y\neq 0$,*

$$\mathbb{P}\big(\ell(t,y)>s\big)=\left(\frac{y^2}{2\pi}\right)^{\frac12}\int_0^t \tau^{-\frac32}e^{-\frac{y^2}{2\tau}}\mathbb{P}\big(\ell(t-\tau,0)>s\big)\,d\tau.$$

In addition,

$$\mathbb{P}\big(\ell(t,0)>0\big)=1\quad\textit{for all } t>0,\tag{3.3.6}$$

and so

$$\mathbb{P}\big(\ell(t,y)>0\big)=2\mathbb{P}\big(B(t)>|y|\big).$$

Proof. First observe that the distribution of $\ell(\,\cdot\,,-y)$ is the same as that of $\ell(\,\cdot\,,y)$. Thus we will assume that $y>0$ throughout.

Next set $\zeta_y=\inf\{t\geq 0: B(t)\geq y\}$. Then

$$\begin{aligned} L\big(t,[y,y+\delta]\big) &= \int_{t\wedge\zeta_y}^{t} \mathbf{1}_{[y,y+\delta]}\big(B(\tau)\big)\,d\tau \\ &= \int_0^{t-t\wedge\zeta_y} \mathbf{1}_{[0,\delta]}\big(B(\tau+t\wedge\zeta_y)-B(t\wedge\zeta_y)\big)\,d\tau. \end{aligned}$$

Since (cf. Exercise 2.5)

$$\big(B(\tau+t\wedge\zeta_y)-B(t\wedge\zeta_y),\mathcal{F}_{\tau+t\wedge\zeta_y},\mathbb{P}\big)$$

is a Brownian motion that is independent of $\mathcal{F}_{t\wedge\zeta_y}$, it follows that, for any bounded, continuous $f\colon[0,\infty)\longrightarrow\mathbb{R}$ with $f(0)=0$,

$$\begin{aligned} &\mathbb{E}^{\mathbb{P}}\Big[f\Big(\frac{1}{\delta}L\big(t,[y,y+\delta]\big)\Big)\Big] \\ &= \int_{\{\zeta_y<t\}} \Big(\int f\Big(\frac{1}{\delta}L\big(t-\zeta_y(\omega_1),[0,\delta]\big)(\omega_2)\Big)\,\mathbb{P}(d\omega_2)\Big)\,\mathbb{P}(d\omega_1). \end{aligned}$$

Hence, after letting $\delta\searrow 0$, we have

$$\mathbb{E}^{\mathbb{P}}\big[f\big(\ell(t,y)\big)\big] = \int_{\{\zeta_y<t\}} \Big(\int f\Big(\ell\big(t-\zeta_y(\omega_1),0\big)(\omega_2)\Big)\,\mathbb{P}(d\omega_2)\Big)\,\mathbb{P}(d\omega_1),$$

from which it follows that

$$\mathbb{P}\big(\ell(t,y)>s\big) = \int_{\{\zeta_y<t\}} \mathbb{P}\big(\ell(t-\zeta_y(\omega),0)>s\big)\,\mathbb{P}(d\omega).$$

Starting from here and using the fact (cf. parts (iv) and (v) in Exercise 2.5) that $\mathbb{P}(\zeta_y\le\tau)=2\mathbb{P}\big(B(\tau)>y\big)$, it is easy to verify the first assertion.

To prove (3.3.6), note that, because $\ell(t,0)=|B(t)|-\widetilde{B}(t)$ and $-\widetilde{B}(\,\cdot\,)$ is a Brownian motion, all that we have to do is note that

$$\mathbb{P}\big(\exists\tau\in(0,t)\ B(\tau)>0\big) = \lim_{y\searrow 0}\mathbb{P}\big(\zeta_y<t\big) = 2\lim_{y\searrow 0}\mathbb{P}\big(B(t)>y\big)=1. \qquad\square$$

We will close this discussion with another of Lévy's remarkable insights. What Lévy wanted to do is "count" the number of times that a Brownian path visits 0 by time t. Of course, due to its fuzzy nature, one suspects it visits infinitely often, and so one has to be careful about what one means. With this in mind, Lévy considered the number $N_\epsilon(t)$ of times before t that $B(\,\cdot\,)$ returns to 0 after leaving $(-\epsilon,\epsilon)$. That is, $N_\epsilon(t)\ge n$ if and only if there exist $0<\tau_1<\cdots<\tau_{2n}\le t$ such that $|B(\tau_{2m-1})|\ge\epsilon$ and $B(\tau_{2m})=0$ for $1\le m\le n$. He then showed that

$$\mathbb{P}\Big(\lim_{\epsilon\searrow 0}\|\epsilon N_\epsilon(\,\cdot\,)-\ell(\,\cdot\,,0)\|_{[0,t]}=0\Big)=1 \quad\text{for all } t>0. \tag{3.3.7}$$

One way to prove (3.3.7) is to use (3.3.5). Indeed, if $\zeta_0 = 0$ and, for $m \geq 1$,

$$\zeta_{2m-1} = \inf\{t \geq \zeta_{2m-2} : |B(t)| \geq \epsilon\} \quad \text{and} \quad \zeta_{2m} = \inf\{t \geq \zeta_{2m-1} : B(t) = 0\},$$

then, because $B(\cdot)$ is continuous, $\zeta_m \geq t$ for all but a finite number of m's, and

$$\sum_{m\geq 1} \big(|B(t\wedge\zeta_{2m-1})| - |B(t\wedge\zeta_{2m})|\big) = \epsilon N_\epsilon(t) + \big(\epsilon - |B(t)|\big) \sum_{m\geq 1} \mathbf{1}_{(\zeta_{2m-1},\zeta_{2m})}(t).$$

On the other hand, because $B(\tau) \neq 0$ for $\tau \in (\zeta_{2m-1}, \zeta_{2m})$, $\ell(\zeta_{2m-1}, 0) = \ell(\zeta_{2m}, 0)$, and therefore

$$\begin{aligned}
\sum_{m\geq 1} \big(|B(t\wedge\zeta_{2m-1})| - |B(t\wedge\zeta_{2m})|\big) &= \sum_{m\geq 1} \big(\widetilde{B}(t\wedge\zeta_{2m-1}) - \widetilde{B}(t\wedge\zeta_{2m})\big) \\
&= -\widetilde{B}(t) - \sum_{m\geq 0} \big(\widetilde{B}(t\wedge\zeta_{2m}) - \widetilde{B}(t\wedge\zeta_{2m+1})\big) \\
&= -|B(t)| + \ell(t,0) + \int_0^t \xi(\tau)\, dB(\tau),
\end{aligned}$$

where

$$\xi(\tau) = \sum_{m\geq 0} \operatorname{sgn}\big(B(\tau)\big)\mathbf{1}_{[\zeta_{2m},\zeta_{2m+1})}(\tau).$$

After combining these, we see that

$$\begin{aligned}
&\epsilon N_\epsilon(t) - \ell(t,0) \\
&\quad = -|B(t)| \sum_{m\geq 0} \mathbf{1}_{[\zeta_{2m},\zeta_{2m+1}]}(t) - \epsilon \sum_{m\geq 1} \mathbf{1}_{(\zeta_{2m-1},\zeta_{2m})}(t) + \int_0^t \xi(\tau)\, dB(\tau).
\end{aligned}$$

Since $|B(t)| \leq \epsilon$ for $t \in [\zeta_{2m}, \zeta_{2m+1}]$, the absolute value of the sum of the first two terms on the right is no larger than ϵ and $|\xi(\tau)| \leq \mathbf{1}_{[0,\epsilon]}\big(|B(\tau)|\big)$. Hence,

$$\mathbb{E}^{\mathbb{P}}\big[\|\epsilon N_\epsilon(\cdot) - \ell(\cdot,0)\|_{[0,t]}^2\big]^{\frac{1}{2}} \leq \epsilon + 2\mathbb{E}^{\mathbb{P}}\left[\int_0^t \mathbf{1}_{[0,\epsilon]}\big(|B(\tau)|\big)\, d\tau\right]^{\frac{1}{2}},$$

which means that there is a $C(t) < \infty$ such that

$$\mathbb{E}^{\mathbb{P}}\big[\|\epsilon N_\epsilon(\cdot) - \ell(\cdot,0)\|_{[0,t]}^2\big]^{\frac{1}{2}} \leq C(t)\epsilon^{\frac{1}{2}} \quad \text{for } \epsilon \in (0,1].$$

In particular,

$$\sum_{k\geq 1} \mathbb{E}^{\mathbb{P}}\big[\|k^{-4} N_{k^{-4}}(\cdot) - \ell(\cdot,0)\|_{[0,t]}^2\big]^{\frac{1}{2}} < \infty,$$

and so $\|k^{-4}N_{k^{-4}}(\,\cdot\,) - \ell(\,\cdot\,,0)\|_{[0,t]} \longrightarrow 0$ (a.s., $\mathbb{P}$). Finally, if $(k+1)^{-4} \le \epsilon \le k^{-4}$, then

$$(k+1)^{-4}N_{k^{-4}}(\,\cdot\,) \le \epsilon N_\epsilon(\,\cdot\,) \le k^{-4}N_{(k+1)^{-4}}(\,\cdot\,),$$

and therefore (3.3.7) holds.

In conjunction with Lemma 3.3.1, (3.3.7) gives abundant, quantitative evidence of the dithering nature of Brownian paths. In fact, it says that, for any $s \ge 0$ and $\delta > 0$, with probability 1, $B(\,\cdot\,)$ leaves and revisits $B(s)$ infinitely often before time $s + \delta$.

3.4 Spacial properties of solutions to (3.0.2)

Let $X(\,\cdot\,,\mathbf{x})$ be the solution to (3.0.2), where σ and b are uniformly Lipschitz continuous. As yet we have concentrated on $X(t,\mathbf{x})$ as a function of t for each fixed $\mathbf{x}$. Indeed, because, for each $\mathbf{x}$, $X(\,\cdot\,,\mathbf{x})$ is defined only up to a set of $\mathcal{W}$-measure 0, we can't even talk about it as a function $(t,\mathbf{x})$. In this section we will show that there is a version of $(t,\mathbf{x}) \rightsquigarrow X(t,\mathbf{x})$ which is continuous and, when σ and b are differentiable, this version can be taken to be differentiable as well.

3.4.1 *Spacial continuity*

Recall the Euler approximations $X_n(t,\mathbf{x})$ in (3.0.1). It is evident that, for all $w \in \mathbb{W}(\mathbb{R}^M)$, $(t,\mathbf{x}) \rightsquigarrow X_n(t,\mathbf{x})(w)$ is continuous. Now set $\Delta_n(t,\mathbf{x}) = X(t,\mathbf{x}) - X_n(t,\mathbf{x})$. Using (3.3.2) and arguing as we did when $p=2$, one can show that for each $p \in [2,\infty)$ and $t > 0$ there is a $C_p(t) < \infty$ such that

$$\begin{aligned}
&\mathbb{E}^{\mathcal{W}}\big[\|\Delta_n(\,\cdot\,,\mathbf{x})\|_{[0,t]}^p\big]^{\frac{1}{p}} \le C_p(t)(1+|\mathbf{x}|)2^{\frac{n}{2}},\\
&\mathbb{E}^{\mathcal{W}}\big[\|X_n(\,\cdot\,,\mathbf{x}) - X_n(s,\mathbf{x})\|_{[s,t]}^p\big]^{\frac{1}{p}} \le C_p(t)(1+|\mathbf{x}|)(t-s)^{\frac{1}{2}},\\
&\mathbb{E}^{\mathcal{W}}\big[\|X_n(\,\cdot\,,\mathbf{y}) - X_n(\,\cdot\,,\mathbf{x})\|_{[0,t]}^p\big]^{\frac{1}{p}} \le C_p(t)|\mathbf{y}-\mathbf{x}|
\end{aligned} \tag{3.4.1}$$

for $n \ge 0$, $0 \le s < t$, and $\mathbf{x},\mathbf{y} \in \mathbb{R}^N$. From these it was clear that, for each $\mathbf{x}$, $\|\Delta_n(\,\cdot\,,\mathbf{x})\|_{[0,t]} \longrightarrow 0$ both (a.s., $\mathcal{W}$) and in $L^p(\mathcal{W};\mathbb{R}^N)$, but we will now use Kolmogorov's continuity criterion to show that this convergence is uniform with respect to $\mathbf{x}$ in compact subsets. To this end, note that, for $p \in [2,\infty)$,

$$\begin{aligned}
&|\Delta_n(t,\mathbf{y}) - \Delta_n(t,\mathbf{x})|^p\\
&\quad = \big|\big(X(t,\mathbf{y}) - X(t,\mathbf{x})\big) - \big(X_n(t,\mathbf{y}) - X_n(t,\mathbf{x})\big)\big|^{\frac{p}{2}}|\Delta_n(t,\mathbf{y}) - \Delta_n(t,\mathbf{x})|^{\frac{p}{2}},
\end{aligned}$$

apply Schwarz's inequality to get

$$\begin{aligned}
&\mathbb{E}^{\mathcal{W}}\big[\|\Delta_n(\,\cdot\,,\mathbf{y})-\Delta_n(\,\cdot\,,\mathbf{x})\|_{[0,t]}^p\big]\\
&\quad\le \mathbb{E}^{\mathcal{W}}\Big[\big\|\big(X(\,\cdot\,,\mathbf{y})-X(\,\cdot\,,\mathbf{x})\big)-\big(X_n(\,\cdot\,,\mathbf{y})-X_n(\,\cdot\,,\mathbf{x})\big)\big\|_{[0,t]}^p\Big]^{\frac{1}{2}}\\
&\qquad\times \mathbb{E}^{\mathcal{W}}\big[\|\Delta_n(\,\cdot\,,\mathbf{y})-\Delta_n(\,\cdot\,,\mathbf{x})\|_{[0,t]}^p\big]^{\frac{1}{2}},
\end{aligned}$$

and, after combining these with the first and third estimates in (3.4.1), conclude that, for each $R>0$, there is a $K_p(t,R)<\infty$ such that

$$\mathbb{E}^{\mathcal{W}}\big[\|\Delta_n(\,\cdot\,,\mathbf{y})-\Delta_n(\,\cdot\,,\mathbf{x})\|_{[0,t]}^p\big]^{\frac{1}{p}}\le K_p(t,R)2^{-\frac{n}{4}}|\mathbf{y}-\mathbf{x}|^{\frac{1}{2}}\text{ for }\mathbf{x},\mathbf{y}\in[-R,R]^N.$$

Hence, by taking $p>2N$, we can apply Theorem 2.1.2 to see that

$$\sup_{n\ge 0}2^{\frac{n}{4}}\mathbb{E}^{\mathcal{W}}\big[\|\Delta_n(\,\cdot\,,\,\cdot\,\cdot\,)\|_{[0,t]\times[-R,R]^N}^p\big]^{\frac{1}{p}}<\infty$$

and therefore that $(t,\mathbf{x})\rightsquigarrow X(t,\mathbf{x})$ can be chosen so that it is continuous and, $\mathcal{W}$-almost surely, $\|X_n(\,\cdot\,,\mathbf{x})-X(\,\cdot\,,\mathbf{x})\|_{[0,t]}\longrightarrow 0$ uniformly for $\mathbf{x}$ in compact subsets of $\mathbb{R}^N$.

3.4.2 *Spacial derivatives*

In preparation for proving the differentiability of $X(t,\,\cdot\,)$, we will need the following lemma, one that is harder to state than it is to prove.

Lemma 3.4.1. *Assume that σ and b are continuously differentiable and that their first derivatives are bounded. Define $\sigma_k^{(1)}:\mathbb{R}^N\longrightarrow\mathrm{Hom}(\mathbb{R}^N;\mathbb{R}^N)$ for $1\le k\le M$ and $b^{(1)}:\mathbb{R}^N\longrightarrow\mathrm{Hom}(\mathbb{R}^N;\mathbb{R}^N)$ by*

$$\left(\sigma_k^{(1)}\right)_{ij}=\frac{\partial\sigma_{i,k}}{\partial x_j}\quad\text{and}\quad b_{ij}^{(1)}=\frac{\partial b_i}{\partial x_j}.$$

For each $n\ge 0$, let $\widetilde{Y}_n:[0,\infty)\times\mathbb{R}^N\times\mathbb{W}(\mathbb{R}^M)\longrightarrow\mathrm{Hom}(\mathbb{R}^{\widetilde{N}};\mathbb{R}^N)$ be a Borel measurable map with the properties that $\widetilde{Y}_n(\,\cdot\,,\mathbf{x})$ is continuous and progressively measurable with respect to $\{W_t:t\ge 0\}$, and, for each $p\in[2,\infty)$ and $t>0$,

$$\begin{aligned}
\mathbb{E}^{\mathcal{W}}\big[\|\widetilde{Y}_n(\,\cdot\,,\mathbf{x})\|_{[0,t]}^p\big]^{\frac{1}{p}}&\le C_p(t),\\
\mathbb{E}^{\mathcal{W}}\big[\|\widetilde{Y}_n(\,\cdot\,,\mathbf{y})-\widetilde{Y}_n(\,\cdot\,,\mathbf{x})\|_{[0,t]}^p\big]^{\frac{1}{p}}&\le C_p(t)|\mathbf{y}-\mathbf{x}|,\\
\mathbb{E}^{\mathcal{W}}\big[\|\widetilde{Y}_n(\,\cdot\,,\mathbf{x})-\widetilde{Y}_n(s,\mathbf{x})\|_{[s,t]}\big]^{\frac{1}{p}}&\le C_p(t)(t-s)^{\frac{1}{2}}
\end{aligned}$$

and

$$\mathbb{E}^{\mathcal{W}}\big[\|\widetilde{Y}_{n+1}(\,\cdot\,,\mathbf{x})-\widetilde{Y}_n(\,\cdot\,,\mathbf{x})\|_{[0,t]}^p\big]^{\frac{1}{p}} \le C_p(t)2^{\frac{n}{2}}$$

for some $C_p(t)<\infty$. *If, for each* $n\ge 0$, $X_n(\,\cdot\,,\mathbf{x})$ *is given by* (3.0.1) *and*

$$\widetilde{X}_n\colon [0,\infty)\times\mathbb{R}^N\times\mathbb{W}(\mathbb{R}^M)\longrightarrow \operatorname{Hom}(\mathbb{R}^{\widetilde{N}};\mathbb{R}^N)$$

is determined by

$$\begin{aligned}\widetilde{X}_n(t,\mathbf{x}) = \sum_{k=1}^M \int_0^t \sigma_k^{(1)}\big(X_n(\lfloor\tau\rfloor_n,\mathbf{x})\big)\widetilde{X}_n(\lfloor\tau\rfloor_n,\mathbf{x})\,dw(\tau)_k \\ + \int_0^t b^{(1)}\big(X_n(\lfloor\tau\rfloor_n,\mathbf{x})\big)\widetilde{X}_n(\lfloor\tau\rfloor_n,\mathbf{x})\,d\tau + \widetilde{Y}_n(t,\mathbf{x}),\end{aligned}$$

then, for each $p\in[2,\infty)$, *there is a non-decreasing* $t\rightsquigarrow \widetilde{C}_p(t)$ *such that*

$$\begin{aligned}\mathbb{E}^{\mathcal{W}}\big[\|\widetilde{X}_n(\,\cdot\,,\mathbf{x})\|_{[0,t]}^p\big]^{\frac{1}{p}} &\le \widetilde{C}_p(t),\\ \mathbb{E}^{\mathcal{W}}\big[\|\widetilde{X}_n(\,\cdot\,,\mathbf{x})-\widetilde{X}_n(s,\mathbf{x})\|_{[s,t]}^p\big]^{\frac{1}{p}} &\le \widetilde{C}_p(t)(t-s)^{\frac{1}{2}},\\ \mathbb{E}^{\mathcal{W}}\big[\|\widetilde{X}_n(\,\cdot\,,\mathbf{y})-\widetilde{X}_n(\,\cdot\,,\mathbf{x})\|_{[0,t]}^p\big]^{\frac{1}{p}} &\le \widetilde{C}_p(t)|\mathbf{y}-\mathbf{x}|\\ \mathbb{E}^{\mathcal{W}}\big[\|\widetilde{X}_{n+1}(\,\cdot\,,\mathbf{x})-\widetilde{X}_n(\,\cdot\,,\mathbf{x})\|_{[0,t]}^p\big]^{\frac{1}{p}} &\le \widetilde{C}_p(t)2^{\frac{n}{2}},\end{aligned}$$

for all $n\ge 0$, $0\le s<t$, *and* $\mathbf{x},\mathbf{y}\in\mathbb{R}^N$.

Proof. There are no new ideas required to prove this lemma because each of the conclusions is proved in the same way as the corresponding estimate in (3.4.1). For example, to prove the first estimate, begin by observing that, by (3.3.2),

$$\begin{aligned}&\mathbb{E}^{\mathcal{W}}\big[\|\widetilde{X}_n(\,\cdot\,,\mathbf{x})\|_{[0,t]}^p\big]\\ &\le 3^{p-1}K_p\sum_{k=1}^M \mathbb{E}^{\mathcal{W}}\left[\left(\int_0^t \|\sigma_k^{(1)}\big(X_n(\lfloor\tau\rfloor_n,\mathbf{x})\big)\|_{\mathrm{H.S.}}^2\|\widetilde{X}_n(\lfloor\tau\rfloor_n,\mathbf{x})\|_{\mathrm{H.S.}}^2\,d\tau\right)^{\frac{p}{2}}\right]\\ &\quad + 3^{p-1}\mathbb{E}^{\mathcal{W}}\left[\left(\int_0^t \|b^{(1)}\big(X_n(\lfloor\tau\rfloor_n,\mathbf{x})\big)\|_{\mathrm{H.S.}}\big\|\widetilde{X}_n(\lfloor\tau\rfloor_n,\mathbf{x})\big\|_{\mathrm{H.S.}}\,d\tau\right)^p\right]\\ &\quad + 3^{p-1}\mathbb{E}^{\mathcal{W}}\big[\|\widetilde{Y}_n(t,\mathbf{x})\|_{\mathrm{H.S.}}^p\big].\end{aligned}$$

Thus there is an $A_p(t)<\infty$ such that

$$\mathbb{E}^{\mathcal{W}}\big[\|\widetilde{X}_n(\,\cdot\,,\mathbf{x})\|_{[0,t]}^p\big] \le A_p(t)\int_0^t \mathbb{E}^{\mathcal{W}}\big[\|\widetilde{X}_n(\,\cdot\,,\mathbf{x})\|_{[0,\tau]}^p\big]\,d\tau + 3^{p-1}C_p(t)^p,$$

and so the first estimate follows from Lemma 1.2.4.

The proofs of the other results are left as an exercise. □

Now assume that σ and b are twice differentiable and that their first and second derivatives are bounded. An advantage of the Euler approximation scheme is that, for each $n \geq 0$, $X_n(t,\,\cdot\,)$ is twice continuously differentiable and that its first derivatives satisfy

$$\begin{aligned}\frac{\partial X_n(t,\mathbf{x})_i}{\partial x_j} = \frac{\partial X_n(m2^{-n},\mathbf{x})_i}{\partial x_j} \\ &+ \sum_{\ell=1}^N \sum_{k=1}^M \frac{\partial \sigma\big(X_n(m2^{-n},\mathbf{x})\big)_{i,k}}{\partial x_\ell} \frac{\partial X_n(m2^{-n},\mathbf{x})_\ell}{\partial x_j} \big(w(t) - w(m2^{-n})\big)_k \\ &+ \sum_{\ell=1}^N \frac{\partial b\big(X_n(m2^{-n},\mathbf{x})\big)_i}{\partial x_\ell} \frac{\partial X_n(m2^{-n},\mathbf{x})_\ell}{\partial x_j} (t - m2^{-n})\end{aligned}$$

for $m2^{-n} \leq t \leq (m+1)2^{-n}$. Hence, if $X_n^{(1)}(t,\mathbf{x})$ is the Jacobian matrix of $X_n(t,\mathbf{x})$, then

$$\begin{aligned}X_n^{(1)}(t,\mathbf{x}) = \mathbf{I} + \sum_{k=1}^M \int_0^t \sigma_k^{(1)}\big(X_n(\lfloor\tau\rfloor_n,\mathbf{x})\big) X_n^{(1)}(\tau,\mathbf{x})\, dw(\tau)_k \\ + \int_0^t b^{(1)}\big(X_n(\lfloor\tau\rfloor_n,\mathbf{x})\big) X_n^{(1)}(\tau,\mathbf{x})\, d\tau,\end{aligned}$$

and so Lemma 3.4.1, applied with $\widetilde{N} = N$, $\widetilde{X} = X^{(1)}$, and $\widetilde{Y} = \mathbf{I}$, says that

$$\begin{aligned}\mathbb{E}^{\mathcal{W}}\big[\|X_n^{(1)}(\,\cdot\,,\mathbf{x})\|_{[0,t]}^p\big]^{\frac{1}{p}} &\leq C_p^{(1)}(t), \\ \mathbb{E}^{\mathcal{W}}\big[\|X_n^{(1)}(\,\cdot\,,\mathbf{x}) - X_n^{(1)}(s,\mathbf{x})\|_{[s,t]}^p\big]^{\frac{1}{p}} &\leq C_p^{(1)}(t)(t-s)^{\frac{1}{2}}, \\ \mathbb{E}^{\mathcal{W}}\big[\|X_n^{(1)}(\,\cdot\,,\mathbf{y}) - X_n^{(1)}(\,\cdot\,,\mathbf{x})\|_{[0,t]}^p\big]^{\frac{1}{p}} &\leq C_p^{(1)}(t)|\mathbf{y}-\mathbf{x}| \\ \mathbb{E}^{\mathcal{W}}\big[\|X_{n+1}^{(1)}(\,\cdot\,,\mathbf{x}) - X_n^{(1)}(\,\cdot\,,\mathbf{x})\|_{[0,t]}^p\big]^{\frac{1}{p}} &\leq C_p^{(1)}(t)2^{\frac{n}{2}}.\end{aligned} \tag{3.4.2}$$

Starting from these and proceeding in the same way as we did when we showed in §3.4.1 that $\mathbf{x} \rightsquigarrow X(t,\mathbf{x})$ is continuous, one sees that there is a measurable map

$$X^{(1)}\colon [0,\infty) \times \mathbb{R}^N \times \mathbb{W}(\mathbb{R}^M) \longrightarrow \mathrm{Hom}(\mathbb{R}^N;\mathbb{R}^N)$$

such that $(t,\mathbf{x}) \rightsquigarrow X^{(1)}(t,\mathbf{x})$ is continuous, $X^{(1)}(\,\cdot\,,\mathbf{x})$ is a progressively solution to

$$X^{(1)}(t,\mathbf{x}) = \mathbf{I} + \sum_{k=1}^{M} \int_0^t \sigma_k^{(1)}\big(X(\tau,\mathbf{x})\big)X^{(1)}(\tau,\mathbf{x})\,dw(\tau)_k \\ + \int_0^t b^{(1)}\big(X(\tau,\mathbf{x})\big)X^{(1)}(\tau,\mathbf{x})\,d\tau, \tag{3.4.3}$$

and, $\mathcal{W}$-almost surely, $X_n^{(1)} \longrightarrow X^{(1)}$ uniformly on compact subsets of $[0,\infty)\times\mathbb{R}^N$. As a result, we now know that $X(t,\,\cdot\,)$ is almost surely continuously differentiable and that $X^{(1)}(t,\mathbf{x})$ is $\mathcal{W}$-almost surely equal to its Jacobian matrix. In addition,

$$\begin{aligned} \mathbb{E}^{\mathcal{W}}\big[\|X^{(1)}(\,\cdot\,,\mathbf{x})\|_{[0,t]}\big]^{\frac{1}{p}} &\le C_p^{(1)}(t) \\ \mathbb{E}^{\mathcal{W}}\big[\|X^{(1)}(\,\cdot\,,\mathbf{x}) - X_n^{(1)}(\,\cdot\,,\mathbf{x})\|_{[0,t]}^p\big]^{\frac{1}{p}} &\le C_p^{(1)}(t)2^{\frac{n}{2}}, \\ \mathbb{E}^{\mathcal{W}}\big[\|X^{(1)}(\,\cdot\,,\mathbf{x}) - X^{(1)}(s,\mathbf{x})\|_{[s,t]}^p\big]^{\frac{1}{p}} &\le C_p^{(1)}(t)(t-s)^{\frac{1}{2}}, \\ \mathbb{E}^{\mathcal{W}}\big[\|X^{(1)}(\,\cdot\,,\mathbf{y}) - X^{(1)}(\,\cdot\,,\mathbf{x})\|_{[0,t]}^p\big]^{\frac{1}{p}} &\le C_p^{(1)}(t)|\mathbf{y}-\mathbf{x}|. \end{aligned} \tag{3.4.4}$$

Next assume that σ and b have three continuous derivatives, all of which are bounded. Then the X_n's have three continuous spacial derivatives and

$$X_n^{(2)}(t,\mathbf{x}) = \sum_{k=1}^{M} \int_0^t \sigma_k^{(1)}\big(X_n(\lfloor\tau\rfloor_n,\mathbf{x})\big)X_n^{(2)}(\lfloor\tau\rfloor_n,\mathbf{x})\big)\,dw(\tau)_k \\ + \int_0^t b^{(1)}\big(X_n(\lfloor\tau\rfloor_n,\mathbf{x})\big)X_n^{(2)}(\lfloor\tau\rfloor_n,\mathbf{x})\,d\tau + Y_n^{(2)}(t,\mathbf{x}),$$

where $X_n^{(2)}(t,\mathbf{x})_{i,(j_1,j_2)} = \frac{\partial^2 X_n(t,\mathbf{x})_i}{\partial x_{j_1}\partial x_{j_2}}$ and $Y_n^{(2)}(t,\mathbf{x})_{i,(j_1j_2)}$ equals

$$\sum_{k=1}^{M}\sum_{\ell_1,\ell_2=1}^{N} \int_0^t \frac{\partial^2\sigma_{i,k}}{\partial x_{\ell_1}\partial x_{\ell_2}}\big(X_n(\lfloor\tau\rfloor_n,\mathbf{x})\big) \\ \times X_n^{(1)}(\lfloor\tau\rfloor_n,\mathbf{x})_{\ell_1,j_1}X_n^{(1)}(\lfloor\tau\rfloor_n,\mathbf{x})_{\ell_2,j_2}\,dw(\tau)_k \\ + \sum_{\ell_1,\ell_2=1}^{N}\int_0^t \frac{\partial^2 b_i}{\partial x_{\ell_1}\partial x_{\ell_2}}\big(X_n(\lfloor\tau\rfloor_n,\mathbf{x})\big)X_n^{(1)}(\lfloor\tau\rfloor_n,\mathbf{x})_{\ell_1,j_1}X_n^{(1)}(\lfloor\tau\rfloor_n,\mathbf{x})_{\ell_2,j_2}\,d\tau.$$

Using (3.4.2), one sees that these quantities again satisfy the conditions in Lemma 3.4.1, and therefore that the analogs of the conclusions just drawn about the $X_n^{(1)}$'s hold for the $X_n^{(2)}$'s and lead to the existence of a map

$$X^{(2)}\colon [0,\infty)\times\mathbb{R}^N\times\mathbb{W}(\mathbb{R}^M) \longrightarrow \mathrm{Hom}\big(\mathrm{Hom}(\mathbb{R}^N;\mathbb{R}^N);\mathbb{R}^N\big)$$

with the properties that $X^{(2)}(\,\cdot\,,\,\cdot\,\cdot\,)$ is continuous,

$$\frac{\partial^2 X(t,\mathbf{x})_i}{\partial x_{j_1}\partial x_{j_2}} = X^{(2)}(t,\mathbf{x})_{i,(j_1 j_2)} \quad (\text{a.s.}, \mathcal{W}),$$

$X^{(2)}(\,\cdot\,,\mathbf{x})$ is progressively measurable for each $\mathbf{x} \in \mathbb{R}^N$, and

$$\begin{aligned}
&X^{(2)}(t,\mathbf{x})_{i,(j_1,j_2)} \\
&= \sum_{k=1}^{M}\sum_{\ell=1}^{N} \int_0^t \sigma_k^{(1)}\big(X(\tau,\mathbf{x})\big)_{i\ell} X^{(2)}(\tau,\mathbf{x})_{\ell,(j_1,j_2)}\, dw(\tau)_k \\
&+ \sum_{\ell=1}^{N} \int_0^t b^{(1)}\big(X(\tau,\mathbf{x})\big)_{i1ll} X^{(2)}(\tau,\mathbf{x})_{\ell,(j_1,j_2)}\, d\tau \\
&+ \sum_{k=1}^{M}\sum_{\ell_1,\ell_2=1}^{N} \int_0^t \frac{\partial^2 \sigma_{i,k}}{\partial x_{\ell_1}\partial x_{\ell_2}}\big(X(\tau,\mathbf{x})\big) X^{(1)}(\tau,\mathbf{x})_{\ell_1,j_1} X^{(1)}(\tau,\mathbf{x})_{\ell_2,j_2}\, dw(\tau)_k \\
&+ \sum_{\ell_1,\ell_2=1}^{N} \int_0^t \frac{\partial^2 b_i}{\partial x_{\ell_1}\partial x_{\ell_2}}\big(X(\tau,\mathbf{x})\big) X^{(1)}(\tau,\mathbf{x})_{\ell_1,j_1} X^{(1)}(\tau,\mathbf{x})_{\ell_2,j_2}\, d\tau.
\end{aligned}$$

Hence, $X(t,\,\cdot\,)$ has two continuous derivatives and, for each $p \in [2,\infty)$, there exists a $C_p^{(2)}(t) < \infty$ such that the analogs of (3.4.2) and (3.4.4) hold. In particular,

$$\sum_{i,j_1,j_2=1}^{N} \mathbb{E}^{\mathcal{W}}\Big[\big\|X^{(2)}(\tau,\,\cdot\,)_{i,(j_1,j_2)}\big\|_{[0,t]}^p\Big] \le C_p^{(2)}(t). \tag{3.4.5}$$

It should be clear that, by continuing in the same way, one can prove that, for any $n \ge 1$, $X(t,\,\cdot\,)$ has n continuous derivatives if σ and b have $(n+1)$ bounded continuous ones. In fact, an examination of the argument reveals that we could have afforded the derivatives of order greater than one to have polynomial growth and that the existence of the $(n+1)$st order derivatives could have been replaced by a modulus of continuity assumption on the nth derivatives.

3.4.3 *An application to Kolmogorov's backward equation*

Astute readers will have noticed that, since it was introduced in §1.2, nothing has been said about Kolmogorov's backward equation (1.2.4). The reason for its absence is that, until now, we did not know how to produce solutions to it.

Recall the transition probability function $P(t,\mathbf{x},\,\cdot\,)$ in §2.2.1, and set

$$u_\varphi(t,\mathbf{x}) = \int_{\mathbb{R}^N} \varphi(\mathbf{y})\, P(t,\mathbf{x},d\mathbf{y}).$$

If $u_\varphi(t,\,\cdot\,) \in C^2_{\mathrm{b}}(\mathbb{R}^N;\mathbb{C})$, then by (1.2.2) and (1.2.3)

$$\begin{aligned} &u_\varphi(t+h,\mathbf{x}) - u_\varphi(t,\mathbf{x}) \\ &\quad = \langle u_\varphi(t,\,\cdot\,), P(h,\mathbf{x},\,\cdot\,)\rangle - u_\varphi(t,\mathbf{x}) = \int_0^h \langle Lu_\varphi(t,\,\cdot\,), P(\tau,\mathbf{x},\,\cdot\,)\rangle\, d\tau, \end{aligned}$$

and so, after dividing by h and letting $h \searrow 0$, one sees that $\partial_t u_\varphi(t,\mathbf{x}) = Lu_\varphi(t,\mathbf{x})$. Now assume that σ and b have three bounded derivatives, and let $\varphi \in C^2_{\mathrm{b}}(\mathbb{R}^N;\mathbb{R})$ be given. Then $u_\varphi(t,\mathbf{x}) = \mathbb{E}^{\mathcal{W}}\big[\varphi(X(t,\mathbf{x})\big]$, and therefore, if $\mathbf{e}_i$ is the element of $\mathbb{S}^{N-1}$ for which $(\mathbf{e}_i)_j = \delta_{i,j}$, then

$$u_\varphi(t,\mathbf{x}+h\mathbf{e}_i) - u_\varphi(t,\mathbf{x}) = \mathbb{E}^{\mathcal{W}}\big[\varphi\big(X(t,\mathbf{x}+h\mathbf{e}_i)\big) - \varphi\big(X(t,\mathbf{x})\big)\big].$$

Since

$$\varphi\big(X(t,\mathbf{x}+h\mathbf{e}_i)\big) - \varphi\big(X(t,\mathbf{x})\big) = \sum_{j=1}^N \int_0^h \partial_{x_j}\varphi(t,\mathbf{x}+\xi\mathbf{e}_i) X^{(1)}(t,\mathbf{x}+\xi\mathbf{e}_i)_{j,i}\, d\xi,$$

we can use (3.4.4) to see that $\partial_{x_i} u_\varphi(t,\,\cdot\,)$ exists, is continuous, and is given by

$$\partial_{x_i} u_\varphi(t,\mathbf{x}) = \sum_{j=1}^N \mathbb{E}^{\mathcal{W}}\big[\partial_{x_j}\varphi\big((t,\mathbf{x})\big) X^{(1)}(t,\mathbf{x})_{j,i}\big].$$

Similarly, using (3.4.5), one sees that $\partial_{x_{i_1}}\partial_{x_{i_2}} u_\varphi(t,\mathbf{x})$ exists, is continuous, and is given by

$$\begin{aligned} &\sum_{j_1,j_2=1}^N \mathbb{E}^{\mathcal{W}}\big[\partial_{x_{j_1}}\partial_{x_{j_2}}\varphi\big(X(t,\mathbf{x})\big) X^{(1)}(t,\mathbf{x})_{j_1,i_1} X^{(1)}(t,\mathbf{x})_{j_2,i_2}\big] \\ &\qquad + \sum_{j=1}^N \mathbb{E}^{\mathcal{W}}\big[\partial_{x_j}\varphi\big(X(t,\mathbf{x})\big) X^{(2)}(t,\mathbf{x})_{j,(i_1,i_2)}\big]. \end{aligned}$$

Hence $u_\varphi \in C^{1,2}_{\mathrm{b}}\big([0,T] \times \mathbb{R}^N;\mathbb{R}\big)$ for all $T > 0$, and therefore it satisfies (1.2.4).

It turns out that one can do much better. In fact, using no probability theory and only clever applications of the weak minimum principle, O. Olenik proved (cf. §3.2 in [22]) that, when $\varphi \in C^2_{\mathrm{b}}(\mathbb{R}^N;\mathbb{R})$, Kolmogorov's backward equation can be solved if a and b are bounded and have two, bounded, continuous derivatives. Her result highlights a basic weakness of the Itô's theory: having to find a smooth σ is a serious drawback.

3.5 Wiener's spaces of homogeneous chaos

Wiener spent a lot of time thinking about *noise* and how to separate it from signals. One of his most profound ideas on the subject was that of decomposing a random variable into components of uniform orders of randomness, or, what, with his usual flare for language, he called components of **homogeneous chaos**. From a mathematical standpoint, what he was doing is write $L^2(\mathcal{W};\mathbb{R})$ as the direct sum of mutually orthogonal subspaces consisting of functions that could be reasonably thought of as having a uniform order of randomness. Wiener's own treatment of this subject is fraught with difficulties, all of which were resolved by Itô. Thus, once again, we will be guided by Itô.

To explain Wiener's idea, we must define what we will mean by **multiple stochastic integrals**. That is, if for $m \geq 1$ and $t \in [0,\infty]$, $\square^{(m)}(t) := [0,t)^m$ and $\square^{(m)} := \square^{(m)}(\infty)$, we want to assign a meaning to expressions like

$$\tilde{I}_F^{(m)}(t) = \int_{\square^{(m)}(t)} \big(F(\vec{\tau}), d\vec{w}(\vec{\tau})\big)_{(\mathbb{R}^M)^m}$$

when $F \in L^2\big(\square^{(m)};(\mathbb{R}^M)^m\big)$. With this goal in mind, when $m = 1$ and $F = f \in L^2\big([0,\infty);\mathbb{R}^M\big)$, take $I_F^{(1)}(t) = I_f(t)$, where $I_f(t)$ is the Paley–Wiener integral of f. When $m \geq 2$ and $F = f_1\otimes\cdots\otimes f_m$ for some $f_1,\dots,f_m \in L^2\big([0,\infty);\mathbb{R}^M\big)$,[7] use induction to define $I_F^{(m)}(t)$ so that

$$I^{(m)}_{f_1\otimes\cdots\otimes f_m}(t) = \int_0^t I^{(m-1)}_{f_1\otimes\cdots\otimes f_{m-1}}(\tau)\big(f_m(\tau), dw(\tau)\big)_{\mathbb{R}^M}, \tag{3.5.1}$$

where now we need Itô's integral. Of course, we are obliged to check that $\tau \rightsquigarrow f_m(\tau)I^{(m-1)}_{f_1\otimes\cdots\otimes f_{m-1}}(\tau)$ is square integrable. But, assuming that $I^{(m-1)}_{f_1\otimes\cdots\otimes f_{m-1}}$ is well defined, we have that

$$\begin{aligned}\mathbb{E}^{\mathcal{W}}&\left[\int_0^T \big|f_m(\tau)I^{(m-1)}_{f_1\otimes\cdots\otimes f_{m-1}}(\tau)\big|^2\,d\tau\right]\\ &= \int_0^T |f_m(\tau)|^2\mathbb{E}^{\mathcal{W}}\Big[\big|I^{(m-1)}_{f_1\otimes\cdots\otimes f_{m-1}}(\tau)\big|^2\Big]\,d\tau.\end{aligned}$$

Hence, at each step in our induction procedure, we can check that

$$\mathbb{E}^{\mathcal{W}}\Big[\big|I^{(m)}_{f_1\otimes\cdots\otimes f_m}(T)\big|^2\Big] = \int_{\Delta^{(m)}(T)} \big|f_1(\tau_1)\big|^2\cdots\big|f_m(\tau_m)\big|^2\,d\tau_1\cdots d\tau_m,$$

[7] Here $f_1\otimes\cdots\otimes f_m$ denotes the $(\mathbb{R}^M)^m$-valued function F on $[0,\infty)^m$ such that $\big(\Xi, F(t_1,\dots,t_m)\big)_{(\mathbb{R}^M)^m} = \big(\boldsymbol{\xi}_1, f_1(t_1)\big)_{\mathbb{R}^M}\cdots\big(\boldsymbol{\xi}_m, f_m(t_m)\big)_{\mathbb{R}^M}$ for $\Xi = (\boldsymbol{\xi}_1,\dots,\boldsymbol{\xi}_m) \in (\mathbb{R}^M)^m$.

where $\Delta^{(m)}(t) := \{(t_1,\dots,t_m) \in \square^{(m)} : 0 \le t_1 < \cdots < t_m < t\}$; and so, after polarization, we arrive at

$$\begin{aligned}\mathbb{E}^{\mathcal{W}}&\Big[I^{(m)}_{f_1\otimes\cdots\otimes f_m}(T)I^{(m)}_{f_1'\otimes\cdots\otimes f_m'}(T)\Big]\\ &= \int_{\Delta^{(m)}(T)} \big(f_1(\tau_1), f_1'(\tau_1)\big)_{\mathbb{R}^M}\cdots\big(f_m(\tau_m), f_m'(\tau_m)\big)_{\mathbb{R}^M}\,d\tau_1\cdots d\tau_m.\end{aligned}$$

So far our integrals are over the simplices $\Delta^{(m)}(T)$ and not the rectangle $\square^{(m)}(T)$. To remedy this, we next introduce

$$\tilde{I}^{(m)}_{f_1\otimes\cdots\otimes f_m}(t) := \sum_{\pi\in\Pi_m} I^{(m)}_{f_{\pi(1)}\otimes\cdots\otimes f_{\pi(m)}}(t), \tag{3.5.2}$$

where Π_m is the symmetric group (i.e., the group of permutations) on $\{1,\dots,m\}$. By the preceding, one sees that

$$\begin{aligned}\mathbb{E}^{\mathcal{W}}&\Big[\tilde{I}^{(m)}_{f_1\otimes\cdots\otimes f_m}(T)\tilde{I}^{(m)}_{f_1'\otimes\cdots\otimes f_m'}(T)\Big]\\ &= \sum_{\pi,\pi'\in\Pi_m}\int_{\Delta^{(m)}(T)}\prod_{\ell=1}^m\big(f_{\pi(\ell)}(\tau_\ell), f'_{\pi'(\ell)}(\tau_\ell)\big)_{\mathbb{R}^M}\,d\tau_1\cdots d\tau_m\\ &= \sum_{\pi\in\Pi_m}\sum_{\pi'\in\Pi_m}\int_{\Delta^{(m)}(T)}\prod_{\ell=1}^m\big(f_{\pi(\ell)}(\tau_{\pi'(\ell)}), f'_\ell(\tau_{\pi'(\ell)})\big)_{\mathbb{R}^M}\,d\tau_1\cdots d\tau_\ell\\ &= \sum_{\pi\in\Pi_m}\int_{\square^{(m)}(T)}\prod_{\ell=1}^m\big(f_{\pi(\ell)}(\tau_\ell), f'_\ell(\tau_\ell)\big)_{\mathbb{R}^M}\,d\tau_1\cdots d\tau_m\\ &= \sum_{\pi\in\Pi_m}\prod_{\ell=1}^m\big(f_{\pi(\ell)}, f'_\ell\big)_{L^2([0,T];\mathbb{R}^M)}.\end{aligned}$$

In preparation for the next step, let $\{g_j : j \ge 1\}$ be an orthonormal basis in $L^2([0,\infty);\mathbb{R}^M)$, and note that $\{g_{j_1}\otimes\cdots\otimes g_{j_m} : (j_1,\dots,j_m) \in (\mathbb{Z}^+)^m\}$ is an orthonormal basis in $L^2\big(\square^{(m)};(\mathbb{R}^M)^m\big)$. Next let $\mathcal{A}$ denote the set of $\boldsymbol{\alpha} \in \mathbb{N}^{\mathbb{Z}^+}$ for which $\|\boldsymbol{\alpha}\| := \sum_1^\infty \alpha_j < \infty$. Finally, given $\boldsymbol{\alpha} \in \mathcal{A}$, set

$$S(\boldsymbol{\alpha}) = \{j \in \mathbb{Z}^+ : \alpha_j \ge 1\} \quad\text{and}\quad G_{\boldsymbol{\alpha}} = \bigotimes_{j\in S(\boldsymbol{\alpha})} g_j^{\otimes\alpha_j}.$$

Then, $G_{\boldsymbol{\alpha}} \in L^2\big(\square^{(\|\boldsymbol{\alpha}\|)};(\mathbb{R}^M)^{\|\boldsymbol{\alpha}\|}\big)$, and, for $\boldsymbol{\alpha},\boldsymbol{\beta} \in \mathcal{A}$ with $\|\boldsymbol{\alpha}\| = \|\boldsymbol{\beta}\| = m$, the preceding calculation shows that

$$\big(\tilde{I}^{(m)}_{G_{\boldsymbol{\alpha}}}, \tilde{I}^{(m)}_{G_{\boldsymbol{\beta}}}\big)_{L^2(\mathcal{W};\mathbb{R})} = \delta_{\boldsymbol{\alpha},\boldsymbol{\beta}}\boldsymbol{\alpha}!, \tag{3.5.3}$$

where $\boldsymbol{\alpha}! = \prod_{j\in S(\boldsymbol{\alpha})}\alpha_j!$.

Now, given $F \in L^2\big(\square^{(m)}; (\mathbb{R}^M)^m\big)$, set

$$\widetilde{F}(t_1, \dots, t_m) := \sum_{\pi \in \Pi_m} F\big(t_{\pi(1)}, \dots, t_{\pi(m)}\big),$$

and observe that

$$\begin{aligned}\widetilde{F} &= \sum_{\mathbf{j} \in (\mathbb{Z}^+)^m} \big(\widetilde{F}, g_{j_1} \otimes \cdots \otimes g_{j_m}\big)_{L^2(\square^{(m)}; (\mathbb{R}^M)^m)} g_{j_1} \otimes \cdots \otimes g_{j_m} \\ &= \sum_{\|\boldsymbol{\alpha}\| = m} \big(\widetilde{F}, G_{\boldsymbol{\alpha}}\big)_{L^2(\square^{(m)}; (\mathbb{R}^M)^m)} \frac{\widetilde{G}_{\boldsymbol{\alpha}}}{\boldsymbol{\alpha}!},\end{aligned}$$

and therefore

$$\|\widetilde{F}\|^2_{L^2(\square^{(m)}; (\mathbb{R}^M)^m)} = \sum_{\|\boldsymbol{\alpha}\| = m} \binom{m}{\boldsymbol{\alpha}!} \big(\widetilde{F}, G_{\boldsymbol{\alpha}}\big)^2_{L^2(\square^{(m)}; (\mathbb{R}^M)^m)},$$

where $\binom{m}{\boldsymbol{\alpha}}$ is the multinomial coefficient $\frac{m!}{\boldsymbol{\alpha}!}$. Hence, after combining this with calculation in (3.5.3), we have that

$$\mathbb{E}^{\mathcal{W}}\left[\left|\sum_{\|\boldsymbol{\alpha}\| = m} \frac{\big(\widetilde{F}, G_{\boldsymbol{\alpha}}\big)_{L^2(\square^{(m)}; (\mathbb{R}^M)^m)}}{\boldsymbol{\alpha}!} \tilde{I}^{(m)}_{G_{\boldsymbol{\alpha}}}(\infty)\right|^2\right] = \frac{1}{m!} \|\widetilde{F}\|^2_{L^2(\square^{(m)}; (\mathbb{R}^M)^m)}.$$

With these considerations, we have proved the following.

Theorem 3.5.1. *For each $m \geq 1$, there is a unique linear map*

$$F \in L^2\big(\square^{(m)}; (\mathbb{R}^M)^m\big) \longmapsto \tilde{I}^{(m)}_F \in M^2(\mathcal{W}; \mathbb{R})$$

such that $\tilde{I}^{(m)}_{f_1 \otimes \cdots \otimes f_m}$ is given as in (3.5.2) and

$$\mathbb{E}^{\mathcal{W}}\Big[\tilde{I}^{(m)}_F(\infty) \tilde{I}^{(m)}_{F'}(\infty)\Big] = \frac{1}{m!} \big(\widetilde{F}, \widetilde{F}'\big)_{L^2(\square^{(m)}; (\mathbb{R}^M)^m)}.$$

In fact,

$$\tilde{I}^{(m)}_F = \sum_{\|\boldsymbol{\alpha}\| = m} \frac{\big(\widetilde{F}, G_{\boldsymbol{\alpha}}\big)_{L^2(\square^{(m)}; (\mathbb{R}^M)^m)}}{\boldsymbol{\alpha}!} \tilde{I}_{G_{\boldsymbol{\alpha}}},$$

where the convergence is in $L^2(\mathcal{W}; \mathbb{R})$.

Although it is somewhat questionable to do so, as indicated at the beginning of this section, it is tempting to think of $\tilde{I}^{(m)}_F(t)$ as

$$\int_{\square^{(m)}(t)} \big(F(\vec{\tau}), d\vec{w}(\vec{\tau})\big)_{(\mathbb{R}^M)^m} \quad \text{where } d\vec{w}(\vec{\tau}) = \big(dw(\tau_1), \dots, dw(\tau_m)\big).$$

The reason why this notation is questionable is that, although it is suggestive, it may suggest the wrong thing. Specifically, in order to avoid stochastic integrals with non-progressively measurable integrands, our definition of $\tilde{I}_F^{(m)}$ carefully avoided integration across diagonals, whereas the preceding notation gives no hint of that fact.

Take $Z^{(0)}$ to be the subspace of $L^2(\mathcal{W};\mathbb{R})$ consisting of the constant functions, and, for $m \geq 1$, set

$$Z^{(m)} = \big\{\tilde{I}_F^{(m)}(\infty) : F \in L^2\big(\square^{(m)};(\mathbb{R}^M)^m\big)\big\}. \tag{3.5.4}$$

Clearly, each $Z^{(m)}$ is a linear subspace of $L^2(\mathcal{W};\mathbb{R})$. Furthermore, if

$$\{F_k : k \geq 1\} \subseteq L^2\big(\square^{(m)};(\mathbb{R}^M)^m\big)$$

and $\big\{\tilde{I}_{F_k}^{(m)}(\infty) : k \geq 1\big\}$ converges in $L^2(\mathcal{W};\mathbb{R})$, then $\big\{\widetilde{F}_k : k \geq 1\big\}$ converges in $L^2\big(\square^{(m)};(\mathbb{R}^M)^m\big)$ to some symmetric function G. Hence, since $\widetilde{G} = m!G$, we see that $\tilde{I}_{F_k}^{(m)}(\infty) \longrightarrow \tilde{I}_F^{(m)}(\infty)$ in $L^2(\mathcal{W};\mathbb{R})$ where $F = \frac{1}{m!}G$. Therefore, each $Z^{(m)}$ is a closed linear subspace of $L^2(\mathcal{W};\mathbb{R})$. Finally, $Z^{(m)} \perp Z^{(m')}$ when $m' \neq m$. This is completely obvious if either m or m' is 0. Thus, suppose that $1 \leq m < m'$. Then

$$\begin{aligned}
&\mathbb{E}^{\mathcal{W}}\Big[I_{f_1\otimes\cdots\otimes f_m}^{(m)}(\infty) I_{f_1'\otimes\cdots\otimes f_{m'}'}^{(m')}(\infty)\Big] \\
&\quad = \int_{\Delta^{(m'-m)}} \prod_{\ell=0}^{m-1} \big(f_{m-\ell}(\tau_\ell), f_{m'-\ell}'(\tau_\ell)\big)_{\mathbb{R}^M} \\
&\qquad\qquad \times \mathbb{E}^{\mathcal{W}}\big[I_{f_1'\otimes\cdots\otimes f_{m'-m}'}^{(m'-m)}(\tau_{m'-m})\big]\, d\tau_1 \cdots d\tau_{m'-m} = 0,
\end{aligned}$$

which completes the proof.

The space $Z^{(m)}$ is the space of **mth order homogeneous chaos**. (See Exercise 5.8 for another characterization of $Z^{(m)}$.) The reason why elements of $Z^{(0)}$ are said to be of 0th order chaos is clear: constants are non-random. To understand why $\tilde{I}_F^{(m)}(\infty)$ is of mth order chaos when $m \geq 1$, it is helpful to replace $dw(\tau)$ by the much more ambiguous $\dot{w}(\tau)\,d\tau$ and write

$$\tilde{I}_F^{(m)}(\infty) = \int_{\square^{(m)}\setminus D^{(m)}} \Big(F(\tau_1,\dots,\tau_m), \big(\dot{w}(\tau_1),\dots,\dot{w}(\tau_m)\big)\Big)_{(\mathbb{R}^M)^m} d\tau_1 \cdots d\tau_m,$$

where $D^{(m)} := \{(\tau_1, \ldots, \tau_m) : \tau_k = \tau_\ell \text{ for some } 1 \le k < \ell \le m\}$. In the world of engineering and physics, $\tau \rightsquigarrow \dot{w}(\tau)$ is called *white noise*.[8] Thus, $Z^{(m)}$ is the space built out of homogeneous mth order polynomials in white noises evaluated at different times. In other words, the order of chaos is the order of the white noise polynomial.

The result of Wiener, alluded to at the beginning of this section, now becomes the assertion that

$$L^2(\mathcal{W};\mathbb{R}) = \bigoplus_{m=0}^{\infty} Z^{(m)}. \tag{3.5.5}$$

The key to Itô's proof of (3.5.5) is found in the following.

Lemma 3.5.2. *If $f \in L^2\big([0,\infty);\mathbb{R}^M\big)$ and $I_f^{(0)} := 1$, then, for each $\lambda \in \mathbb{C}$,*

$$e^{\lambda I_f(\infty) - \frac{\lambda^2}{2}\|f\|^2_{L^2([0,\infty);\mathbb{R}^M)}} = \sum_{k=0}^{\infty} \frac{\lambda^k}{k!} \tilde{I}^{(k)}_{f^{\otimes k}}(\infty) := \lim_{m\to\infty} \sum_{k=0}^{m} \frac{\lambda^k}{k!} \tilde{I}^{(k)}_{f^{\otimes k}}(\infty)$$

$\mathcal{W}$-almost surely and in $L^2(\mathcal{W};\mathbb{R})$. In fact, if

$$R_f^m(\infty,\lambda) := e^{\lambda I_f(\infty) - \frac{\lambda^2}{2}\|f\|^2_{L^2([0,\infty);\mathbb{R}^M)}} - \sum_{0\le k<m} \frac{\lambda^k}{k!} \tilde{I}^{(k)}_{f^{\otimes k}}(\infty),$$

then

$$\begin{aligned}
&\mathbb{E}^{\mathcal{W}}\Big[\big|R_f^m(\infty,\lambda)\big|^2\Big] \\
&\quad = e^{|\lambda|^2\|f\|^2_{L^2([0,\infty);\mathbb{R}^M)}} - \sum_{k=0}^{m-1} \frac{\big(|\lambda|\|f\|_{L^2([0,\infty);\mathbb{R}^M)}\big)^{2k}}{k!} \\
&\quad \le \frac{\big(|\lambda|\|f\|_{L^2([0,\infty);\mathbb{R}^M)}\big)^{2m}}{m!} e^{|\lambda|^2\|f\|^2_{L^2([0,\infty);\mathbb{R}^M)}}.
\end{aligned}$$

Proof. Obviously, the first assertion follows from the second one.

Set $E(t,\lambda) = e^{\lambda I_f(t) - \frac{\lambda^2}{2}\|\mathbf{1}_{[0,t)} f\|^2_{L^2([0,\infty);\mathbb{R}^M)}}$. Then, by Itô's formula,

$$E(t,\lambda) = 1 + \lambda \int_0^t E(\tau,\lambda)\big(f(\tau), dw(\tau)\big)_{\mathbb{R}^M}.$$

Thus, if (cf. (3.5.1)) $I^{(m)}(t) := I^{(m)}_{f^{\otimes m}}(t) = \frac{1}{m!}\tilde{I}^{(m)}_{f^{\otimes m}}$ and

[8] The terminology comes from the observation that, no matter how one interprets $t \rightsquigarrow \dot{w}(t)$, it is a stationary, centered Gaussian process whose covariance is the Dirac delta function times the identity, and, as such, it is totally uncorrelated in the sense that $\dot{w}(t_1)$ is independent of $\dot{w}(t_2)$ when $t_1 \ne t_2$. Thus the Fourier transform of its covariance is constant, and so, from the time series point of view, its spectrum is *white* and therefore contains no information.

$$R^0(t,\lambda) := E(t,\lambda) \quad \text{and} \quad R^{m+1}(t,\lambda) := \lambda \int_0^t R^m(\tau,\lambda)\big(f(\tau), dw(\tau)\big)_{\mathbb{R}^M}$$

for $m \geq 0$, then, by induction, one sees that

$$E(t,\lambda) = 1 + \sum_{k=1}^{m} \lambda^k I^{(k)}(t) + R^{m+1}(t,\lambda)$$

for all $m \geq 0$. Finally, if $A(t) = \int_0^t |f(\tau)|^2\,d\tau$, then

$$\begin{aligned}&\mathbb{E}^{\mathcal{W}}\big[|R^0(t,\lambda)|^2\big]\\ &\quad = \exp\Big(\big(2\,\mathfrak{Re}(\lambda)^2 - \mathfrak{Re}(\lambda^2)\big)A(t)\Big)\mathbb{E}^{\mathcal{W}}\big[e^{2\,\mathfrak{Re}(\lambda) I_f(t) - 2\,\mathfrak{Re}(\lambda)^2 A(t)}\big] = e^{|\lambda|^2 A(t)}\end{aligned}$$

and

$$\mathbb{E}^{\mathcal{W}}\big[|R^{m+1}(t,\lambda)|^2\big] = |\lambda|^2 \int_0^t \mathbb{E}^{\mathcal{W}}\big[|R^m(\tau,\lambda)|^2\big]\dot{A}(\tau)\,d\tau.$$

Hence, the desired conclusion follows by induction on $m \geq 0$. □

Theorem 3.5.3. *The span of*

$$\mathbb{R} \oplus \big\{\tilde{I}^{(m)}_{f^{\otimes m}}(\infty) : m \geq 1 \;\&\; f \in L^2\big([0,\infty);\mathbb{R}^M\big)\big\}$$

is dense in $L^2(\mathcal{W};\mathbb{R})$*. In particular,* (3.5.5) *holds.*

Proof. Let $\mathbb{H}$ denote the smallest closed subspace of $L^2(\mathcal{W};\mathbb{R})$ containing constants and $\tilde{I}^{(m)}_{f^{\otimes m}}(\infty)$ for all $f \in L^2([0,\infty);\mathbb{R}^N)$. By Lemma 3.5.2, we know that $\cos \circ I_f(\infty)$ and $\sin \circ I_f(\infty)$ are in $\mathbb{H}$ for all $f \in L^2\big([0,\infty);\mathbb{R}^M\big)$.

Next, observe that the space of functions $\Phi\colon C\big([0,\infty);\mathbb{R}^M\big) \longrightarrow \mathbb{R}$ which have the form

$$\Phi = F\big(I_{f_1}(\infty), \dots, I_{f_L}(\infty)\big)$$

for some $L \geq 1$, $F \in \mathscr{S}(\mathbb{R}^L;\mathbb{R})$, and $f_1, \dots, f_L \in L^2\big([0,\infty);\mathbb{R}^M\big)$ is dense in $L^2(\mathcal{W};\mathbb{R}^M)$. Indeed, this follows immediately from the density in $L^2(\mathcal{W};\mathbb{R})$ of the space of functions of the form

$$w \rightsquigarrow F\big(w(t_1), \dots, w(t_L)\big),$$

where $L \geq 1$, $F \in \mathscr{S}(\mathbb{R}^L;\mathbb{R})$, and $0 \leq t_0 < \cdots < t_L$.

Now suppose that $F \in \mathscr{S}(\mathbb{R}^L;\mathbb{R})$ is given, and let $\widehat{F}$ denote its Fourier transform. Then, by elementary Fourier analysis,

$$F_n(\mathbf{x}) := \big(2(4^n+1)\pi\big)^{-L} \sum_{\|\mathbf{m}\|_\infty \leq 4^n} e^{-\sqrt{-1}\,(2^{-n}\mathbf{m},\mathbf{x})_{\mathbb{R}^L}}\,\widehat{F}(\mathbf{m}2^{-n}) \longrightarrow F(\mathbf{x}),$$

both uniformly and boundedly, where $\mathbf{m} = (m_1, \dots, m_L) \in \mathbb{Z}^L$ and $\|\mathbf{m}\|_\infty = \max_{1 \leq \ell \leq L} |m_\ell|$. Finally, since $\overline{\widehat{F}(\boldsymbol{\xi})} = \widehat{F}(-\boldsymbol{\xi})$ for all $\boldsymbol{\xi} \in \mathbb{R}^L$, we can write

$$
\begin{aligned}
&\big(2(4^n+1)\pi\big)^L F_n\big(I_{f_1}(\infty),\dots,I_{f_L}(\infty)\big)\\
&\quad=\widehat{F}(\mathbf{0})+2\sum_{\substack{\mathbf{m}\in\mathbb{N}^L\\ 1\le\|\mathbf{m}\|_\infty\le 4^n}}\Big(\mathfrak{Re}\big(\widehat{F}(\mathbf{m}2^{-n})\big)\cos\circ I_{\mathbf{m},n}(\infty)\\
&\qquad\qquad+\mathfrak{Im}\big(\widehat{F}(\mathbf{m}2^{-n})\big)\sin\circ I_{\mathbf{m},n}(\infty)\Big)\in\mathbb{H},
\end{aligned}
$$

where $I_{\mathbf{m},n}:=2^{-n}I_{f_{\mathbf{m}}}$ with $f_{\mathbf{m}}=\sum_{\ell=1}^L m_\ell f_\ell$. □

The following corollary is an observation made by Itô after he cleaned up Wiener's treatment of (3.5.5). It is often called **Itô's representation theorem** and turns out to play an important role in applications of stochastic analysis to, of all things, models of financial markets.[9]

Corollary 3.5.4. *The map*

$$
(x,\boldsymbol{\xi})\in\mathbb{R}\times PM^2(\mathbb{R}^M)\longmapsto x+\int_0^\infty\big(\boldsymbol{\xi}(\tau),dw(\tau)\big)_{\mathbb{R}^M}\in L^2(\mathcal{W};\mathbb{R})
$$

is a linear, isometric surjection. Hence, for each $\Phi\in L^2(\mathcal{W};\mathbb{R})$ there is a $\mathcal{W}$-almost surely unique $\boldsymbol{\xi}\in PM^2(\mathbb{R}^M)$ such that

$$
\Phi=\mathbb{E}^{\mathcal{W}}[\Phi]+\int_0^\infty\big(\boldsymbol{\xi}(\tau),dw(\tau)\big)_{\mathbb{R}^M}\quad \mathcal{W}\text{-almost surely.}
$$

In particular,

$$
\mathbb{E}^{\mathcal{W}}[\Phi\mid \mathcal{W}_t]=\mathbb{E}^{\mathcal{W}}[\Phi]+\int_0^t\big(\boldsymbol{\xi}(\tau),dw(\tau)\big)_{\mathbb{R}^M}\ \mathcal{W}\text{-almost surely}\quad\text{for each } t\ge 0.
$$

Finally, for any $\Phi\in L^1(\mathcal{W};\mathbb{R})$, there is a version of $(t,w)\rightsquigarrow\mathbb{E}^{\mathcal{W}}\big[\Phi\mid\mathcal{W}_t\big](w)$ that is continuous as function of $t\ge 0$.

Proof. Since it is clear that the map is linear and isometric, the first assertion will be proved once we check that the map is onto. But, because it is a linear isometry, we know that its image is a closed subspace, and so we need only show that its image contains a set whose span is dense. However, for each $f\in L^2\big([0,\infty);\mathbb{R}^M\big)$ and $m\ge 1$,

$$
\tilde{I}^{(m)}_{f^{\otimes m}}(\infty)=m\int_0^\infty\tilde{I}^{(m-1)}_{f^{\otimes(m-1)}}(\tau)\big(f(\tau),dw(\tau)\big)_{\mathbb{R}^M},
$$

and so, by the first part of Theorem 3.5.3, we are done.

Given the first assertion, the second assertion follows and shows that $(t,w)\rightsquigarrow E^{\mathcal{W}}[\Phi\mid\mathcal{W}_t](w)$ can be chosen so that it is continuous with respect

[9] In fact, Corollary 3.5.4 along with Itô's formula are responsible for the widespread misconception in the financial community that Itô was an economist.

to t for any $\Phi \in L^2(\mathcal{W};\mathbb{R})$. Finally, if $\Phi \in L^1(\mathcal{W};\mathbb{R})$, choose $\{\Phi_k : k \geq 1\} \subseteq L^2(\mathcal{W};\mathbb{R})$ so that $\Phi_k \longrightarrow \Phi$ in $L^1(\mathcal{W};\mathbb{R})$, and let $(t,w) \rightsquigarrow X_k(t,w)$ be a version of $(t,w) \rightsquigarrow E^{\mathcal{W}}[\Phi_k \mid W_t](\omega)$ which is continuous in t. Then, by Doob's inequality, for all $t > 0$,

$$\sup_{\ell>k} \mathcal{W}\big(\|X_\ell(\,\cdot\,) - X_k(\,\cdot\,)\|_{[0,t]} \geq \epsilon\big) \leq \sup_{\ell>k} \frac{\|\Phi_\ell - \Phi_k\|_{L^1(W;\mathbb{R})}}{\epsilon} \longrightarrow 0$$

as $k \to \infty$. Hence there exists a progressively measurable $X\colon [0,\infty)\times\Omega\longrightarrow\mathbb{R}$ to which a subsequence of $\{X_k(\,\cdot\,) : k \geq 1\}$ converges $\mathcal{W}$-almost surely uniformly on compacts, and, since, for each $t \geq 0$,

$$\lim_{k\to\infty} \big\|X_k(t) - E^{\mathcal{W}}[\Phi \mid W_t]\big\|_{L^1(\mathcal{W};\mathbb{R})} = 0,$$

it follows that $X(t)$ is a version of $\mathbb{E}^{\mathcal{W}}[\Phi \mid W_t]$. □

3.6 Exercises

Exercise 3.1. Let $\big(B(t), \mathcal{F}_t, \mathbb{P}\big)$ be an $\mathbb{R}^N$-valued Brownian motion and $\eta \in PM^2_{\text{loc}}(\mathbb{R}^N)$. Given stopping times $\zeta_1 \leq \zeta_2 \leq T$ and a bounded $\mathcal{F}_{\zeta_1}$-measurable function α, show that

$$(t,\omega) \rightsquigarrow \alpha(\omega)\mathbf{1}_{[\zeta_1(\omega),\zeta_2(\omega))}(t)$$

is progressively measurable and that

$$\begin{aligned}\int_{\zeta_1}^{\zeta_2} \alpha\big(\eta(\tau), dB(\tau)\big)_{\mathbb{R}^N} &:= \int_0^T \alpha\mathbf{1}_{[\zeta_1,\zeta_2)}(\tau)\big(\eta(\tau), dB(\tau)\big)_{\mathbb{R}^N} \\ &= \alpha \int_{\zeta_1}^{\zeta_2} \big(\eta(\tau), dB(\tau)\big)_{\mathbb{R}^N}.\end{aligned}$$

Exercise 3.2. Let $\big(B(t), \mathcal{F}_t, \mathbb{P}\big)$ be an $\mathbb{R}^N$-valued Brownian motion, and let η_1 and η_2 be elements of $L^2([0,\infty);\mathbb{R}^N)$. Show that $\big(I_{\eta_1}(\infty), I_{\eta_2}(\infty)\big)$ is a centered $\mathbb{R}^2$-valued Gaussian random variable with covariance

$$\begin{pmatrix} (\eta_1,\eta_1)_{L^2([0,\infty);\mathbb{R}^N)} & (\eta_1,\eta_2)_{L^2([0,\infty);\mathbb{R}^N)} \\ (\eta_1,\eta_2)_{L^2([0,\infty);\mathbb{R}^N)} & (\eta_2,\eta_2)_{L^2([0,\infty);\mathbb{R}^N)} \end{pmatrix}.$$

Next, set $\boldsymbol{\zeta} = \eta_1 + i\eta_2$, and define $I_{\boldsymbol{\zeta}}(t) = I_{\eta_1}(t) + iI_{\eta_2}(t)$ for $t \in [0,\infty]$. Show that $\mathbb{E}^{\mathbb{P}}\big[e^{I_{\boldsymbol{\zeta}}(\infty)}\big]$ equals

$$\exp\left(\frac{\|\eta_1\|^2_{L^2([0,\infty);\mathbb{R}^N)} - \|\eta_2\|^2_{L^2([0,\infty);\mathbb{R}^N)}}{2} + i(\eta_1,\eta_2)_{L^2([0,\infty);\mathbb{R}^N)}\right).$$

Exercise 3.3. Let $\mathfrak{G}$ be an open subset of $\mathbb{R} \times \mathbb{R}^N$ and, given $(s, \mathbf{x}) \in \mathfrak{G}$, define $\zeta_{s,\mathbf{x}} : \mathbb{W}(\mathbb{R}^N) \longrightarrow [0, \infty]$ by

$$\zeta_{s,\mathbf{x}}(w) = \inf\{t \geq 0 : (s + t, \mathbf{x} + w(t)) \notin \mathfrak{G}\}.$$

If $u \in C_{\mathrm{b}}^{1,2}(\mathfrak{G}; \mathbb{R}) \cap C(\overline{\mathfrak{G}}; \mathbb{R})$, show that

$$u(s + t \wedge \zeta_{s.\mathbf{x}}, \mathbf{x} + w(t \wedge \zeta_{s,\mathbf{x}})) - \int_0^{t\wedge\zeta_{s,\mathbf{x}}} (\partial_\tau + \tfrac{1}{2}\Delta)u(s + \tau, \mathbf{x} + w(\tau))\, d\tau$$

is a $\mathcal{W}$-martingale relative to $\{W_t : t \geq 0\}$.

Hint: First assume that $u \in C_{\mathrm{b}}^{1,2}(\mathbb{R} \times \mathbb{R}^N; \mathbb{R})$, and use Doob's stopping time theorem. Next, use a bump function combined with stopping times to reduce to this case.

Exercise 3.4. The **elliptic strong minimum principle** states that if $G \ni \mathbf{0}$ is a connected, open subset of $\mathbb{R}^N$ and $u \in C^2(G; \mathbb{R})$ is a harmonic function on G that achieves is minimum value at $\mathbf{0}$, then $u = u(\mathbf{0})$ on G. This is of course a special case of the parabolic strong minimum principle, but it has much simpler proofs. For example, use part (vii) in Exercise 2.5 and Exercise 3.3 to prove the **mean value property** that, for any $\mathbf{x} \in G$,

$$u(\mathbf{x}) = \frac{1}{\omega_{N-1}} \int_{\mathbb{S}^{N-1}} u(\mathbf{x} + r\omega)\, \lambda_{\mathbb{S}^{N-1}}(d\omega)$$

and therefore, if $\Omega_N = \lambda_{\mathbb{R}^N}\big(B(\mathbf{0}, 1)\big) = \frac{\omega_{M-1}}{N}$,

$$u(\mathbf{x}) = \frac{1}{\Omega_N r^N} \int_{B(x,r)} u(\mathbf{y})\, d\mathbf{y}$$

as long as $\overline{B(\mathbf{x}, r)} \subseteq G$. From this conclude that the set of $\mathbf{x} \in G$ at which u equals $u(\mathbf{0})$ is both open and closed.

Exercise 3.5. The goal of this exercise is to construct a typical example of a continuous local martingale that is not a martingale. Set $f(z) = e^{\frac{z^2}{2}}$ for $z \in \mathbb{C}$ and

$$F(x_1, x_2) = \exp\left(\frac{x_1^2 - x_2^2}{2} + i x_1 x_2\right).$$

Using the fact f is analytic, show that $\Delta F = 0$, and therefore, by Itô's formula, that $\big(F(w(t)), W_t, \mathcal{W}\big)$ is a continuous local $\mathbb{C}$-valued martingale, where $\mathcal{W}$ is Wiener measure on $\mathbb{W}(\mathbb{R}^2)$. On the other hand, show that $\mathbb{E}^{\mathcal{W}}\big[|F(w(t))|\big] = \infty$ if $t \geq 1$, and so it is not a martingale. See Exercise 5.3 for a more interesting example.

Exercise 3.6. This exercise presents an entirely different approach to proving estimates of the sort in (3.3.2).

Let $\sigma \in PM^2_{\text{loc}}(\text{Hom}(\mathbb{R}^M;\mathbb{R}^N))$, and set $A(t) = \int_0^t \sigma(\tau)\sigma(\tau)^\top\, d\tau$. For a given $\boldsymbol{\xi} \in \mathbb{R}^N$, apply (3.3.1) to the function

$$(\mathbf{y}, v) \in \mathbb{R}^N \times \mathbb{R} \longmapsto e^{(\boldsymbol{\xi},\mathbf{y})_{\mathbb{R}^N} - \frac{v}{2}} \in \mathbb{R}$$

to see that

$$\left(\exp\Big(\big(\boldsymbol{\xi}, I_\sigma(t)\big)_{\mathbb{R}^N} - \tfrac{1}{2}(\boldsymbol{\xi}, A(t)\boldsymbol{\xi})_{\mathbb{R}^N}\Big), \mathcal{F}_t, \mathbb{P}\right)$$

is a continuous local martingale. Thus, if

$$\zeta_R = \inf\{t \geq 0 : \text{Trace}\big(A(t)\big) \geq R\},$$

show that, for any $\boldsymbol{\xi} \in \mathbb{R}^N$,

$$(*) \qquad \mathbb{E}^{\mathbb{P}}\left[\exp\Big(\big(\boldsymbol{\xi}, I_\sigma(t\wedge\zeta_R)\big)_{\mathbb{R}^N} - \tfrac{1}{2}\big(\boldsymbol{\xi}, A(t\wedge\zeta_R)\boldsymbol{\xi}\big)_{\mathbb{R}^N}\Big)\right] = 1.$$

Equations of this sort allow one to draw interesting conclusions about the relationship between the relative sizes of the terms in the exponential.

(i) Suppose that, for all $\xi \in \mathbb{R}$,

$$\mathbb{E}^{\mathbb{P}}\big[e^{\xi X - \frac{1}{2}\xi^2 A}\big] = 1,$$

where X and A are random variables on some probability space $(\Omega, \mathcal{F}, \mathbb{P})$ and A is bounded and non-negative. By integrating both sides with respect to $\gamma_{0,1}(d\xi)$, show that

$$\mathbb{E}^{\mathbb{P}}\big[e^{\frac{X^2}{2(1+\|A\|_{\mathrm{u}})}}\big] \leq (1 + \|A\|_{\mathrm{u}})^{\frac{1}{2}}.$$

In particular, $\mathbb{E}^{\mathbb{P}}\big[e^{\alpha|X|}\big] < \infty$ for all $\alpha > 0$.

(ii) Define the **Hermite polynomials** $\{H_m : m \geq 0\}$ by

$$H_m(x) = (-1)^m e^{\frac{x^2}{2}} \partial_x^m e^{-\frac{x^2}{2}}.$$

Equivalently, $H_0 = \mathbf{1}$ and $H_{m+1}(x) = (x - \partial_x)H_m(x)$. Show that H_m is an mth order polynomial in which the coefficient of x^m is 1 and the coefficient of x^k is 0 unless the parity of k is the same as that of m. Next use Taylor's theorem to see that

$$e^{\xi x - \frac{\xi^2}{2}} = \sum_{m=0}^{\infty} \frac{\xi^m}{m!} H_m(x),$$

and conclude that $H_{2m}(0) = (-1)^m \frac{(2m)!}{2^m m!} = (-1)^m \prod_{k=1}^m (2k-1)$. Hence,

$$H_{2m}(x) = \sum_{k=0}^{m} c_{2m,2k} x^{2k} \quad \text{and} \quad H_{2m+1} = \sum_{k=0}^{m} c_{2m+1,2k+1} x^{2k+1},$$

where $c_{m,m} = 1$ and $c_{2m,0} = (-1)^m \prod_{k=1}^m (2k-1)$.

(iii) Given $a \geq 0$, set

$$H_{2m}(x,a) = \sum_{k=0}^{m} c_{2m,2k} a^{m-k} x^{2k} \text{ and } H_{2m+1}(x,a) = \sum_{k=0}^{m} c_{2m+1,2k+1} a^{m-k} x^{2k+1},$$

and check that

$$e^{\xi x - \frac{\xi^2}{2} a} = \sum_{m=0}^{\infty} \frac{\xi^m}{m!} H_m(x,a) \quad \text{for all } a \geq 0.$$

As a consequence, show that

$$\sum_{m=0}^{\infty} \frac{\xi^m}{m!} \mathbb{E}^{\mathbb{P}}\big[H_m(X,A)\big] = \mathbb{E}^{\mathbb{P}}\big[e^{\xi X - \frac{1}{2}\xi^2 A}\big] = 1,$$

and therefore that $\mathbb{E}^{\mathbb{P}}\big[H_m(X,A)\big] = 0$ for all $m \geq 1$. In particular, since $H_1(x,a) = x$ and $H_2(x,a) = x^2 - a$, this proves that $\mathbb{E}^{\mathbb{P}}[X] = 0$ and $\mathbb{E}^{\mathbb{P}}\big[X^2\big] = \mathbb{E}^{\mathbb{P}}[A]$. For $m \geq 2$, use Hölder's inequality and the fact that

$$a^\theta b^{1-\theta} \leq \theta t^{\frac{1}{\theta}} a + (1-\theta) t^{-\frac{1}{1-\theta}} b$$

for any $a, b \geq 0$, $\theta \in (0,1)$, and $t > 0$, to show that

$$\mathbb{E}[X^{2m}] \leq (-1)^{m+1} \frac{(2m)!}{2^m m!} \mathbb{E}[A^m] + f_m(t)\mathbb{E}[X^{2m}] + g_m(t)\mathbb{E}[A^m],$$

where

$$f_m(t) = \sum_{k=1}^{m-1} \frac{k}{m} |c_{2m,2k}| t^{\frac{m}{k}} \quad \text{and} \quad g_m(t) = \sum_{k=1}^{m-1} \frac{m-k}{m} |c_{2m,2k}| t^{-\frac{m}{m-k}}.$$

Thus if $s_m > 0$ is determined by $f_m(s_m) = \frac{1}{2}$, then

$$\mathbb{E}[X^{2m}] \leq 2\left((-1)^{m+1} \frac{(2m)!}{2^m m!} + g_m(s_m) \right) \mathbb{E}[A^m].$$

Continuing to assume that $m \geq 2$, determine $t_m > 0$ by $g_m(t_m) = \frac{(2m)!}{2^{m+1} m!}$, and conclude that

$$\mathbb{E}[A^m] \leq \frac{2^{m+1} m!}{(2m)!} \Big((-1)^{m+1} + f_m(t_m) \Big) \mathbb{E}[X^{2m}].$$

Thus there exists a $\kappa_{2m} \in [1, \infty)$ such that

$$\kappa_{2m}^{-1} \mathbb{E}^{\mathbb{P}}\big[A^m\big] \leq \mathbb{E}^{\mathbb{P}}\big[X^{2m}\big] \leq \kappa_{2m} \mathbb{E}^{\mathbb{P}}\big[A^m\big].$$

(iv) In conjunction with the final estimate in **(iii)**, $(*)$ implies that

$$\kappa_{2m}^{-1}\mathbb{E}^{\mathbb{P}}\left[\left(\boldsymbol{\xi}, A(t\wedge\zeta_R)\boldsymbol{\xi}\right)_{\mathbb{R}^N}^m\right] \le \mathbb{E}^{\mathbb{P}}\left[\left|\left(\boldsymbol{\xi}, I_\sigma(t\wedge\zeta_R)\right)_{\mathbb{R}^N}\right|^{2m}\right] \le \kappa_{2m}\mathbb{E}^{\mathbb{P}}\left[\left(\boldsymbol{\xi}, A(t\wedge\zeta_R)\boldsymbol{\xi}\right)_{\mathbb{R}^N}^m\right].$$

Using the fact that $R \rightsquigarrow \left(\boldsymbol{\xi}, A(t\wedge\zeta_R)\boldsymbol{\xi}\right)_{\mathbb{R}^N}$ is non-decreasing and the monotone convergence theorem together with Fatou's lemma conclude that

$$\mathbb{E}^{\mathbb{P}}\left[\left|\left(\boldsymbol{\xi}, I_\sigma(t)\right)_{\mathbb{R}^N}\right|^{2m}\right] \le \kappa_{2m}\mathbb{E}^{\mathbb{P}}\left[\left(\boldsymbol{\xi}, A(t)\boldsymbol{\xi}\right)_{\mathbb{R}^N}^m\right].$$

Hence,

$$\mathbb{E}^{\mathbb{P}}\left[|I_\sigma(t)|^{2m}\right] \le \kappa_{2m}N^{m-1}\mathbb{E}^{\mathbb{P}}\left[\mathrm{Trace}\left(A(t)\right)^m\right].$$

(v) Show that

$$\mathbb{E}^{\mathbb{P}}\left[\mathrm{Trace}\left(A(t)\right)^m\right] \le \kappa_{2m}N^{m-1}\mathbb{E}^{\mathbb{P}}\left[\|I_\sigma(\,\cdot\,)\|_{[0,t]}^{2m}\right].$$

Exercise 3.7. Let $\left(B(t), \mathcal{F}_t, \mathbb{P}\right)$ be an $\mathbb{R}^N$-valued Brownian motion and $\sigma\colon [0,\infty) \longrightarrow \mathrm{Hom}(\mathbb{R}^N;\mathbb{R}^N)$ a Borel measurable function for which there exists a $\kappa \in (0,1]$ such that $\kappa^2 \le \left(\mathbf{e}, \sigma(t)\sigma(t)^\top\mathbf{e}\right) \le \kappa^{-2}$ for all $t \ge 0$ and $\mathbf{e} \in \mathbb{S}^{N-1}$. Next, suppose that $\beta\colon [0,\infty)\times\Omega \longrightarrow \mathbb{R}^N$ is a $\mathcal{B}_{[0,\infty)}\times\mathcal{F}$-measurable function with the properties that $\sigma\left(\{\beta(t) : t \ge 0\}\right)$ is independent of $\sigma\left(\{B(t) : t \ge 0\}\right)$ and

$$\int_0^T \mathbb{E}^{\mathbb{P}}\left[|\beta(t)|^2\right]dt < \infty \quad \text{for all } T \ge 0.$$

Finally, set

$$Z(t) = \int_0^t \sigma(\tau)\,dB(\tau) + \int_0^t \beta(\tau)\,d\tau,$$

and define

$$\int_0^T \left(f(t), dZ(t)\right)_{\mathbb{R}^N} = \int_0^T \left(\sigma(t)^\top f(t), dB(t)\right)_{\mathbb{R}^N} + \int_0^T \left(f(t), \beta(t)\right)_{\mathbb{R}^N}dt$$

for $T \in [0,\infty]$ and $f \in L^2\left([0,T];\mathbb{R}^N\right)$.

(i) For each $T > 0$, show that there is an $\epsilon_T \in (0,1]$ such that

$$\epsilon_T\|f\|_{L^2([0,T];\mathbb{R}^N)} \le \mathbb{E}^{\mathbb{P}}\left[\left(\int_0^T \left(f, dZ_t\right)_{\mathbb{R}^N}\right)^2\right]^{\frac{1}{2}} \le \epsilon_T^{-1}\|f\|_{L^2([0,T];\mathbb{R}^N)}$$

for all $f \in L^2\left([0,T];\mathbb{R}^N\right)$.

(ii) Let $T > 0$, and set

$$L_T = \left\{ \int_0^T \big(f(t), dZ(t)\big)_{\mathbb{R}^N} : f \in L^2\big([0,T];\mathbb{R}^N\big) \right\}.$$

Show that L_T is smallest closed linear subspace of $L^2(\mathbb{P};\mathbb{R})$ that contains $\big(\boldsymbol{\xi}, Z(t)\big)_{\mathbb{R}^N}$ for all $t \in [0,T]$ and $\boldsymbol{\xi} \in \mathbb{R}^N$.

(iii) Assume that there is a closed Gaussian family $\mathfrak{G}$ containing $\big(\boldsymbol{\xi}, B(t)\big)_{\mathbb{R}^N}$ and $\big(\boldsymbol{\xi}, \beta(t)\big)_{\mathbb{R}^N}$ for all $t \geq 0$ and $\boldsymbol{\xi} \in \mathbb{R}^N$. Show that $\big(\boldsymbol{\xi}, Z(t)\big)_{\mathbb{R}^N} \in \mathfrak{G}$ for all $t \geq 0$ and $\boldsymbol{\xi} \in \mathbb{R}^N$, and, using (ii) and Exercise 2.1, conclude that for each $X \in \mathfrak{G}$ and $T > 0$ there is an $m_{T,X} \in \mathbb{R}$ and an $f_{T,X} \in L^2\big([0,T];\mathbb{R}^N\big)$ such that

$$\mathbb{E}^{\mathbb{P}}\big[X \mid \sigma\big(\{Z(t) : t \in [0,T]\}\big)\big] = m_{T,X} + \int_0^T \big(f_{T,X}(t), dZ(t)\big)_{\mathbb{R}^N}.$$

Exercise 3.8. Show that if $\boldsymbol{\eta} \in PM^2_{\text{loc}}(\mathbb{R}^M)$ is a function of locally bounded variation, then

$$(t,w) \in [0,\infty) \times \mathbb{W}(\mathbb{R}^M) \longrightarrow \int_0^t \big(\boldsymbol{\eta}(\tau), dw(\tau)\big)_{\mathbb{R}^M} \in \mathbb{R}$$

is a continuous function. On the other hand, in general, there is no reason to think that Itô stochastic integrals will be continuous functions of the driving Brownian motion unless the integrand has locally bounded variation. Nonetheless, use (3.3.1) to show that if $F \in C^1_{\text{b}}(\mathbb{R}^M;\mathbb{R}^M)$ is exact in the sense that $\partial_{x_i} F_j = \partial_{x_j} F_i$ for $1 \leq i < j \leq M$, then one can take

$$(t,w) \in [0,\infty) \times \mathbb{W}(\mathbb{R}^M) \longmapsto \int_0^t \big(F\big(w(\tau)\big), dw(\tau)\big)_{\mathbb{R}^M} \in \mathbb{R}$$

to be a continuous function. Next, given a square integrable function $f : [0,\infty) \longrightarrow \mathbb{R}^M$ of locally bounded variation and $m \geq 1$, show (cf. (3.5.2)) that

$$\tilde{I}^{(m)}_{f^{\otimes m}}(t) = H_m\big(I_f(t), \|f\|^2_{L^2([0,\infty);\mathbb{R}^M)}\big),$$

where $H_m(x,a)$ is defined as in (ii) of Exercise 3.6, and conclude that

$$(t,w) \in [0,\infty) \times \mathbb{W}(\mathbb{R}^M) \longmapsto \tilde{I}^{(m)}_{f^{\otimes m}}(t)(w) \in \mathbb{R}$$

can be chosen so that it is continuous. More generally, suppose that $f_1, \dots, f_m$ are functions of locally bounded variation on $[0,\infty)$, and set $f_0 = \mathbf{1}$. If $m \geq 2$, show that

$$\begin{aligned} I_{f_m}(t) I^{(m-1)}_{f_1 \otimes \cdots \otimes f_{m-1}}(t) &= I^{(m)}_{f_1 \otimes \cdots \otimes f_m}(t) + I^{(m)}_{f_m \otimes f_1 \otimes \cdots \otimes f_{m-1}}(t) \\ &\quad + \int_0^t f_m(\tau) f_{m-1}(\tau) I^{(m-2)}_{f_1 \otimes \cdots \otimes f_{m-2}}(\tau)\, d\tau, \end{aligned}$$

and use this to see that

$$\tilde{I}^{(m)}_{f_1\otimes\cdots\otimes f_m}(t) = \frac{1}{2}\left(\sum_{\pi\in S_{m-1}}\left(I_{f_{\pi(m)}}(t)I^{(m-1)}_{f_{\pi(1)}\otimes\cdots\otimes f_{\pi(m-1)}}(t)\right.\right.$$
$$\left.\left.-\int_0^t f_{\pi(m)}(\tau)f_{\pi(m-1)}(\tau)I^{(m-2)}_{f_{\pi(1)}\otimes\cdots\otimes f_{\pi(m-2)}}(\tau)\,d\tau\right)\right),$$

where S_m is the symmetric group of permutations of $\{1,\ldots,m\}$. Proceeding by induction on m, conclude that

$$(t,w) \rightsquigarrow \tilde{I}^{(m)}_{f_1\otimes\cdots\otimes f_m}(t,w)$$

can be chosen to be continuous.

Chapter 4
Other Theories of Stochastic Integration

Doob's presentation of Itô's theory in his book [3] indicates that he understood that one can apply Itô's ideas to any continuous, square integrable martingale $\big(M(t), \mathcal{F}_t, \mathbb{P}\big)$ for which one knows that there is a progressively measurable function $t \rightsquigarrow A(t)$ which is continuous and non-decreasing in t and for which $\big(M(t)^2 - A(t), \mathcal{F}_t, \mathbb{P}\big)$ is a martingale. At the time, Doob did not know what is now called the Doob–Meyer decomposition theorem, a special case of which guarantees that such an $A(\,\cdot\,)$ always exists. In this chapter, we will first prove this existence result and then, following Kunita and Watanabe (cf. [10]), develop a version of the theory that Doob had in mind.

4.1 The Doob–Meyer decomposition theorem

Doob noticed that if $\big(X_n, \mathcal{F}_n, \mathbb{P}\big)$ is a discrete parameter, integrable submartingale, then there is a unique sequence $\{A_n : n \geq 0\}$ such that $A_0 = 0$, A_n is $\mathcal{F}_{n-1}$-measurable and $A_{n-1} \leq A_n$ for $n \geq 1$, and $\big(X_n - A_n, \mathcal{F}_n, \mathbb{P}\big)$ is a martingale. To see that such A_n's exist, simply set $A_0 = 0$ and $A_n = A_{n-1} + \mathbb{E}^{\mathbb{P}}[X_n - X_{n-1} \mid \mathcal{F}_{n-1}]$ for $n \geq 1$, and check that $\big(X_n - A_n, \mathcal{F}_n, \mathbb{P}\big)$ is a martingale. To prove uniqueness, suppose that $\{B_n : n \geq 0\}$ is a sequence of random variables such that $B_0 = 0$, B_n is $\mathcal{F}_{n-1}$-measurable for $n \geq 1$, and $\big(X_n - B_n, \mathcal{F}_n, \mathbb{P}\big)$ is a martingale. Then $B_0 = A_0$ and, since $B_n - B_{n-1}$ is $\mathcal{F}_{n-1}$-measurable,

$$B_n - B_{n-1} = \mathbb{E}^{\mathbb{P}}[X_n - X_{n-1} \mid \mathcal{F}_{n-1}] = A_n - A_{n-1}$$

for $n \geq 1$.

Trivial as Doob's observation is, it greatly simplified proofs of results like his stopping time and convergence theorems. However, even formulating, much less proving, a continuous parameter analog was a non-trivial challenge. Indeed, wholly aside from a proof of existence, in a continuous param-

D. W. Stroock, *Elements of Stochastic Calculus and Analysis*,
CRM Short Courses, https://doi.org/10.1007/978-3-319-77038-3_4

eter context it was not obvious what should replace the condition that A_n be $\mathcal{F}_{n-1}$-measurable. The person who figured out how to carry out this program was P.-A. Meyer, who, in the process, launched a program that led to a deep *théorie générale* of stochastic processes. I know no elementary proof of Meyer's full theorem, and so it is fortunate that we need only the particularly simple, special case covered by Theorem 4.1.2 below.

Lemma 4.1.1. *If $\big(X(t),\mathcal{F}_t,\mathbb{P}\big)$ is a continuous, $\mathbb{R}$-valued local martingale and $X(\,\cdot\,)$ has locally bounded variation, then $X(\,\cdot\,)$ is $\mathbb{P}$-almost surely constant.*

Proof. By using stopping times, one can reduce to the case when $\big(X(t),\mathcal{F}_t,\mathbb{P}\big)$ is a bounded martingale. Thus, without loss in generality, we will assume that $X(\,\cdot\,)$ is bounded and that $X(0)=0$.

Because $X(\,\cdot\,)$ is continuous and has locally bounded variation, $X(t)^2 = 2\int_0^t X(\tau)\,dX(\tau)$, where the integral is taken in the sense of Riemann–Stieltjes. Next, let $|X|(t) = \mathrm{var}_{[0,t]}\big(X(\,\cdot\,)\big)$, and set $X_R(t) = X(t\wedge\zeta_R)$ where $\zeta_R = \inf\{t\ge 0 : |X|(t)\ge R\}$. Then

$$
\begin{aligned}
\mathbb{E}^{\mathbb{P}}\big[X_R(t)^2\big] &= \sum_{m<2^n t} \mathbb{E}^{\mathbb{P}}\big[X_R(t\wedge(m+1)2^{-n})^2 - X_R(m2^{-n})^2\big]\\
&= \sum_{m<2^n t} \mathbb{E}^{\mathbb{P}}\Big[X_R\big(t\wedge(m+1)2^{-n}\big)\Big(X_R\big(t\wedge(m+1)2^{-n}\big) - X_R(m2^{-n})\Big)\Big]\\
&\longrightarrow \mathbb{E}^{\mathbb{P}}\left[\int_0^t X_R(\tau)\,dX_R(\tau)\right] \quad \text{as } n\to\infty.
\end{aligned}
$$

Since this means that

$$
\mathbb{E}^{\mathbb{P}}\left[\int_0^t X_R(\tau)\,dX_R(\tau)\right] = 2\mathbb{E}^{\mathbb{P}}\left[\int_0^t X_R(\tau)\,dX_R(\tau)\right],
$$

it follows that $\mathbb{E}^{\mathbb{P}}\big[X(t\wedge\zeta_R)^2\big] = 2\mathbb{E}^{\mathbb{P}}\big[X(t\wedge\zeta_R)^2\big]$, and so, $\mathbb{P}$-almost surely, $X(t)=0$ for $t\in[0,\zeta_R)$. Because $\zeta_R\nearrow\infty$ as $R\to\infty$, the proof is complete. □

Theorem 4.1.2. *If $\big(M(t),\mathcal{F}_t,\mathbb{P}\big)$ is a continuous, $\mathbb{R}$-valued local martingale and $M(0)=0$, then there is a $\mathbb{P}$-almost surely unique continuous, non-decreasing, progressively measurable function $t\rightsquigarrow A(t)$ such that $A(0)=0$ and*

$$
\big(M(t)^2 - A(t),\mathcal{F}_t,\mathbb{P}\big) \text{ is a local martingale.}
$$

Proof.[1] To prove the uniqueness assertion, suppose that both $A(\,\cdot\,)$ and $B(\,\cdot\,)$ are continuous, non-decreasing, progressively measurable functions for which $A(0)=0=B(0)$ and both $\big(M(t)^2-A(t),\mathcal{F}_t,\mathbb{P}\big)$ and $\big(M(t)^2-B(t),\mathcal{F}_t,\mathbb{P}\big)$ are

[1] I learned the basic idea for this proof of existence during a conversation with Itô.

local martingales. If $X(t) = B(t) - A(t)$, then $\big(X(t), \mathcal{F}_t, \mathbb{P}\big)$ is a continuous local martingale, $X(0) = 0$, and $X(\,\cdot\,)$ has locally bounded variation. Hence, by Lemma 4.1.1, $X(\,\cdot\,) = 0$ (a.s., $\mathbb{P}$).

Next, note that if M is uniformly bounded and $A(\,\cdot\,)$ exists, then

$$\big(M(t)^2 - A(t), \mathcal{F}_t, \mathbb{P}\big) \text{ is a martingale}$$

and $\mathbb{E}^{\mathbb{P}}[A(t)] = E^{\mathbb{P}}[M(t)^2]$. In fact, if $\{\zeta_m : m \geq 1\}$ are stopping times for the local martingale $\big(M(t)^2 - A(t), \mathcal{F}_t, \mathbb{P}\big)$, then $\mathbb{E}^{\mathbb{P}}[A(t \wedge \zeta_m)] = \mathbb{E}^{\mathbb{P}}[M(t \wedge \zeta_m)^2]$, $A(t \wedge \zeta_m) \nearrow A(t)$, and, as a consequence,

$$M(t \wedge \zeta_m)^2 - A(t \wedge \zeta_m) \longrightarrow M(t)^2 - A(t) \text{ in } L^1(\mathbb{P}; \mathbb{R}),$$

and therefore $\big(M(t)^2 - A(t), \mathcal{F}_t, \mathbb{P}\big)$ is a martingale.

We will now show how to reduce to the case when M is bounded. To this end, assume that existence is known in that case, and introduce the stopping times $\zeta_m = \inf\{t \geq 0 : |M(t)| \geq m\}$. Then, for each $m \geq 1$, there is a unique $A_m(\,\cdot\,)$ for $t \rightsquigarrow M(t \wedge \zeta_m)$. Further, by uniqueness, we can assume that, for all $m \geq 1$, $A_{m+1}(t) = A_m(t)$ when $t \in [0, \zeta_m)$. Thus, if $A(t) := A_m(t)$ for $t \in [0, \zeta_m)$, then $\big(M(t \wedge \zeta_m)^2 - A(t \wedge \zeta_m), \mathcal{F}_t, \mathbb{P}\big)$ is a martingale for all $m \geq 1$, and so $\big(M(t)^2 - A(t), \mathcal{F}_t, \mathbb{P}\big)$ is a local martingale. Having made this observation, from now on we will assume that $M(0) = 0$ and $M(\,\cdot\,)$ is uniformly bounded.

Set $\zeta_{0,n} = 0$ for $n \geq 0$ and $\zeta_{k,0} = k$ for $k \geq 1$. Next, assuming that $\{\zeta_{m,n} : m \geq 0\}$ has been chosen so that $\zeta_{m,n} \nearrow \infty$ as $m \to \infty$, define $\zeta_{m+1,n+1}$ to be

$$\zeta_{\ell,n} \wedge \inf\{t \geq \zeta_{m,n+1} : |M(t) - M(\zeta_{m,n})| \geq 2^{-n}\}$$
$$\text{where } \ell \text{ is the element of } \mathbb{Z}^+ \text{ for which } \zeta_{\ell-1,n} \leq \zeta_{m,n+1} < \zeta_{\ell,n}.$$

Clearly, for each $n \geq 0$, $\{\zeta_{m,n} : m \geq 0\}$ is non-decreasing sequence of bounded stopping times that tend to ∞. Further, these sequences are nested in the sense that $\{\zeta_{m,n} : m \geq 0\} \subseteq \{\zeta_{m,n+1} : m \geq 0\}$. For $n \geq 1$, set

$$M_{m,n} = M(\zeta_{m,n}) \quad \text{and} \quad \Delta_{m,n}(t) = M(t \wedge \zeta_{m+1,n}) - M(t \wedge \zeta_{m,n}),$$

and observe that $M(t)^2 = 2Y_n(t) + A_n(t)$, where

$$Y_n(t) = \sum_{m=0}^{\infty} M_{m,n} \Delta_{m,n}(t) \quad \text{and} \quad A_n(t) = \sum_{m=0}^{\infty} \Delta_{m,n}(t)^2.$$

In addition, $\big(Y_n(t), \mathcal{F}_t, \mathbb{P}\big)$ is a square integrable martingale, $A_n(0) = 0$, $A_n(\,\cdot\,)$ is continuous, and $A_n(t) + 4^{-n} \geq A_n(s)$ for $0 \leq s < t$. Thus, if, for each $T > 0$, we show that $\{A_n(\,\cdot\,) \restriction [0,T] : n \geq 0\}$ converges in $L^2\big(\mathbb{P}; C([0,T]; \mathbb{R})\big)$ or, equivalently, that $\{Y_n(\,\cdot\,) \restriction [0,T] : n \geq 0\}$ does, then we will be done. For that purpose, define

$$\widetilde{M}_{m,n+1} = M_{\ell,n} \quad \text{if}\, \zeta_{\ell,n} \le \zeta_{m,n+1} < \zeta_{\ell+1,n},$$

and observe that $|M_{m,n+1} - \widetilde{M}_{m,n+1}| \le 2^{-n}$ and

$$Y_{n+1}(t) - Y_n(t) = \sum_{m=0}^{\infty} \big(M_{m,n+1} - \widetilde{M}_{m,n+1}\big)\Delta_{m,n+1}(t).$$

Since $M_{m,n+1} - \widetilde{M}_{m,n+1}$ is $\mathcal{F}_{\zeta_{m,n+1}}$-measurable, one can use Hunt's stopping time theorem to see that the terms in this series are orthogonal in $L^2(\mathbb{P};\mathbb{R})$, and therefore, by Doob's inequality,

$$\begin{aligned}\mathbb{E}^{\mathbb{P}}\big[\|Y_{n+1}(\,\cdot\,) - Y_n(\,\cdot\,)\|^2_{[0,t]}\big] &\le 4^{1-n}\sum_{m=0}^{\infty}\mathbb{E}^{\mathbb{P}}\big[\Delta_{m,n+1}(t)^2\big] \\ = 2^{1-2n}\sum_{m=0}^{\infty}\big[M(t\wedge\zeta_{m+1,n+1})^2 &- M(t\wedge\zeta_{m,n+1})^2\big] = 4^{1-n}\mathbb{E}^{\mathbb{P}}\big[M(t)^2\big],\end{aligned}$$

and the desired convergence follows from this. □

One should appreciate how much more subtle than the preceding Meyer's reasoning was. Perhaps the most subtle aspect of his theory is the one having to do with uniqueness. In the discrete setting, uniqueness relied on Doob's insistence that A_n be $\mathcal{F}_{n-1}$-measurable, but what is the analog of that requirement in the continuous parameter context? Loosely speaking, Meyer's answer is that $A(\,\cdot\,)$ should have the property that $A(t)^2 = 2\int_0^t A(\tau)\,dA(\tau)$. An example that illustrates the importance of this point is the simple Poisson process $t \rightsquigarrow N(t)$, the one that starts at 0, waits a unit exponential holding time before jumping to 1, waits there for a second, independent unit exponential holding time before jumping to 2, etc. Since it is non-decreasing, $N(t)$ is a submartingale with respect to any filtration $\{\mathcal{F}_t : t \ge 0\}$ for which $N(\,\cdot\,)$ is progressively measurable. Further, there is no doubt that $N(t) - A(t)$ is a martingale if one takes $A(t) = N(t)$. But this is not the choice that Meyer's theory makes. Instead, his theory chooses $A(t) = t$, because $N(t) - t$ is also a martingale and this choice satisfies $A(t)^2 = 2\int_0^t A(\tau)\,dA(\tau)$. More generally, if a continuous $A(\,\cdot\,)$ exists, his theory would choose it. However, in general, there is no continuous choice of $A(\,\cdot\,)$, and to deal with those cases Meyer had to introduce a raft of new ideas.

From now on, $\langle\!\langle M\rangle\!\rangle(\,\cdot\,)$ will be used to denote the function $A(\,\cdot\,)$ in Theorem 4.1.2. In addition, given a stopping time ζ, M^ζ will be the local martingale $t \rightsquigarrow M(t\wedge\zeta)$. The following lemma contains a few elementary facts that will be used in the next section.

Lemma 4.1.3. *Let $\big(M(t),\mathcal{F}_t,\mathbb{P}\big)$ be a continuous local martingale with $M(0) = 0$ and ζ a stopping time. Then $\langle\!\langle M^\zeta\rangle\!\rangle(t) = \langle\!\langle M\rangle\!\rangle(t\wedge\zeta)$. Furthermore, if $\mathbb{E}^{\mathbb{P}}\big[\langle\!\langle M\rangle\!\rangle(\zeta)\big] < \infty$, then*

$$\left(M^\zeta(t), \mathcal{F}_t, \mathbb{P}\right) \quad \textit{and} \quad \left(M^\zeta(t)^2 - \langle\!\langle M^\zeta \rangle\!\rangle(t), \mathcal{F}_t, \mathbb{P}\right)$$

are martingales.

Proof. To prove the first assertion, simply observe that, by Doob's stopping time theorem, $\left(M^\zeta(t)^2 - \langle\!\langle M \rangle\!\rangle(t \wedge \zeta), \mathcal{F}_t, \mathbb{P}\right)$ is a local martingale, and therefore, by uniqueness, $\langle\!\langle M^\zeta \rangle\!\rangle(t) = \langle\!\langle M \rangle\!\rangle(t \wedge \zeta)$.

Now assume that $\mathbb{E}^{\mathbb{P}}\left[\langle\!\langle M \rangle\!\rangle(\zeta)\right] < \infty$, and set

$$\zeta_m = \inf\{t \geq 0 : |M(t)| \vee \langle\!\langle M \rangle\!\rangle(t) \geq m\}.$$

Then

$$\left(M^{\zeta_m \wedge \zeta}(t), \mathcal{F}_t, \mathbb{P}\right) \quad \text{and} \quad \left(M^{\zeta_m \wedge \zeta}(t)^2 - \langle\!\langle M \rangle\!\rangle(t \wedge \zeta_m \wedge \zeta), \mathcal{F}_t, \mathbb{P}\right)$$

are martingales, and so, by Doob's inequality,

$$\mathbb{E}^{\mathbb{P}}\left[\|M^{\zeta_m \wedge \zeta}\|_{[0,t]}^2\right] \leq 4\mathbb{E}^{\mathbb{P}}\left[\langle\!\langle M \rangle\!\rangle(t \wedge \zeta_m \wedge \zeta)\right] \leq 4\mathbb{E}^{\mathbb{P}}\left[\langle\!\langle M \rangle\!\rangle(\zeta)\right],$$

and therefore $\mathbb{E}^{\mathbb{P}}\left[\|M^\zeta\|_{[0,\zeta)}^2\right] < \infty$. Since this means that $M^\zeta(t \wedge \zeta_m) \longrightarrow M^\zeta(t)$ in $L^2(\mathbb{P};\mathbb{R})$, and, because $\langle\!\langle M^\zeta \rangle\!\rangle(t \wedge \zeta_m) \nearrow \langle\!\langle M^\zeta \rangle\!\rangle(t)$, there is nothing more to do. □

4.2 Kunita and Watanabe's integral

As mentioned earlier, Doob already understood that Itô's integration theory could be extended to martingales other than Brownian motion once one had a result like the one in Theorem 4.1.2. However, it was Kunita and Watanabe who not only carried out the program that Doob had in mind but did so in a particularly elegant fashion.

Let $(\Omega, \mathcal{F}, \mathbb{P})$ be a complete probability space and $\{\mathcal{F}_t : t \geq 0\}$ a non-decreasing filtration of complete sub σ-algebras, use $M^2(\mathbb{P};\mathbb{R})$ and $M_{\text{loc}}(\mathbb{P};\mathbb{R})$, respectively, to denote the set of all square integrable $\mathbb{R}$-valued continuous martingales and continuous local martingales under $\mathbb{P}$ relative to $\{\mathcal{F}_t : t \geq 0\}$ with $M(0) = 0$. Clearly, both $M^2(\mathbb{P};\mathbb{R})$ and $M_{\text{loc}}(\mathbb{P};\mathbb{R})$ are vector spaces over $\mathbb{R}$. Next, given $M_1, M_2 \in M_{\text{loc}}(\mathbb{P};\mathbb{R})$, set

$$\langle M_1, M_2 \rangle = \frac{\langle\!\langle M_1 + M_2 \rangle\!\rangle - \langle\!\langle M_1 - M_2 \rangle\!\rangle}{4},$$

and note that

$$\left(M_1(t)M_2(t) - \langle M_1, M_2 \rangle(t), \mathcal{F}_t, \mathbb{P}\right)$$

is a continuous local martingale. Moreover, by Lemma 4.1.1, one sees that $\langle M_1, M_2 \rangle$ is the only progressively measurable, continuous function $A(\,\cdot\,)$ such that $A(0) = 0$, $A(\,\cdot\,) \restriction [0,t]$ has bounded variation for each $t \geq 0$, and

$\big(M_1(t)M_2(t) - A(t), \mathcal{F}_t, \mathbb{P}\big)$ is a continuous local martingale.

Lemma 4.2.1. *The map* $(M_1, M_2) \rightsquigarrow \langle M_1.M_2\rangle$ *is symmetric and bilinear on* $M_{\mathrm{loc}}(\mathbb{P};\mathbb{R})^2$ *in the sense that* $\langle M_1, M_2\rangle = \langle M_2, M_1\rangle$ *and, for all* $\alpha_1, \alpha_2 \in \mathbb{R}$,

$$\langle \alpha_1 M_1 + \alpha_2 M_2, M_3\rangle = \alpha_1\langle M_1, M_3\rangle + \alpha_2\langle M_2, M_3\rangle$$

$\mathbb{P}$*-almost surely. Furthermore,* $\langle M, M\rangle = \langle\!\langle M\rangle\!\rangle \ge 0$ *and, for* $0 \le s < t$,

$$\begin{aligned} &\big|\langle M_1, M_2\rangle(t) - \langle M_1, M_2\rangle(s)\big| \\ &\quad \le \sqrt{\langle\!\langle M_1\rangle\!\rangle(t) - \langle\!\langle M_1\rangle\!\rangle(s)}\sqrt{\langle\!\langle M_2\rangle\!\rangle(t) - \langle\!\langle M_2\rangle\!\rangle(s)} \end{aligned} \tag{4.2.1}$$

$\mathbb{P}$*-almost surely. In particular,*

$$\Big\|\sqrt{\langle\!\langle M_2\rangle\!\rangle} - \sqrt{\langle\!\langle M_1\rangle\!\rangle}\Big\|_{[0,t]} \le \sqrt{\langle\!\langle M_2 - M_1\rangle\!\rangle(t)} \quad \text{(a.s., }\mathbb{P}),$$

and for any stopping time ζ,

$$\langle M_1^\zeta, M_2\rangle(t) = \langle M_1, M_2\rangle(t\wedge\zeta).$$

Proof. The first assertions are all easy applications of uniqueness. To prove the Schwarz type inequality in (4.2.1), one uses the same reasoning as usual. That is,

$$0 \le \langle\!\langle \alpha M_1 \pm \alpha^{-1} M_2\rangle\!\rangle = \alpha^2\langle\!\langle M_1\rangle\!\rangle \pm 2\langle M_1, M_2\rangle + \alpha^{-2}\langle\!\langle M_2\rangle\!\rangle$$

$\mathbb{P}$-almost surely, first for each $\alpha \in \mathbb{R}$ and then for all α simultaneously. Thus, $\mathbb{P}$-almost surely,

$$2\big|\langle M_1, M_2\rangle\big| \le \alpha^2\langle\!\langle M_1\rangle\!\rangle + \alpha^{-2}\langle\!\langle M_2\rangle\!\rangle,$$

and so, after minimizing with respect to α, one sees that

$$|\langle M_1, M_2\rangle| \le \sqrt{\langle\!\langle M_1\rangle\!\rangle}\sqrt{\langle\!\langle M_2\rangle\!\rangle} \quad \text{(a.s., }\mathbb{P}).$$

Finally, by applying this with $M_i(t) - M_i(t\wedge s)$ in place of $M_i(t)$, one arrives at (4.2.1).

Once one has (4.2.1), the first of the concluding assertions follows in the same way as the triangle inequality follows from Schwarz's inequality. To prove the second, note that, by (4.2.1),

$$\big|\langle M_1^\zeta, M_2\rangle(t) - \langle M_1^\zeta, M_2\rangle(t\wedge\zeta)\big| \le \sqrt{\langle\!\langle M_1^\zeta\rangle\!\rangle(t) - \langle\!\langle M_1^\zeta\rangle\!\rangle(t\wedge\zeta)}\sqrt{\langle\!\langle M_2\rangle\!\rangle(t)} = 0$$

and

$$\big|\langle M_1^\zeta, M_2\rangle(t\wedge\zeta) - \langle M_1, M_2\rangle(t\wedge\zeta)\big| \le \sqrt{\langle\!\langle M_1^\zeta - M_1\rangle\!\rangle(t\wedge\zeta)}\sqrt{\langle\!\langle M_2\rangle\!\rangle(t)} = 0,$$

since

$$\langle\!\langle M_1^\zeta - M_1 \rangle\!\rangle(t \wedge \zeta) = \langle\!\langle (M_1^\zeta - M_1)^\zeta \rangle\!\rangle(t)$$

and, by Lemma 4.1.3, $(M_1^\zeta - M_1)^\zeta = M_1^\zeta - M_1^\zeta = 0$. □

Because $\langle\!\langle M \rangle\!\rangle(t)$ is continuous and non-decreasing, it determines a non-atomic, locally finite Borel measure $\langle\!\langle M \rangle\!\rangle(dt)$ on $[0,\infty)$, and because, $\langle M_1, M_2\rangle$ is continuous and has locally bounded variation, it determines a non-atomic, locally finite Borel signed measure $\langle M_1, M_2\rangle(dt)$ there. Starting from (4.2.1) and using Riemann sum approximations, one sees that, for all $\alpha > 0$, $T > 0$, and $\varphi \in C([0,T];\mathbb{R})$,

$$2\left|\int_0^T \varphi(\tau)\,\langle M_1, M_2\rangle(d\tau)\right| \le \alpha^2 \int_0^T \varphi(\tau)^2\,\langle\!\langle M_1\rangle\!\rangle(d\tau) + \alpha^{-2}\langle\!\langle M_2\rangle\!\rangle(T),$$

and therefore that

$$\left|\int_0^T \varphi(\tau)\,\langle M_1, M_2\rangle(d\tau)\right| \le \sqrt{\int_0^T \varphi(\tau)^2 \langle\!\langle M_1\rangle\!\rangle(d\tau)}\sqrt{\langle\!\langle M_2\rangle\!\rangle(T)}.$$

Because this inequality holds for all continuous φ's on $[0,T]$, it also holds for all bounded Borel measurable ones, and therefore we know that, for all Borel measurable $\varphi : [0,T] \longrightarrow \mathbb{R}$,

$$\left|\int_0^T |\varphi(\tau)|\,|\langle M_1, M_2\rangle|(d\tau)\right| \le \sqrt{\int_0^T \varphi(\tau)^2 \langle\!\langle M_1\rangle\!\rangle(d\tau)}\sqrt{\langle\!\langle M_2\rangle\!\rangle(T)}, \qquad (4.2.2)$$

where $|\langle M_1, M_2\rangle|(dt)$ is the variation measure determined by $\langle M_1, M_2\rangle(dt)$.

With these preparations, we can say how Kunita and Watanabe defined stochastic integrals with respect to any element of $M_{\text{loc}}(\mathbb{P};\mathbb{R})$. Given $M \in M_{\text{loc}}(\mathbb{P};\mathbb{R})$, use $PM^2_{\text{loc}}(M;\mathbb{R})$ to denote the space of progressively measurable functions ξ such that

$$\int_0^t |\xi(\tau)|^2\,\langle\!\langle M\rangle\!\rangle(d\tau) < \infty \quad \text{for all}\, t > 0.$$

Noting that, by (4.2.2), every $\xi \in PM^2_{\text{loc}}(M;\mathbb{R})$ is locally integrable with respect to $|\langle M, M'\rangle|(dt)$ for all $M' \in M^2_{\text{loc}}(\mathbb{P};\mathbb{R})$, define the stochastic integral I_ξ^M of ξ with respect to M to be the element of $M_{\text{loc}}(\mathbb{P};\mathbb{R})$ such that $I_\xi^M(0){=}0$ and $\langle I_\xi^M, M'\rangle(dt) = \xi(t)\langle M, M'\rangle(dt)$ for all $M' \in M_{\text{loc}}(\mathbb{P};\mathbb{R})$. It is obvious that this definition uniquely determines I_ξ^M since, if I and J were two such elements of $M_{\text{loc}}(\mathbb{P};\mathbb{R})$, then $\langle I - J, I - J\rangle(dt) = 0$. Thus, before adopting this definition, all that we have to do is prove that such an element exists.

Lemma 4.2.2. *Let ζ be a finite stopping time and α a bounded, $\mathcal{F}_\zeta$-measurable function, and set $\xi(t) = \alpha\mathbf{1}_{[\zeta,\infty)}(t)$. Then $\xi \in PM^2_{\text{loc}}(M;\mathbb{R})$ and*

$$\alpha\big(M(t) - M^\zeta(t)\big) = I_\xi^M(t).$$

Proof. Without loss in generality, assume that $M(\,\cdot\,)$ is bounded.

Because $t \rightsquigarrow \xi(t)$ is adapted and right-continuous, ξ is progressively measurable.

Set $\widetilde{M}(t) = \alpha\big(M(t) - M^\zeta(t)\big)$, and let $s < t$ and $\Gamma \in \mathcal{F}_s$ be given. Then

$$\mathbb{E}^{\mathbb{P}}\big[\widetilde{M}(t), \Gamma\big] = \mathbb{E}^{\mathbb{P}}\big[\widetilde{M}(t), \Gamma \cap \{\zeta \le s\}\big] + \mathbb{E}^{\mathbb{P}}\big[\widetilde{M}(t), \Gamma \cap \{s < \zeta \le t\}\big].$$

Because $\Gamma \cap \{\zeta \le s\} \in \mathcal{F}_s$ and $\alpha \mathbf{1}_{\Gamma \cap \{\zeta \le s\}}$ is $\mathcal{F}_s$-measurable,

$$\begin{aligned}\mathbb{E}^{\mathbb{P}}\big[\widetilde{M}(t), \Gamma \cap \{\zeta \le s\}\big] &= \mathbb{E}^{\mathbb{P}}\big[\alpha\big(M(t) - M(t \wedge \zeta)\big), \Gamma \cap \{\zeta \le s\}\big] \\ &= \mathbb{E}^{\mathbb{P}}\big[\alpha\big(M(s) - M(s \wedge \zeta)\big), \Gamma \cap \{\zeta \le s\}\big] = \mathbb{E}^{\mathbb{P}}\big[\widetilde{M}(s), \Gamma\big].\end{aligned}$$

At the same time, $\alpha \mathbf{1}_{\Gamma \cap \{s < \zeta \le t\}}$ is $\mathcal{F}_{t\wedge\zeta}$-measurable, and so, by Hunt's stopping time theorem,

$$\mathbb{E}^{\mathbb{P}}\big[\alpha M(t), \Gamma \cap \{s < \zeta \le t\}\big] = \mathbb{E}^{\mathbb{P}}\big[\alpha M(t \wedge \zeta), \Gamma \cap \{s < \zeta \le t\}\big],$$

which means that $\mathbb{E}^{\mathbb{P}}\big[\widetilde{M}(t), \Gamma \cap \{s < \zeta \le t\}\big] = 0$. Thus $\big(\widetilde{M}(t), \mathcal{F}_t, \mathbb{P}\big)$ is a martingale.

To show that $\widetilde{M} = I_\xi^M$, let $M' \in M_{\mathrm{loc}}(\mathbb{P}, \mathbb{R})$, and set $X(t) = M(t)M'(t) - \langle M, M'\rangle(t)$. Then, by the preceding

$$\Big(\alpha\big(X(t) - X(t \wedge \zeta)\big), \mathcal{F}_t, \mathbb{P}\Big) \quad \text{and} \quad \Big(\alpha M(\zeta)\big(M(t) - M(t \wedge \zeta)\big), \mathcal{F}_t, \mathbb{P}\Big)$$

are continuous, local martingales. Thus, since

$$\begin{aligned}&\widetilde{M}(t)M'(t) - \int_0^t \xi(\tau)\,\langle M, M'\rangle(d\tau) \\ &\qquad = \widetilde{M}(t)M'(t) - \alpha\big(\langle M, M'\rangle(t) - \langle M, M'\rangle(t \wedge \zeta)\big) \\ &\qquad = \alpha\big(X(t) - X(t \wedge \zeta)\big) - \alpha M(\zeta)\big(M'(t) - M'(t \wedge \zeta)\big),\end{aligned}$$

we know that

$$\left(\widetilde{M}(t)M'(t) - \int_0^t \xi(\tau)\,\langle M, M'\rangle(d\tau), \mathcal{F}_t, \mathbb{P}\right)$$

is a continuous local martingale and therefore that $\widetilde{M} = I_\xi^M$. □

Lemma 4.2.3. *If I_ξ^M exists and ζ is a stopping time, then $I_\xi^{M^\zeta}$ and $I^M_{\mathbf{1}_{[0,\zeta)}\xi}$ exist and both are equal to $(I_\xi^M)^\zeta$. Next suppose that $\{\xi_n : n \ge 0\} \cup \{\xi\} \subseteq PM^2_{\mathrm{loc}}(M; \mathbb{R})$ and that*

$$\lim_{n\to\infty}\int_0^t \left(\xi_n(\tau)-\xi(\tau)\right)^2\langle\!\langle M\rangle\!\rangle(d\tau)=0 \quad (\text{a.s.}, \mathbb{P})$$

for all $t \geq 0$. If $I_{\xi_n}^M$ exists for each $n \geq 0$, then I_ξ^M also exits and, for all $t \geq 0$, $I_{\xi_n}^M(t) \longrightarrow I_\xi^M(t)$ in $\mathbb{P}$-probability.

Proof. The first assertion is an easy application of the last part of Lemma 4.2.1. Indeed,

$$\begin{aligned}\langle (I_\xi^M)^\zeta, M'\rangle(t) &= \langle I_\xi^M, M'\rangle(t\wedge\zeta)\\ &= \int_0^{t\wedge\zeta}\xi(\tau)\langle M,M'\rangle(d\tau) = \int_0^t \mathbf{1}_{[0,\zeta)}(\tau)\xi(\tau)\langle M,M'\rangle(d\tau),\end{aligned}$$

and $\langle M^\zeta, M'\rangle(d\tau) = \mathbf{1}_{[0,\zeta)}(\tau)\langle M,M'\rangle(d\tau)$.

Turning to the second assertion, begin by assuming that

$$\|M\|_{[0,t]} \quad\text{and}\quad \sup_{n\geq 0}\int_0^t \xi_n(\tau)^2\langle\!\langle M\rangle\!\rangle(d\tau)$$

are uniformly bounded for each $t \geq 0$, and set $I_n = I_{\xi_n}^M$. Then

$$\left(\left(I_n(t)-I_m(t)\right)^2 - \int_0^t \left(\xi_n(\tau)-\xi_m(\tau)\right)^2\langle\!\langle M\rangle\!\rangle(d\tau), \mathcal{F}_t, \mathbb{P}\right)$$

is a martingale, and so

$$\mathbb{E}^{\mathbb{P}}\left[\left(I_n(\,\cdot\,)-I_m(\,\cdot\,)\right)^2\right] = \mathbb{E}^{\mathbb{P}}\left[\int_0^t \left(\xi_n(\tau)-\xi_m(\tau)\right)^2\langle\!\langle M\rangle\!\rangle(d\tau)\right],$$

which, by Doob's inequality, means that

$$\lim_{m\to\infty}\sup_{n>m}\mathbb{E}^{\mathbb{P}}\left[\|I_n(\,\cdot\,)-I_m(\,\cdot\,)\|_{[0,t]}^2\right]=0.$$

Hence there exists a continuous, square integrable martingale I such that

$$\|I-I_m\|_{[0,t]} \longrightarrow 0 \text{ in } L^2(\mathbb{P};\mathbb{R}) \quad \text{for all } t \geq 0.$$

In addition, if $M' \in M_{\text{loc}}(\mathbb{P};\mathbb{R})$ and $\langle\!\langle M'\rangle\!\rangle(t)$ is bounded for all $t \geq 0$, then, for $0 \leq s < t$,

$$\begin{aligned}\mathbb{E}^{\mathbb{P}}\left[I(t)M'(t)-I(s)M'(s)\,\middle|\,\mathcal{F}_s\right] &= \lim_{n\to\infty}\mathbb{E}^{\mathbb{P}}\left[I_n(t)M'(t)-I_n(s)M'(s)\,\middle|\,\mathcal{F}_s\right]\\ &= \lim_{n\to\infty}\mathbb{E}^{\mathbb{P}}\left[\int_s^t \xi_n(\tau)\langle M,M'\rangle(d\tau)\,\middle|\,\mathcal{F}_s\right]\end{aligned}$$

and, by (4.2.2),

$$\int_s^t |\xi(\tau) - \xi_n(\tau)|\,|\langle M, M'\rangle|(d\tau) \le \sqrt{\int_0^t \big(\xi(\tau) - \xi_n(\tau)\big)^2 \langle\!\langle M \rangle\!\rangle(d\tau)}\sqrt{\langle\!\langle M' \rangle\!\rangle(t)},$$

which tends to 0 $\mathbb{P}$-almost surely. Hence, $\langle I, M'\rangle(dt) = \xi(t)\langle M, M'\rangle(dt)$ (a.s., $\mathbb{P}$) when $\langle\!\langle M' \rangle\!\rangle(t)$ is bounded for all $t \ge 0$. To prove that the same equality for general M', take $\zeta_k = \inf\{t \ge 0 : \langle\!\langle M' \rangle\!\rangle(t) \ge k\}$. Then, using Lemma 4.2.1, one sees that

$$\langle I, M'\rangle(t \wedge \zeta_k) = \langle I, (M')^{\zeta_k}\rangle(t) = \int_0^{t\wedge\zeta_k} \xi(\tau)\langle M, M'\rangle(d\tau)$$

$\mathbb{P}$-almost surely for all $k \ge 1$. Since $\zeta_k \nearrow \infty$ as $k \to \infty$, it follows that $I = I_\xi^M$.

To remove the boundedness assumptions on M and $\int_0^t \xi_n(\tau)^2 \langle\!\langle M \rangle\!\rangle(d\tau)$, define

$$\zeta_k = \inf\left\{t \ge 0 : |M(t)| \vee \sup_{n\ge 0} \int_0^t \xi_n(\tau)^2 \,\langle\!\langle M \rangle\!\rangle(t) \ge k\right\},$$

and set $M_k = M^{\zeta_k}$. By the preceding, we know that $I_k := I_\xi^{M_k}$ exists, and, by the first part of this lemma, $I_{k+1}(t\wedge\zeta_k) = I_k(t\wedge\zeta_k)$ (a.s., $\mathbb{P}$) for all $k \ge 1$. Hence, if we define $I \in M_{\mathrm{loc}}(\mathbb{P};\mathbb{R})$ so that $I(t \wedge \zeta_k) = I_k(t \wedge \zeta_k)$ (a.s., $\mathbb{P}$) for all $k \ge 1$, then $\langle I, M'\rangle(dt) = \xi(t)\langle M, M'\rangle(dt)$ for all $M' \in M_{\mathrm{loc}}(\mathbb{P};\mathbb{R})$. □

In view of the preceding, what remains is to find a sufficiently rich set of ξ's for which we can show that I_ξ^M exists, and, as was the case when M was a Brownian motion, a good guess is that ξ's of locally bounded variation should be the place to look.

Lemma 4.2.4. *Suppose $M \in M_{\mathrm{loc}}(\mathbb{P};\mathbb{R})$ and ξ is an element of $PM^2_{\mathrm{loc}}(M;\mathbb{R})$ for which $\xi(\,\cdot\,)$ has locally bounded variation. Then I_ξ^M exists and $I_\xi^M(t)$ is equal to the Riemann–Stieltjes integral $\int_0^t \xi(\tau)\,dM(\tau)$.*

Proof. Begin with the assumption that ξ is uniformly bounded. Set $I(t) = \int_0^t \xi(\tau)\,dM(\tau)$, where the integral is taken in the sense of Riemann–Stieltjes. Then $I(t) = \lim_{n\to\infty} I_n(t)$, where, by Lemma 4.2.2,

$$I_n(t) := \sum_{m<2^n t} \xi(m2^{-n})\big(M(t\wedge(m+1)2^{-n}) - M(t\wedge m2^{-n})\big) = I_{\xi_n}(t)$$

when $\xi_n(t) = \xi(\lfloor t\rfloor_n)$. At the same time, because $\langle\!\langle M \rangle\!\rangle(dt)$ is non-atomic and ξ has at most a countable number of discontinuities,

$$\lim_{n\to\infty} \int_0^t \big(\xi(\tau) - \xi_n(\tau)\big)^2 \,\langle\!\langle M \rangle\!\rangle(d\tau) = 0,$$

Hence, by Lemma 4.2.3, $I = I_\xi^M$.

To remove the boundedness assumption on ξ, for each $n \geq 1$, define $\xi_n(t) = \xi(t)$ if $|\xi(t)| \leq n$ and $\xi_n(t) = \pm n$ if $\pm\xi(t) > n$. Then $\mathrm{var}_{[0,t]}(\xi_n) \leq \mathrm{var}_{[0,t]}(\xi)$, $|\xi_n(t)| \leq n \wedge |\xi(t)|$, and $\int_0^t \big(\xi(\tau) - \xi_n(\tau)\big)^2 \langle\!\langle M \rangle\!\rangle(d\tau) \longrightarrow 0$. Hence, by the preceding combined with Lemmas 3.2.1 and 4.2.3, I_ξ^M exists,

$$I_{\xi_n}^M(t) \longrightarrow \int_0^t \xi(\tau)\, dM(\tau), \quad \text{and} \quad I_{\xi_n}^M(t) \longrightarrow I_\xi^M(t). \qquad \square$$

To complete the program, we have to show that if $\xi \in PM^2_{\mathrm{loc}}(M;\mathbb{R})$, then there exist $\{\xi_n : n \geq 1\} \subseteq PM^2_{\mathrm{loc}}(M;\mathbb{R})$ such that each ξ_n has locally bounded variation and $\int_0^t \big(\xi(\tau) - \xi_n(\tau)\big)^2 \langle\!\langle M \rangle\!\rangle(d\tau) \longrightarrow 0$. The construction of such a sequence is a little trickier than it was in the Brownian case and makes use of the following lemma.

Lemma 4.2.5. *Let $F : [0,\infty) \longrightarrow \mathbb{R}$ be a continuous, non-decreasing function with $F(0) = 0$, and define*

$$F^{-1}(t) = \inf\{\tau \geq 0 : F(\tau) \geq t\} \quad \textit{for } t \geq 0.$$

Then F^{-1} is left-continuous and non-decreasing on $\big[0, F(\infty)\big)$. Further,

$$F \circ F^{-1}(t) = t, \quad F^{-1} \circ F(\tau) \leq \tau, \quad \textit{and} \quad D := \{\tau : F^{-1} \circ F(\tau) < \tau\}$$

is either empty or the at most countable union of mutually disjoint intervals of the form $(a,b]$ with the properties that $0 \leq a < b \leq \infty$, $F(a) = F(b)$, $F(\tau) < F(a)$ for $\tau < a$, and $F(\tau) > F(b)$ for $\tau > b$. Next, if $F(d\tau)$ is the Borel measure determined by F, then $F(D) = 0$ and

$$\int_0^{F(T)} f \circ F^{-1}(t)\, dt = \int_0^T f(\tau)\, F(d\tau)$$

for non-negative, Borel measurable f on $[0,\infty)$. Finally, for $\epsilon > 0$, set $\rho_\epsilon(t) = \epsilon^{-1}\rho(\epsilon^{-1}t)$ where $\rho \in C^\infty\big(\mathbb{R};[0,\infty)\big)$ vanishes off of $(0,1)$ and $\int \rho(t)\, dt = 1$, and, given $f \in L^2\big(F;[0,\infty)\big)$, set

$$f_\epsilon(\tau) = \int_0^{F(\infty)} \rho_\epsilon\big(F(\tau) - \sigma\big) f \circ F^{-1}(\sigma)\, d\sigma.$$

Then f_ϵ is a continuous function of locally bounded variation, $\|f_\epsilon\|_{L^2(F;\mathbb{R})} \leq \|f\|_{L^2(F;\mathbb{R})}$, and $\|f - f_\epsilon\|_{L^2(F;\mathbb{R})} \longrightarrow 0$ as $\epsilon \searrow 0$.

Proof. There is no problem checking that F^{-1} is left-continuous, non-decreasing, and satisfies $F \circ F^{-1}(\tau) = \tau$ and $F^{-1} \circ F(\tau) \leq \tau$. In addition, if $F^{-1} \circ F(\tau) < \tau$, then $\tau \in (a,b]$ where

$$a = \inf\{\sigma : F(\sigma) = F(\tau)\} \quad \text{and} \quad b = \sup\{\sigma : F(\sigma) = F(\tau)\}.$$

Thus, if $D \neq \varnothing$, then it is the union of mutually disjoint intervals of the described sort, and there can be at most a countable number of such intervals. Hence, since $F\big((a,b]\big) = F(a) - F(b)$, it follows that $F(D) = 0$.

Next, by the standard change of variables formula for Riemann–Stieltjes integrals,

$$\int_0^T f \circ F(\tau)\, F(d\tau) = \int_0^T f \circ F(\tau)\, dF(\tau) = \int_0^{F(T)} f(t)\, dt$$

for $f \in C\big([0,\infty);\mathbb{R}\big)$, and so the same is true for all non-negative Borel measurable f's. Hence, because $F(D) = 0$,

$$\int_0^{F(T)} f \circ F^{-1}(t)\, dt = \int_0^T (f \circ F^{-1} \circ F)(\tau)\, F(d\tau) = \int_0^T f(\tau)\, F(d\tau).$$

As the composition of a smooth function with a continuous, monotone one, it is clear that f_ϵ is a continuous function of locally bounded variation. In addition, by the preceding,

$$\begin{aligned}\int_0^\infty f_\epsilon(\tau)^2\, F(d\tau) &= \int_0^{F(\infty)} \big(\rho_\epsilon * (f \circ F^{-1})\big)(t)^2\, dt \\ &\le \int_0^{F(\infty)} f \circ F^{-1}(t)^2\, dt = \int_0^\infty f(\tau)^2\, F(d\tau),\end{aligned}$$

and similarly

$$\|f - f_\epsilon\|^2_{L^2(F;\mathbb{R})} = \int_0^{F(\infty)} \Big(f \circ F^{-1}(t) - \big(\rho_\epsilon * (f \circ F^{-1})\big)(t)\Big)^2\, dt \longrightarrow 0. \qquad \square$$

Theorem 4.2.6. *For each $M \in M_{\text{loc}}(\mathbb{P};\mathbb{R})$ and $\xi \in PM^2_{\text{loc}}(M;\mathbb{R})$ there exists a unique $I^M_\xi \in M_{\text{loc}}(\mathbb{P};\mathbb{R})$ such that $\langle I^M_\xi, M'\rangle(dt) = \xi(t)\langle M, M'\rangle(dt)$ for all $M' \in M_{\text{loc}}(\mathbb{P};\mathbb{R})$. Furthermore, if $\xi(\,\cdot\,)$ has locally bounded variation, then $I^M_\xi(t)$ equals the Riemann–Stieltjes integral on $[0,t]$ of ξ with respect to M.*

Proof. All that remains is to prove the existence assertion when ξ isn't of locally bounded variation. For that purpose, define ξ_ϵ from ξ by the prescription in Lemma 4.2.5 with $F = \langle\!\langle M\rangle\!\rangle$ and $f = \xi$. Then ξ_ϵ has locally bounded variation and

$$\lim_{\epsilon \searrow 0} \int_0^t \big(\xi(\tau) - \xi_\epsilon(\tau)\big)^2 \langle\!\langle M\rangle\!\rangle(d\tau) = 0.$$

Thus, by Lemmas 4.2.3 and 4.2.4, we are done. $\square$

In the future, we will call I^M_ξ the **stochastic integral** of ξ with respect to M and will usually use $\int_0^t \xi(\tau)\, dM(\tau)$ to denote $I^M_\xi(t)$. Notice that if ζ_1 and ζ_2 are stopping times and $\zeta_1 \le \zeta_2$, then

$$\begin{aligned}\int_{t\wedge\zeta_1}^{t\wedge\zeta_2} \xi(\tau)\,dM(\tau) &:= I_\xi^M(t\wedge\zeta_2) - I_\xi^M(\zeta_1) \\ &= \int_0^t \mathbf{1}_{[\zeta_1,\zeta_2)}(\tau)\xi(\tau)\,dM(\tau).\end{aligned} \tag{4.2.3}$$

Also, if $\xi \in PM^2_{\text{loc}}(M;\mathbb{R})$ and $\eta \in PM^2_{\text{loc}}(I_\xi^M;\mathbb{R})$, then $\xi\eta \in PM_{\text{loc}}(M;\mathbb{R})$ and

$$\int_0^t \eta(\tau)\,dI_\xi^M(\tau) = \int_0^t \xi(\tau)\eta(\tau)\,dM(\tau). \tag{4.2.4}$$

To check this, set $J(t) = \int_0^t \eta(\tau)\,dI_\xi^M(\tau)$. Then, for any $M' \in M_{\text{loc}}(\mathbb{P};\mathbb{R})$,

$$\langle J, M'\rangle(dt) = \eta(t)\langle I_\xi^M, M'\rangle(dt) = \xi(t)\eta(t)\langle M, M'\rangle(dt).$$

Finally, suppose that $M(t) = \big(M_1(t),\dots,M_N(t)\big)$, where $M_i \in M_{\text{loc}}(\mathbb{P};\mathbb{R})$ for each $1 \le i \le N$, set

$$A(t) = \big(\big(\langle M_i, M_j\rangle(t)\big)\big)_{1\le i,j\le N},$$

let $PM^2_{\text{loc}}(M;\mathbb{R}^N)$ denote the space of $\mathbb{R}^N$-valued, progressively measurable $\boldsymbol{\eta}$ such that

$$\int_0^t \big(\boldsymbol{\eta}(\tau), dA(\tau)\boldsymbol{\eta}(\tau)\big)_{\mathbb{R}^N} < \infty \quad \text{for all } t \ge 0,$$

and define

$$I_{\boldsymbol{\eta}}^M(t) = \int_0^t \big(\boldsymbol{\eta}(\tau), dM(\tau)\big)_{\mathbb{R}^N} := \sum_{i=1}^N \int_0^t \eta_i(\tau)\,dM_i(\tau).$$

Then it is easy to check that, for $\boldsymbol{\eta},\boldsymbol{\xi} \in PM^2_{\text{loc}}(M;\mathbb{R}^N)$,

$$\left(I_{\boldsymbol{\xi}}^M(t)I_{\boldsymbol{\eta}}^M(t) - \int_0^t \big(\boldsymbol{\xi}(\tau), dA(\tau)\boldsymbol{\eta}(\tau)\big)_{\mathbb{R}^N}, \mathcal{F}_t, \mathbb{P}\right)$$

is a local martingale. Next, let $PM^2_{\text{loc}}\big(M;\text{Hom}(\mathbb{R}^N;\mathbb{R}^{N'})\big)$ be the space of progressively measurable, $\text{Hom}(\mathbb{R}^N;\mathbb{R}^{N'})$-valued functions σ with the property that

$$\int_0^t \text{Trace}\big(\sigma(\tau)dA(\tau)\sigma(\tau)^\top\big) < \infty \quad \text{for all } t \ge 0.$$

Then, for any $\boldsymbol{\xi} \in \mathbb{R}^{N'}$, $\sigma(\,\cdot\,)^\top\boldsymbol{\xi} \in PM^2_{\text{loc}}(M;\mathbb{R}^N)$, and so we can define I_σ^M to be the $\mathbb{R}^{N'}$-valued progressively measurable function such that

$$\big(\boldsymbol{\xi}, I_\sigma^M(t)\big)_{\mathbb{R}^{N'}} = \int_0^t \big(\sigma(\tau)^\top\boldsymbol{\xi}, dM(\tau)\big)_{\mathbb{R}^N} \quad \text{for all } \boldsymbol{\xi} \in \mathbb{R}^{N'},$$

in which case

$$\left(I_\sigma^M(t), \mathcal{F}_t, \mathbb{P}\right) \quad \text{and} \quad \left(I_\sigma^M(t) \otimes I_\sigma^M(t) - \int_0^t \sigma(\tau)^\top dA(\tau)\sigma(\tau)\, \mathcal{F}_t, \mathbb{P}\right)$$

are, respectively, $\mathbb{R}^{N'}$- and $\mathrm{Hom}(\mathbb{R}^N;\mathbb{R}^{N'})$-valued continuous local martingales.

4.3 Itô's formula again

Having developed the theory of stochastic integration for general continuous, local martingales, it is only reasonable to see what Itô's formula looks like in that context. What follows is Kunita and Watanabe's version of his formula.

Theorem 4.3.1. *For each $1 \le i \le N_1$, let V_i be a continuous, progressively measurable $\mathbb{R}$-valued function of locally bounded variation, and for each $1 \le j \le N_2$ let $\left(M_j(t), \mathcal{F}_t, \mathbb{P}\right)$ be a continuous local martingale. Set $\mathbf{V}(t) = \left(V_1, \dots, V_{N_1}\right)$, $\mathbf{M}(t) = \left(M_1(t), \dots, M_{N_2}(t)\right)$, and*

$$A(t) = \left(\left(\langle M_i, M_j\rangle\right)\right)_{1\le i,j\le N_2}.$$

If $\varphi \in C^{1,2}(\mathbb{R}^{N_1} \times \mathbb{R}^{N_2};\mathbb{C})$, then

$$\begin{aligned}\varphi\left(\mathbf{V}(t), \mathbf{M}(t)\right) &= \varphi\left(\mathbf{V}(0), \mathbf{M}(0)\right)\\ &\quad + \int_0^t \left(\nabla_{(1)}\varphi\left(\mathbf{V}(\tau), \mathbf{M}(\tau)\right), d\mathbf{V}(\tau)\right)_{\mathbb{R}^{N_1}}\\ &\quad + \int_0^t \left(\nabla_{(2)}\varphi\left(\mathbf{V}(\tau), \mathbf{M}(\tau)\right), d\mathbf{M}(\tau)\right)_{\mathbb{R}^{N_2}}\\ &\quad + \frac{1}{2}\int_0^t \mathrm{Trace}\left(\nabla^2_{(2)}\varphi\left(\mathbf{V}(\tau), \mathbf{M}(\tau)\right)dA(\tau)\right).\end{aligned}$$

Proof. Without loss in generality, we will assume that $\mathbf{M}(0) = \mathbf{0}$ and $\mathbf{V}(0) = \mathbf{0}$.

Begin by observing that, for $0 \le s < t$, $A(t) - A(s)$ is non-negative definite and symmetric, and therefore $\|A(t) - A(s)\|_{\mathrm{op}} \le \mathrm{Trace}\left(A(t) - A(s)\right)$.

Using bump functions and stopping times, one can reduce to the case when $\varphi \in C_c^\infty(\mathbb{R}^{N_1} \times \mathbb{R}^{N_2};\mathbb{R})$ and $\mathrm{var}_{[0,\infty)}(\mathbf{V})$, $\|\mathbf{M}\|_{[0,\infty]}$, and $\sup_{t\ge 0}\|A(t)\|_{\mathrm{op}}$ are all uniformly bounded. Therefore we will proceed under these assumptions. In particular, this means that $\left(\mathbf{M}(t), \mathcal{F}_t, \mathbb{P}\right)$ is a bounded, continuous martingale.

The difference between the proof of this general case and the Brownian one is that we now have control only on the second moment of the $|\mathbf{M}(t) - \mathbf{M}(s)|$. Thus we must rely more heavily on continuity, and the way to do that is to control the increments by using stopping times. With this in mind, for a given $T > 0$, define $\zeta_{0,n} = 0$ for all $n \ge 0$ and, for $m \ge 1$, define

$$\zeta_{m,n} = \inf\{t \geq \zeta_{m-1,n} : |\mathbf{V}(t) - \mathbf{V}(\zeta_{m-1,n})| \vee |\mathbf{M}(t) - \mathbf{M}(\zeta_{m-1,n})| \\ \vee \operatorname{Trace}\big(A(t) - A(\zeta_{m-1,n})\big) \geq 2^{-n}\big\} \wedge T.$$

By continuity, $\zeta_{m,n} = T$ for all but a finite number of m's, and so

$$\begin{aligned}&\varphi\big(\mathbf{V}(T), \mathbf{M}(T)\big) - \varphi\big(\mathbf{V}(0), \mathbf{0}\big)\\ &= \sum_{m=0}^{\infty} \int_{\zeta_{m,n}}^{\zeta_{m+1,n}} \Big(\nabla_{(1)}\varphi\big(\mathbf{V}(\tau), \mathbf{M}(\zeta_{m+1,n})\big), d\mathbf{V}(\tau)\Big)_{\mathbb{R}^{N_2}}\\ &\quad + \sum_{m=0}^{\infty} \Big(\varphi\big(\mathbf{V}(\zeta_{m,n}), \mathbf{M}(\zeta_{m+1,n})\big) - \varphi\big(\mathbf{V}(\zeta_{m,n}), \mathbf{M}(\zeta_{m,n})\big)\Big).\end{aligned}$$

Clearly the first sum on the right tends to $\int_0^T \big(\nabla_{(1)}\varphi(\mathbf{V}(\tau), \mathbf{M}(\tau), d\mathbf{V}(\tau)\big)_{R^{N_1}}$ as $n \to \infty$. To handle the second sum, use Taylor's theorem to write

$$\begin{aligned}&\varphi\big(\mathbf{V}(\zeta_{m,n}), \mathbf{M}(\zeta_{m+1,n})\big) - \varphi\big(\mathbf{V}(\zeta_{m,n}), \mathbf{M}(\zeta_{m,n})\big)\\ &= \Big(\nabla_{(2)}\varphi(\mathbf{X}_{m,n}), \Delta_{m,n}\Big)_{\mathbb{R}^{N_2}} + \tfrac{1}{2}\operatorname{Trace}\Big(\nabla^2_{(2)}\varphi(\mathbf{X}_{m,n})\Delta_{m,n} \otimes \Delta_{m,n}\Big) + E_{m,n},\end{aligned}$$

where $\mathbf{X}_{m,n} = \big(\mathbf{V}(\zeta_{m,n}), \mathbf{M}(\zeta_{m,n})\big)$, $\Delta_{m,n} = \mathbf{M}(\zeta_{m+1,n}) - \mathbf{M}(\zeta_{m,n})$, and $|E_{m,n}| \leq C|\Delta_{m,n}|^3$ for some $C < \infty$. Using Lemma 4.2.2 and (4.2.3), one sees that

$$\sum_{m=0}^{\infty} \Big(\nabla_{(2)}\varphi(\mathbf{X}_{m,n}), \Delta_{m,n}\Big)_{\mathbb{R}^{N_2}} = \int_0^T (\boldsymbol{\xi}_n(\tau), dM(\tau),$$

where $\boldsymbol{\xi}_n(\tau) = \nabla_{(2)}\varphi(\mathbf{X}_{m,n})$ for $\tau \in [\zeta_{m,n}, \zeta_{m+1})$, and therefore, since

$$\mathbb{E}^{\mathbb{P}}\left[\int_0^T \big|\boldsymbol{\xi}_n(\tau) - \nabla_{(2)}\varphi\big(\mathbf{V}(\tau), \mathbf{M}(\tau)\big)\big|^2 \langle\!\langle M \rangle\!\rangle(d\tau)\right] \longrightarrow 0,$$

$$\lim_{n\to\infty} \sum_{m=0}^{\infty} \Big(\nabla_{(2)}\varphi(\mathbf{X}_{m,n}), \Delta_{m,n}\Big)_{\mathbb{R}^{N_2}} = \int_0^T \Big(\nabla_{(2)}\varphi\big(\mathbf{V}(\tau), d\mathbf{M}(\tau)\big)\Big)_{\mathbb{R}^{N_2}}$$

in $L^2(\mathbb{P};\mathbb{R})$. At the same time, by Hunt's stopping time theorem,

$$\sum_{m=0}^{\infty} \mathbb{E}^{\mathbb{P}}\big[|\Delta_{m,n}|^2\big] = \sum_{m=0}^{\infty} \mathbb{E}^{\mathbb{P}}\big[|\mathbf{M}(\zeta_{m+1,n})|^2 - |\mathbf{M}(\zeta_{m,n})|^2\big] = \mathbb{E}^{\mathbb{P}}\big[|\mathbf{M}(T)|^2\big],$$

which means that

$$\sum_{m=0}^{\infty} \mathbb{E}^{\mathbb{P}}\big[|E_{m,n}|\big] \leq 2^{-n} C \mathbb{E}^{\mathbb{P}}\big[|\mathbf{M}(T)|^2\big].$$

Finally, set $D_{m,n} = \Delta_{m,n} \otimes \Delta_{m,n} - \big(A(\zeta_{m_1,n}) - A(\zeta_{m,n})\big)$. Then

$$\sum_{m=0}^{\infty} \operatorname{Trace}\Big(\nabla^2_{(2)}\varphi(\mathbf{X}_{m,n})\Delta_{m,n} \otimes \Delta_{m,n}\Big)$$
$$= \sum_{m=0}^{\infty} \operatorname{Trace}\Big(\nabla^2_{(2)}\varphi(\mathbf{X}_{m,n})\big(A(\zeta_{m+1,n}) - A(\zeta_{m,n})\big)\Big)$$
$$+ \sum_{m=0}^{\infty} \operatorname{Trace}\Big(\nabla^2_{(2)}\varphi(\mathbf{X}_{m,n})D_{m,n}\Big),$$

and the first sum on the right tends to

$$\int_0^T \operatorname{Trace}\Big(\nabla_{(2)}\varphi\big(\mathbf{V}(\tau), \mathbf{M}(\tau)\big)dA(\tau)\Big)$$

as $n \to \infty$. At the same time, by Hunt's stopping time theorem, the terms in the second sum are orthogonal in $L^2(\mathbb{P};\mathbb{R})$, and therefore the second moment of that sum is dominated by a constant times

$$\sum_{m=0}^{\infty} \mathbb{E}^{\mathbb{P}}\big[\|D_{m,n}\|^2_{\text{H.S.}}\big].$$

Since $\|D_{m,n}\|^2_{\text{H.S.}} \leq C\big(|\Delta_{m,n}|^4 + \|A(\zeta_{m+1,n}) - A(\zeta_{m,n})\|^2_{\text{op}}\big)$ for some $C < \infty$, the preceding sum is dominated by a constant times

$$2^{-n}\mathbb{E}^{\mathbb{P}}\big[|\mathbf{M}(T)|^2 + \operatorname{Trace}\big(A(T)\big)\big]. \qquad \square$$

An important consequence of Theorem 4.3.1 is that, by exactly the same argument as was used to derive (3.3.2) from Theorem (3.3.1), one can show that for any $\mathbb{R}^N$-valued local martingale $\big(M(t), \mathcal{F}_t, \mathbb{P}\big)$ with $M(0) = 0$ and $p \in [2, \infty)$,

$$\mathbb{E}^{\mathbb{P}}\big[\|M(\,\cdot\,)\|^p_{[0,t]}\big] \leq K_p \mathbb{E}^{\mathbb{P}}\left[\left(\sum_{i=1}^{N} \langle\!\langle M_i \rangle\!\rangle(t)\right)^{\frac{p}{2}}\right], \tag{4.3.1}$$

where K_p is the same constant as in (3.3.2).

A particularly striking application of Theorem 4.3.1 is Kunita and Watanabe's derivation of Lévy's characterization of Brownian motion.

Corollary 4.3.2. *If $\big(M(t), \mathcal{F}_t, \mathbb{P}\big)$ is an $\mathbb{R}^N$-valued, continuous local martingale, then it is a Brownian motion if and only if $M(0) = \mathbf{0}$ and $\langle M_i, M_j\rangle(t) = t\delta_{i,j}$ for $t \geq 0$ and $1 \leq i, j \leq N$.*

Proof. The necessity is obvious. To prove the sufficiency, first observe that, because $\langle\!\langle M_i \rangle\!\rangle(t)$ is bounded for all $1 \leq i \leq N$ and $t \geq 0$, $\big(M(t), \mathcal{F}_t, \mathbb{P}\big)$ is a martingale. Next, given $\boldsymbol{\xi} \in \mathbb{R}^N$, apply Theorem 4.3.1 to show that

$$E_{i\boldsymbol{\xi}}(t) := e^{i(\boldsymbol{\xi}, M(t))_{\mathbb{R}^N} + \frac{|\boldsymbol{\xi}|^2 t}{2}} = 1 + i\int_0^t e^{i(\boldsymbol{\xi}, M(\tau))_{\mathbb{R}^N} + \frac{|\boldsymbol{\xi}|^2 \tau}{2}} \big(\boldsymbol{\xi}, dM(\tau)\big)_{\mathbb{R}^N}.$$

Finally, if $X(t)$ and $Y(t)$ denote the real and imaginary parts of the preceding stochastic integral, check that $\langle\langle X\rangle\rangle(t) + \langle\langle Y\rangle\rangle(t) \le e^{|\boldsymbol{\xi}|^2 t} - 1$, and therefore that $\big(E_{i\boldsymbol{\xi}}(t), \mathcal{F}_t, \mathbb{P}\big)$ is a martingale. Since this means that

$$\mathbb{E}^{\mathbb{P}}\Big[e^{i(\boldsymbol{\xi}, M(t)-M(s))_{\mathbb{R}^N}} \,\Big|\, \mathcal{F}_s\Big] = e^{-\frac{|\boldsymbol{\xi}|^2(t-s)}{2}} \quad \text{for } 0 \le s < t,$$

the proof is complete. □

4.3.1 *Semi-martingales*

Another important consequence of Theorem 4.3.1 is what it says about the way local martingales transform under smooth maps. It is clear that linear maps preserve the martingale property and equally clear that non-linear ones need not. Nonetheless, Theorem 4.3.1 says that the image of a continuous local martingale under a twice continuously differentiable map is the sum of a local martingale and a progressively measurable function of locally bounded variation. In fact, it shows that such a sum is transformed into another such sum, and so it is reasonable to introduce terminology for this sort of stochastic processes. Thus, given an $\mathbb{R}$-valued progressively measurable function $t \rightsquigarrow X(t)$, one says that $\big(X(t), \mathcal{F}_t, \mathbb{P}\big)$ is a continuous local **semi-martingale** if $X = M + V$, where $\big(M(t), \mathcal{F}_t, \mathbb{P}\big)$ is a continuous local martingale with $M(0) = 0$ and V is a progressively measurable function for which $t \rightsquigarrow V(t)$ is a continuous function of locally bounded variation. By Lemma 4.1.1, one sees that, up to a $\mathbb{P}$-null set, M and V are uniquely determined. Therefore we can unambiguously talk about the martingale part M and bounded variation part V of a continuous semi-martingale $X = M + V$, and so we can define $\langle X_1, X_2\rangle = \langle M_1, M_2\rangle$ if $X_1 = M_1 + V_1$ and $X_2 = M_2 + V_2$. Notice that $\langle X_1, X_2\rangle = 0$ if either $M_1 = 0$ or $M_2 = 0$. Finally, if $X = M + V$ and ξ is an $\mathbb{R}$-valued, continuous, progressively measurable function, define

$$\int_0^t \xi(\tau)\, dX(\tau) = \int_0^t \xi(\tau)\, dM(\tau) + \int_0^t \xi(\tau)\, dV(\tau),$$

where the first integral on the right is a stochastic integral and the second is a Riemann–Stieltjes one. Obviously, such integrals are again semi-martingales, and, using (4.2.4) and the properties of Riemann–Stieltjes integrals, one sees that

$$\int_0^t \eta(\tau)\, d\left(\int_0^\tau \xi(\sigma)\, dX(\sigma)\right) = \int_0^t \xi(\tau)\eta(\tau)\, dX(\tau)$$

if η is a second continuous, progressively measurable function.

The following statement is an immediate consequence of Theorem 4.3.1.

Corollary 4.3.3. *Suppose* $\mathbf{X}(t) = \big(X_1(t), \dots, X_N(t)\big)$, *where* $\big(X_j(t), \mathcal{F}_t, \mathbb{P}\big)$ *is a continuous local semi-martingale for each* $1 \le j \le N$, *and set*

$$A(t) = \big(\!\big(\langle X_i(t), X_j(t)\rangle\big)\!\big)_{1\le i,j\le N}.$$

If $\varphi \in C^2(\mathbb{R}^N;\mathbb{C})$, *then*

$$\varphi(\mathbf{X}(t)) - \varphi(\mathbf{X}(0)) = \int_0^t \Big(\nabla\varphi(\mathbf{X}(\tau)), d\mathbf{X}(\tau)\Big)_{\mathbb{R}^N} + \frac{1}{2}\int_0^t \mathrm{Trace}\Big(\nabla^2\varphi(\mathbf{X}(\tau))dA(\tau)\Big).$$

Thus, for any continuous, local submartingale Y,

$$\langle \varphi\circ\mathbf{X}, Y\rangle(dt) = \sum_{j=1}^N \big(\partial_{x_j}\varphi\big)\big(\mathbf{X}(t)\big)\langle X_j, Y\rangle(dt). \tag{4.3.2}$$

Corollary 4.3.3 enables us to prove the following generalization of (2.1.2).

Corollary 4.3.4. *If* $\big(X_1(t), \mathcal{F}_t, \mathbb{P}\big)$ *and* $\big(X_2(t), \mathcal{F}_t, \mathbb{P}\big)$ *are a pair of* $\mathbb{R}$*-valued continuous local semi-martingales, then, for all* $T > 0$,

$$\lim_{n\to\infty}\sup_{t\in[0,T]}\left|\sum_{m=0}^\infty \big(X_1(t\wedge(m+1)2^{-n}) - X_1(t\wedge m2^{-n})\big) \times \big(X_2(t\wedge(m+1)2^{-n}) - X_2(t\wedge m2^{-n})\big) - \langle X_1, X_2\rangle(t)\right| = 0$$

in $\mathbb{P}$*-probability.*

Proof. First note that, by polarization, it suffices to treat the case when $X_1 = X = X_2$. In addition, we may and will assume that X and $\langle\!\langle X\rangle\!\rangle$ are uniformly bounded.

Observe that, by Corollary 4.3.3,

$$\begin{aligned}
&\sum_{m=0}^\infty \big(X(t\wedge(m+1)2^{-n}) - X(t\wedge m2^{-n})\big)^2 \\
&\quad= \sum_{m=0}^\infty \Big(X(t\wedge(m+1)2^{-n})^2 - X(t\wedge m2^{-n})^2 \\
&\qquad\qquad - 2X(t\wedge m2^{-n})\big(X(t\wedge(m+1)2^{-n}) - X(t\wedge m2^{-n})\big)\Big) \\
&\quad= \langle\!\langle X\rangle\!\rangle(t) + 2\int_0^t \big(X(\tau) - X(\lfloor\tau\rfloor_n)\big)\,dX(\tau).
\end{aligned}$$

Next let M and V denote the martingale and bounded variation parts of X. Clearly

$$\sup_{t\in[0,T]}\left|\int_0^t \big(X(\tau)-X(\lfloor\tau\rfloor_n)\big)\,dV(\tau)\right| \le \int_0^T |X(\tau)-X(\lfloor\tau\rfloor_n)|\,d|V|(\tau)\longrightarrow 0$$

as $n\to\infty$. At the same time,

$$\begin{aligned}&\mathbb{E}^{\mathbb{P}}\left[\sup_{t\in[0,T]}\left(\int_0^t \big(X(\tau)-X(\lfloor\tau\rfloor_n)\big)\,dM(\tau)\right)^2\right]\\&\qquad\le 4\mathbb{E}^{\mathbb{P}}\left[\int_0^T \big(X(\tau)-X(\lfloor\tau\rfloor_n)\big)^2\,\langle\!\langle M\rangle\!\rangle(d\tau)\right]\longrightarrow 0,\end{aligned}$$

and, when combined with the preceding, that completes the proof. □

4.4 Representing continuous local martingales

Suppose that $\big(M(t),\mathcal{F}_t,\mathbb{P}\big)$ is an $\mathbb{R}^N$-valued continuous local martingale with $M(0)=\mathbf{0}$ on some complete probability space $(\Omega,\mathcal{F},\mathbb{P})$, and assume that

$$A(t):=\big(\!\big(\langle M_i,M_j\rangle(t)\big)\!\big)_{1\le i,j\le N}=\int_0^t a(\tau)\,d\tau$$

for some progressively measurable, symmetric, non-negative definite-valued $a:[0,\infty)\times\Omega\longrightarrow \mathrm{Hom}(\mathbb{R}^N;\mathbb{R}^N)$. From (2.2.7), it is clear that $\big(\epsilon I+a\big)^{\frac12}$ is progressively measurable for all $\epsilon>0$, and so $\sigma:=a^{\frac12}=\lim_{\epsilon\searrow 0}\big(\epsilon I+a\big)^{\frac12}$ is also. Furthermore,

$$\int_0^t \sigma^2(\tau)\,d\tau = A(t),$$

and so one can integrate σ with respect to a Brownian motion.

Now assume that a, and therefore σ, are strictly positive definite. Then σ^{-1} is progressively measurable. In addition,

$$\int_0^t \big(\sigma^{-1}(\tau)\boldsymbol{\xi},dA(\tau)\sigma^{-1}(\tau)\boldsymbol{\eta}\big)_{\mathbb{R}^N}=t(\boldsymbol{\xi},\boldsymbol{\eta})_{\mathbb{R}^N}\quad\text{for all }\boldsymbol{\xi},\boldsymbol{\eta}\in\mathbb{R}^N,$$

and so, if

$$B(t)=\int_0^t \sigma^{-1}(\tau)\,dM(\tau),$$

then $\langle B_i,B_j\rangle(t)=t\delta_{i,j}$, which, by Corollary 4.3.2, means that $\big(B(t),\mathcal{F}_t,\mathbb{P}\big)$ is an $\mathbb{R}^N$-valued Brownian motion. Moreover, by (4.2.4),

$$\int_0^t \sigma(\tau)\, dB(\tau) = \int_0^t \sigma(\tau)\sigma^{-1}(\tau)\, dM(\tau) = M(t).$$

When a is not strictly positive definite, the preceding argument breaks down. Indeed, if $a = 0$, the sample space can consist of a single point, and there is no way that a one point sample space can support a Brownian motion. Thus, in general one will be able to build only part of a Brownian motion from M. For this reason one has to have a Brownian motion in reserve, ready to be inserted to fill the gaps caused by a becoming degenerate. With this in mind, denote by $N(t)$ the null space of $a(t)$ and by $\Pi(t)$ orthogonal projection onto $N(t)$, and let $\sigma^{-1}(t)$ be the linear map for which $N(t)$ is the null space and $\sigma^{-1}(t) \restriction N(t)^\perp$ is the inverse of $\sigma(t) \restriction N(t)^\perp$. Then $\sigma\sigma^{-1} = \sigma^{-1}\sigma = \Pi^\perp := \mathbf{I} - \Pi$, and, since

$$\Pi(t) = \lim_{\epsilon \searrow 0} a(t)\big(\epsilon\mathbf{I} + a(t)\big)^{-1} \quad \text{and} \quad \sigma^{-1}(t) = \lim_{\epsilon \searrow 0} \sigma(t)\big(\epsilon\mathbf{I} + a(t)\big)^{-1},$$

both these functions are progressively measurable. In particular, for any $\boldsymbol{\xi} \in \mathbb{R}^N$,

$$\int_0^t \big(\sigma^{-1}(\tau)\boldsymbol{\xi}, dA(\tau)\sigma^{-1}\boldsymbol{\xi}\big)_{\mathbb{R}^N} = \int_0^t |\Pi(\tau)^\perp\boldsymbol{\xi}|^2\, d\tau \le t|\boldsymbol{\xi}|^2,$$

and so stochastic integrals of σ^{-1} with respect to M are well defined. Now take $\mathcal{W}$ to be Wiener measure on $\mathbb{W}(\mathbb{R}^N)$, $\widetilde{\mathbb{P}} = \mathbb{P} \times \mathcal{W}$, and $\widetilde{\mathcal{F}}_t$ to be the completion of $\mathcal{F}_t \times \mathcal{W}_t$ with respect to $\widetilde{\mathbb{P}}$, and set $\widetilde{M}(t)(\omega, w) = M(t)(\omega)$, $\widetilde{\Pi}(t)(\omega, w) = \Pi(t)(\omega)$, $\widetilde{\sigma}(t)(\omega, w) = \sigma(t)(\omega)$, and $\widetilde{\sigma}^{-1}(t)(\omega, w) = \sigma^{-1}(t)(\omega)$. It is then an easy matter to check that $\big(\widetilde{M}(t), \widetilde{\mathcal{F}}_t, \widetilde{\mathbb{P}}\big)$ is a local martingale, that $\big(w(t), \widetilde{\mathcal{F}}_t, \widetilde{\mathbb{P}}\big)$ is a Brownian motion, and that (cf. Exercise 4.2) $\langle w_i, \widetilde{M}_j\rangle = 0$.

Define

$$\widetilde{B}(t) := \int_0^t \widetilde{\sigma}^{-1}(\tau)\, d\widetilde{M}(\tau) + \int_0^t \widetilde{\Pi}(\tau)\, dw(\tau),$$

and, using the properties discussed above, check that $\langle \widetilde{B}_i, \widetilde{B}_j\rangle(t) = t\delta_{i,j}$ and therefore that $\big(\widetilde{B}(t), \widetilde{\mathcal{F}}_t, \widetilde{\mathbb{P}}\big)$ is a Brownian motion. Further, by (4.2.4),

$$\begin{aligned}
\int_0^t \widetilde{\sigma}(\tau)\, d\widetilde{B}(\tau) &= \int_0^t \widetilde{\sigma}(\tau)\widetilde{\sigma}^{-1}(\tau)\, d\widetilde{M}(\tau) + \int_0^t \widetilde{\sigma}(\tau)\widetilde{\Pi}(\tau)\, dw(\tau) \\
&= \widetilde{M}(t) - \int_0^t \widetilde{\Pi}(\tau)\, d\widetilde{M}(\tau)
\end{aligned}$$

since $\widetilde{\sigma}(\tau)\widetilde{\Pi}(\tau) = 0$. In addition,

$$\mathbb{E}^{\widetilde{\mathbb{P}}}\left[\left|\int_0^t \widetilde{\Pi}(\tau)\,d\widetilde{M}(\tau)\right|^2\right] = \mathbb{E}^{\mathbb{P}}\left[\left|\int_0^t \Pi(\tau)\,dM(\tau)\right|^2\right]$$
$$= \mathbb{E}^{\mathbb{P}}\left[\int_0^t \mathrm{Trace}\big(\Pi(\tau)a(\tau)\Pi(\tau)\big)\,d\tau\right] = 0,$$

and so $\widetilde{M}(t) = \int_0^t \widetilde{\sigma}(\tau)\,d\widetilde{B}(\tau)$.

We have now proved the following representation theorem. See Exercise 4.4 for an extension of this result.

Theorem 4.4.1. *Let $\big(M(t),\mathcal{F}_t,\mathbb{P}\big)$ be an $\mathbb{R}^N$-valued continuous local martingale on $(\Omega,\mathcal{F},\mathbb{P})$, and assume that*

$$A(t) := \big(\langle\langle X_i, X_j\rangle\rangle\big)_{1\le i,j\le N} = \int_0^t a(\tau)\,d\tau$$

for some progressively measurable, non-negative definite valued function $a : [0,\infty)\times\Omega \longrightarrow \mathrm{Hom}(\mathbb{R}^N;\mathbb{R}^N$. If $\sigma = a^{\frac{1}{2}}$, then there is a probability space $\big(\widetilde{\Omega},\widetilde{\mathcal{F}},\widetilde{\mathbb{P}}\big)$ on which there is a Brownian motion $\big(\widetilde{B}(t),\widetilde{\mathcal{F}}_t,\widetilde{\mathbb{P}}\big)$ and measurable map $F : \widetilde{\Omega} \longrightarrow \Omega$ such that $\mathbb{P} = F_*\widetilde{\mathbb{P}}$ and

$$\widetilde{M}(t) = \widetilde{M}(0) + \int_0^t \widetilde{\sigma}(\tau)\,d\widetilde{B}(\tau)$$

when $\widetilde{M} = M \circ F$ and $\widetilde{\sigma} = \sigma\circ F$.

4.5 Stratonovich integration

When probabilists look at an operator

$$L = \frac{1}{2}\sum_{i,j}^N a_{ij}(\mathbf{x})\partial_{x_i}\partial_{x_j} + \sum_{i=1}^N b_i(\mathbf{x})\partial_{x_i},$$

they are inclined to think of the matrix a as governing the diffusive behavior and b as governing the deterministic behavior of the associated diffusion process, and this is entirely reasonable as long as a and b are constant. However, if what one means is that the diffusive part of the process is the one whose increments during a time interval dt are of order $\sqrt{dt}$ as opposed to the deterministic part whose increments are of order dt, then, as the following example shows, this interpretation of a and b is flawed. Take $N = 2$ and $a(\mathbf{x}) = \begin{pmatrix} x_2^2 & -x_1x_2 \\ -x_1x_3 & x_1^2\end{pmatrix}$ and $b = 0$. Then the prediction is that the associated diffusion is purely diffusive, but, that is not true. Indeed, $a = \sigma\sigma^\top$, where $\sigma(\mathbf{x}) = \begin{pmatrix} -x_2 \\ x_1\end{pmatrix}$, and so an Itô representation of the associated diffusion is

$$\begin{pmatrix} X_1(t,\mathbf{x}) \\ X_2(t,\mathbf{x}) \end{pmatrix} = \begin{pmatrix} x_1 \\ x_2 \end{pmatrix} + \int_0^t \begin{pmatrix} -X_2(\tau,\mathbf{x}) \\ X_1(\tau,\mathbf{x}) \end{pmatrix} dw(\tau),$$

where w is an $\mathbb{R}$-valued Wiener path. The easiest way to solve this equation is to write it in terms of complex variables. That is, set $z = x_1 + ix_2$ and $Z(t,z) = X_1(t,\mathbf{x}) + iX_2(t,\mathbf{x})$. Then

$$Z(t,z) = z + i\int_0^t Z(\tau,z)\,dw(\tau),$$

and so $Z(t,z) = ze^{iw(t)+\frac{t}{2}}$, or, equivalently,

$$X(t,\mathbf{x}) = e^{\frac{t}{2}} \begin{pmatrix} x_1 \cos w(t) - x_2 \sin w(t) \\ x_2 \cos w(t) + x_1 \sin w(t) \end{pmatrix}.$$

In particular, $|X(t,\mathbf{x})| = |\mathbf{x}|e^{\frac{t}{2}}$, which, if $\mathbf{x} \neq \mathbf{0}$, means that the distance of $X(t,\mathbf{x})$ from $\mathbf{0}$ is deterministic and its increments are of order dt.

The preceding example demonstrates why it is too naïve to interpret a and b as governing the diffusive and deterministic parts of the associated diffusion, the point being that it is too coordinate dependent an interpretation. To wit, if one represents the preceding L in terms of polar coordinates, one finds that it is equal to

$$\tfrac{1}{2}\big(\partial_\theta^2 + r\partial_r\big),$$

which makes it clear that, although the angular coordinate of $X(t,\mathbf{x})$ is a Brownian motion, the radial coordinate is deterministic. As this example shows, the absence or presence of the "drift" term b in (1.2.5) depends on the coordinate system in which one is working, whereas the absence or presence of diffusive behavior is coordinate independent. One reason why this flaw was, for the most part, ignored by probabilists is that it doesn't cause any problems when a is uniformly elliptic in the sense that $a \geq \epsilon\mathbf{I}$, in which case, at least over short time intervals, the associated diffusion is diffusive for every choice of b. In fact, as we will show in Corollary 5.2.5, when $a \geq \epsilon\mathbf{I}$, the distributions of the diffusions corresponding to different choices of b are mutually absolutely continuous during finite time intervals, and therefore their almost certain behavior over bounded time intervals will look the same in all coordinate systems.

To address the issues raised above, it is desirable to represent L in a form that looks the same in all coordinate systems, and such a representation was introduced by L. Hörmander. To describe his representation, for a vector field $V \in C(\mathbb{R}^N;\mathbb{R}^N)$, we will use the notation $\mathcal{L}_V$ to denote the directional derivative operator $\sum_{j=1}^N V_j\partial_{x_j}$. To represent an L given by (1.2.5) in Hörmander's form, again write $a = \sigma\sigma^\top$, where $\sigma \in C(\mathbb{R}^N; \mathrm{Hom}(\mathbb{R}^M; \mathbb{R}^N))$, and, for each $1 \le k \le M$, let $V_k(\mathbf{x}) \in \mathbb{R}^N$ be the kth column of $\sigma(\mathbf{x})$. Assuming that σ is continuously differentiable, one then has

$$L = \frac{1}{2}\sum_{k=1}^M \mathcal{L}_{V_k}^2 + \mathcal{L}_{V_0}, \tag{4.5.1}$$

when one takes $V_0 = b - \frac{1}{2}\sum_{k=1}^M \mathcal{L}_{V_k} V_k$. In the preceding example, $M = 1$, $N = 2$, $V_1 = \left(\begin{smallmatrix} -x_2 \\ x_1 \end{smallmatrix}\right)$, and $V_0 = \frac{1}{2}\left(\begin{smallmatrix} x_1 \\ x_2 \end{smallmatrix}\right)$.

The beauty of the representation in (4.5.1) is that it looks the same in all coordinate systems. That is, suppose that F is a diffeomorphism on some open set $G \subseteq \mathbb{R}^N$, and, given a vector field V on G, define the vector field F_*V on $F(G)$ so that $\mathcal{L}_{F_*V}\varphi = \big(\mathcal{L}_V(\varphi \circ F)\big) \circ F^{-1}$ for $\varphi \in C^1\big(F(G); \mathbb{R}\big)$. More explicitly,

$$F_*V = (\mathcal{L}_V F) \circ F^{-1} = (F^{(1)}V) \circ F^{-1} \quad \text{where } F^{(1)} = \big((\partial_{x_j} F_i)\big)_{1 \le i,j \le N}$$

is the Jacobian matrix of F. Then

$$\mathcal{L}_V^2(\varphi \circ F) = \mathcal{L}_V\big((\mathcal{L}_{F_*V}\varphi) \circ F\big) = \big(\mathcal{L}_{F_*V}^2\varphi\big) \circ F,$$

and so, in the coordinate system on $F(G)$ determined by F,

$$L = \frac{1}{2}\sum_{k=1}^M \mathcal{L}_{F_*V_k}^2 + \mathcal{L}_{F_*V_0}.$$

In particular, this means that the presence or absence of the term $\mathcal{L}_{V_0}$ does not depend on the coordinate system, and so it provides a much more reliable prediction of the presence or absence of a deterministic component.

Related to these considerations is the following. In that the Hörmander representation is in terms of vector fields, one suspects that the paths of the associated diffusion associated with the L in (4.5.1) should transform under changes of coordinates the same way as integral curves do. Namely, if $X(\,\cdot\,)$ is an integral curve of the vector field V and F is a diffeomorphism, then $F \circ X(\,\cdot\,)$ is an integral curve of F_*V. Hence, we should expect that if $X(\,\cdot\,,\mathbf{x})$ is the solution to (3.0.2) when $\sigma = \big(V_1, \dots, V_M\big)$ and $b = V_0 + \frac{1}{2}\sum_{k=1}^M \mathcal{L}_{V_k} V_k$, then $F \circ X(\,\cdot\,,\mathbf{x})$ should be solution to (3.0.2) with σ and b replaced by

$$\big(F_*V_1, \dots, F_*V_M\big) \quad \text{and} \quad F_*V_0 + \frac{1}{2}\sum_{k=1}^M \mathcal{L}_{F_*V_k} F_*V_k,$$

and, with sufficient patience, one can check that this is true. However, it would be preferable to have a formulism that made such computations easier and brought out the relationship between $X(\,\cdot\,,\mathbf{x})$ and integral curves of the

V_k's. For that purpose, reconsider the equation

$$\dot{X}(t,\mathbf{x})(w) = \sum_{k=1}^{M} V_k\big(X(t,\mathbf{x})(w)\big)\dot{w}(t)_k + V_0\big(X(t,\mathbf{x})(w)\big) \quad \text{with}\, X(0,\mathbf{x}) = \mathbf{x},$$

where the interpretation now, unlike Itô's interpretation, is that $X(t,\mathbf{x})(w)$ is constructed by taking the limit of integral curves corresponding to mollifications of the Brownian paths w. For instance, in our example,

$$F_1(t,\mathbf{x}) = \begin{pmatrix} x_1 \cos t - x_2 \sin t \\ x_2 \cos t + x_1 \sin t \end{pmatrix} \quad \text{and} \quad F_0(t,\mathbf{x}) = e^{\frac{t}{2}}\mathbf{x}$$

are the integral curves of V_1 and V_0 starting at $\mathbf{x}$, and, for any continuously differentiable w, $F_0\big(t, F_1(w(t),\mathbf{x})\big)$ is the integral curve of $\dot{w}(t)V_1 + V_0$ starting at $\mathbf{x}$. Thus, one might guess that $F_0\big(t, F_1(w(t),\mathbf{x})\big)$ is the diffusion associated with L, as indeed we saw that it is. In Theorem 4.5.3 we will see that this example is as simple as it is because $\mathcal{L}_{V_0}$ commutes with $\mathcal{L}_{V_1}$. Things are more complicated when dealing with non-commuting vector fields.

In view of the preceding, one should wonder whether there is a way to incorporate these ideas into a theory of stochastic integration. Such a theory was introduced by the Russian engineer L. Stratonovich (cf. [16]) and produces what is now called the **Stratonovich integral**. However, Stratonovich's treatment was rather cumbersome, and mathematicians remained skeptical about it until Itô rationalized it. Itô understood that there is no way to define such an integral for all locally square integrable progressively measurable integrands, but he also realized that one could do so if the integrand is a semi-martingale. Namely, given a pair of continuous semimartingales $\big(X(t),\mathcal{F}_t,\mathbb{P}\big)$ and $\big(Y(t),\mathcal{F}_t,\mathbb{P}\big)$, define $X_n(\,\cdot\,)$ and $Y_n(\,\cdot\,)$ to be the polygonal paths such that $X_n(m2^{-n}) = X(m2^{-n})$, $Y_n(m2^{-n}) = Y(m2^{-n})$, and $X_n(\,\cdot\,)$ and $Y_n(\,\cdot\,)$ are linear on $[m2^{-n},(m+1)2^{-n}]$ for each $m \geq 0$. Set $\Delta^X_{m,n} = X((m+1)2^{-n}) - X(m2^{-n})$ and $\Delta^Y_{m,n} = Y((m+1)2^{-n}) - Y(m2^{-n})$. Then, by Corollary 4.3.4,

$$\begin{aligned} \int_0^t X_n(\tau)\,dY_n(\tau) = &\sum_{m<2^n t} \big((m+1)\wedge 2^n t - m\big)X(m2^{-n})\Delta^Y_{m,n} \\ &+ \tfrac{1}{2}\sum_{m<2^n t} \big((m+1)\wedge 2^n t - m\big)^2 \Delta^X_{m,n}\Delta^Y_{m,n} \\ \longrightarrow &\int_0^t X(\tau)\,dY(\tau) + \tfrac{1}{2}\langle X,Y\rangle(t). \end{aligned}$$

(See Exercise 4.5 for more information.) Thus Itô said that the Stratonovich integral of $Y(\,\cdot\,)$ with respect to $X(\,\cdot\,)$ should be[2]

[2] The standard notation is $\circ\, dX(\tau)$ rather than $\bullet\, dX(\tau)$, but, because I use $\circ$ to denote composition, I have chosen to use $\bullet$ to denote Stratonovich integration.

$$\int_0^t Y(\tau) \bullet dX(\tau) := \int_0^t Y(\tau)\, dX(\tau) + \tfrac{1}{2}\langle X, Y\rangle(t).$$

Because, by (4.3.2),

$$\langle ZY, X\rangle(dt) = Z(t)\langle Y, X\rangle(dt) + Y(t)\langle Z, X\rangle(dt),$$

one can use (4.2.4) to check that if $I(t) = \int_0^t Y(\tau) \bullet d(x(\tau)$, then

$$\int_0^t Z(\tau) \bullet dI(\tau) = \int_0^t Z(\tau)Y(\tau) \bullet dX(\tau). \tag{4.5.2}$$

Notice, if one adopts this definition, then the equation in Corollary (4.3.3) becomes

$$\varphi\big(\mathbf{X}(t)\big) = \varphi\big(\mathbf{X}(0)\big) + \sum_{j=1}^{N} \int_0^t \partial_{x_j}\varphi\big(\mathbf{X}(\tau)\big) \bullet dX_j(\tau) \tag{4.5.3}$$

for $\varphi \in C^3(\mathbb{R}^N;\mathbb{R})$. Indeed, by (4.3.2),

$$\sum_{j=1}^{N} \langle \partial_{x_j}\varphi \circ \mathbf{X}, X_j\rangle(dt) = \sum_{i,j=1}^{N} \partial_{x_i}\partial_{x_j}\varphi\big(\mathbf{X}(t)\big)\langle X_i, X_j\rangle(dt).$$

Because the equation in (4.5.3) resembles the fundamental theorem of calculus, at first sight one might think that it is an improvement on the formula in Corollary 4.3.3. However, after a second look, one realizes that the opposite is true. Namely, in Itô's formulation, the integrand in a Stratonovich integral has to be a semi-martingale, and so we had to assume that $\varphi \in C^3(\mathbb{R}^N;\mathbb{R})$ in order to be sure that $\partial_{x_j}\varphi\big(\mathbf{X}(t)\big)$ would be one. Thus we have obtained a pleasing looking formula in which only first order derivatives of φ appear explicitly, but we have done so at the price of requiring that φ have three continuous derivatives. It turns out that in special cases there are ingenious subterfuges that allow one to avoid this objectionable requirement, but the basic fact remains: the fundamental theorem of calculus and martingales are inherently incompatible.

4.5.1 *Stratonovich integral equations*

In spite of the somewhat disparaging comments at the end of the preceding, Stratonovich integration has value. However, to appreciate its value, one has to abandon Itô's clever formulation of it and return to ideas based on its relationship to integral curves. Namely, for each $0 \le k \le M$, let V_k be an element of $C^2(\mathbb{R}^N;\mathbb{R}^N)$ with bounded first and second order derivatives, and consider the stochastic integral equation

$$X(t,\mathbf{x}) = \mathbf{x} + \int_0^t V_0\big(X(\tau,\mathbf{x})\big)\,d\tau + \sum_{k=1}^M V_k\big(X(\tau,\mathbf{x})\big) \bullet dw(\tau)_k. \tag{4.5.4}$$

Using (4.5.3), (4.5.2), and (4.3.2), one sees that

$$\begin{aligned}\varphi\big(X(t,\mathbf{x})\big) = \varphi(\mathbf{x}) &= \sum_{k=1}^M \int_0^t \mathcal{L}_{V_k}\varphi\big(X(\tau,\mathbf{x})\big) \bullet dw_k(\tau)\\ &= \sum_{k=1}^M \int_0^t \mathcal{L}_{V_k}\varphi\big(X(\tau,\mathbf{x})\big)\,dw_k(\tau) + \frac{1}{2}\int_0^t L\varphi\big(X(\tau,\mathbf{x})\big)\,d\tau,\end{aligned}$$

where L is given by (4.5.1). Of course, (4.5.4) is equivalent to (3.0.2) when σ is the $N \times M$-matrix whose kth column, for $1 \le k \le M$, is V_k and $b = V_0 + \frac{1}{2}\sum_{k=1}^M \mathcal{L}_{V_k} V_k$. Thus we know that (4.5.4) has a solution, and so it may seem pointless to replace (3.0.2) by (4.5.4). On the other hand, notice that using the Stratonovich formalism, it is easy to verify that, if F is a diffeomorphism, then $F \circ X(\,\cdot\,,\mathbf{x})$ is the solution to (4.5.4) with the V_k's replaced by F_*V_k's. Indeed, by (4.5.3),

$$F\big(X(t,\mathbf{x})\big) = F(\mathbf{x}) + \int_0^t \mathcal{L}_{V_0}F\big(X(\tau,\mathbf{x})\big)\,d\tau + \int_0^t \mathcal{L}_{V_k}F\big(X(\tau,\mathbf{x})\big) \bullet dw(\tau)_k,$$

and $\mathcal{L}_{V_k}F = (F_*V_k) \circ F$. Further evidence of the virtues of the Stratonovich formalism is provided by the computation in the following theorem.

Theorem 4.5.1. *Assume that, for each $0 \le k \le M$, $V_k \in C^3(\mathbb{R}^N;\mathbb{R}^N)$ has bounded derivatives, and let $V_k^{(1)}$ denote the Jacobian matrix of V_k. If $X(\,\cdot\,,\mathbf{x})$ is the solution to (4.5.4), then $X(t,\,\cdot\,)$ admits a version that is continuously differentiable, and its Jacobian matrix $X^{(1)}(t,\mathbf{x})$ satisfies*

$$\begin{aligned}X^{(1)}(t,\mathbf{x}) = \mathbf{I} &+ \int_0^t V_0^{(1)}\big(X(\tau,\mathbf{x})\big)X^{(1)}(\tau,\mathbf{x})\,d\tau\\ &+ \sum_{k=1}^M \int_0^t V_k^{(1)}\big(X(\tau,\mathbf{x})\big)X^{(1)}(\tau,\mathbf{x}) \bullet dw(\tau)_k.\end{aligned}$$

In addition

$$\begin{aligned}
&\det\big(X^{(1)}(t,\mathbf{x})\big)\\
&= \exp\bigg(\int_0^t \mathrm{div}V_0\big(X(\tau,\mathbf{x})\big)\,d\tau + \sum_{k=1}^M \int_0^t \mathrm{div}V_k\big(X(\tau,\mathbf{x})\big) \bullet dw(\tau)_k\bigg)\\
&= \exp\bigg(\int_0^t \mathrm{div}V_0\big(X(\tau,\mathbf{x})\big)\,d\tau + \sum_{k=1}^M \int_0^t \mathrm{div}V_k\big(X(\tau,\mathbf{x})\big)\,dw(\tau)_k\\
&\qquad + \frac{1}{2}\sum_{k=1}^M \int_0^t \big(\mathcal{L}_{V_k}\mathrm{div}V_k\big)\big(X(\tau,\mathbf{x})\big)\,d\tau\bigg).
\end{aligned}$$

In particular, if $X(\,\cdot\,,\mathbf{x})$ *is the solution to* (3.0.2), V_k *is the* k*th column of* σ, *and* $V_0 = b - \frac{1}{2}\sum_{k=1}^M \mathcal{L}_{V_k} V_k$, *then*

$$\begin{aligned}
\det\big(X^{(1)}(t,\mathbf{x})\big) = \exp\bigg(&\int_0^t \mathrm{div}b\big(X(\tau,\mathbf{x})\big)\,d\tau + \sum_{k=1}^M \int_0^t \mathrm{div}V_k\big(X(\tau,\mathbf{x})\big)\,dw(\tau)_k\\
&- \frac{1}{2}\sum_{k=1}^M \int_0^t \mathrm{Trace}\Big(\big(V_k^{(1)}\big)^2\big(X(\tau,\mathbf{x})\big)\Big)\,d\tau\bigg).
\end{aligned}$$

Proof. The facts that $X(t,\,\cdot\,)$ has a continuously differentiable version and that $X^{(1)}(\,\cdot\,,\mathbf{x})$ satisfies the asserted equation are easily checked by converting (4.5.4) to its equivalent Itô form, applying (3.4.3), and reconverting it to Stratonovich form. Furthermore, in the second assertion, once one knows the first equality, the second and final ones follow when one again uses Itô's prescription for converting Stratonovich integrals to Itô ones.

Given an $N\times N$ matrix $A = \big((a_{ij})\big)_{1\le i,j\le N}$, let $C(A)$ be the matrix whose (i,j)th entry is $(-1)^{i+j}$ times the determinant of the $(N-1)\times(N-1)$ matrix obtained by deleting the ith row and jth column from A. Using Cramer's rule, one sees that $\partial_{a_{ij}}\det(A) = C(A)_{ij}$ and that $AC(A)^\top = \det(A)\mathbf{I}$. Now set $D(t,\mathbf{x}) = \det\big(X^{(1)}(t,\mathbf{x})\big)$. Applying (4.5.3) and the preceding, one sees that

$$\begin{aligned}
D(t,\mathbf{x}) = 1 + &\int_0^t \mathrm{div}V_0\big(X(\tau,\mathbf{x})\big)D(\tau,\mathbf{x})\,d\tau\\
&+ \sum_{m=1}^M \int_0^t \mathrm{div}V_m\big(X(\tau,\mathbf{x})\big)D(\tau,\mathbf{x}) \bullet dw(\tau)_k.
\end{aligned}$$

Since

$$\exp\bigg(\int_0^t \mathrm{div}V_0\big(X(\tau,\mathbf{x})\big)\,d\tau + \sum_{k=1}^M \int_0^t \mathrm{div}V_k\big(X(\tau,\mathbf{x})\big) \bullet dw(\tau)_k\bigg)$$

is the one and only solution to this equation, the proof is complete. □

Of course, one could have carried out the preceding computation without using the Stratonovich formalism, but it would have been a far more tedious exercise. Furthermore, such computations are not the only virtue of the expression in (4.5.4). Indeed, although we know how to solve (4.5.4) by converting it to an Itô stochastic integral equation, there is another way to go about solving it. Namely, for $\boldsymbol{\xi} = (\xi_0, \dots, \xi_M) \in \mathbb{R}\times\mathbb{R}^M$, set $V_{\boldsymbol{\xi}} = \sum_{k=0}^M \xi_k V_k$, and consider the the ordinary differential equation

$$\dot{F}_{\boldsymbol{\xi}}(t,\mathbf{x}) = V_{\boldsymbol{\xi}}\big(F_{\boldsymbol{\xi}}(t,\mathbf{x})\big) \quad \text{with } F_{\boldsymbol{\xi}}(\mathbf{0},\mathbf{x}) = \mathbf{x}.$$

In other words, $t \rightsquigarrow F_{\boldsymbol{\xi}}(t,\mathbf{x})$ is the integral curve of $V_{\boldsymbol{\xi}}$ that passes through $\mathbf{x}$ at time 0. Next define $E : (\mathbb{R}\times\mathbb{R}^M)\times\mathbb{R}^N \longrightarrow \mathbb{R}^N$ by $E(\boldsymbol{\xi},\mathbf{x}) = F_{\boldsymbol{\xi}}(1,\mathbf{x})$. Note that $E(t\boldsymbol{\xi},\mathbf{x}) = F_{\boldsymbol{\xi}}(t,\mathbf{x})$ for all $t \in \mathbb{R}$. Furthermore, standard results (cf. [1]) from the theory of ordinary differential equations show that $E(\boldsymbol{\xi},\,\cdot\,)$ is a diffeomorphism from $\mathbb{R}^N$ onto itself, $E(-\boldsymbol{\xi},\,\cdot\,)$ is the inverse of $E(\boldsymbol{\xi},\,\cdot\,)$, and there exist constants $C < \infty$ and $\nu \in [0,\infty)$ such that

$$\begin{aligned} &|\partial_{\xi_k} E(\boldsymbol{\xi},\mathbf{x})| \vee |\partial_{x_j} E(\boldsymbol{\xi},\mathbf{x})| \vee |\partial_{\xi_k}\partial_{\xi_\ell} E(\boldsymbol{\xi},\mathbf{x})| \\ &\qquad \vee |\partial_{\xi_k}\partial_{x_j} E(\boldsymbol{\xi},\mathbf{x})| \vee |\partial_{x_i}\partial_{x_j} E(\boldsymbol{\xi},\mathbf{x})| \le C(1+|\mathbf{x}|)e^{\nu|\boldsymbol{\xi}|} \end{aligned} \tag{4.5.5}$$

for all $1 \le i,j \le N$ and $0 \le k,\ell \le M$.

For $V, W \in C^1(\mathbb{R}^N;\mathbb{R}^N)$, define the **commutator** $[V,W] = \mathcal{L}_V W - \mathcal{L}_W V$. Equivalently, $\mathcal{L}_{[V,W]} = \mathcal{L}_V \circ \mathcal{L}_W - \mathcal{L}_W \circ \mathcal{L}_V$.

Lemma 4.5.2. *Assume that $V, W \in C^1(\mathbb{R}^N;\mathbb{R}^N)$ have bounded first order derivatives, and determine $t \rightsquigarrow F(t,\mathbf{x})$ and $t \rightsquigarrow G(t,\mathbf{x})$ by*

$$\dot{F}(t,\mathbf{x}) = V\big(F(t,\mathbf{x})\big) \quad \textit{and} \quad \dot{G}(t,\mathbf{x}) = W\big(G(t,\mathbf{x})\big) \textit{ with } F(0,\mathbf{x}) = G(0,\mathbf{x}) = \mathbf{x}.$$

If $[V,W] = 0$, then $F\big(s,G(t,\mathbf{x})\big) = G\big(t,F(s,\mathbf{x})\big)$ for all $(s,t) \in \mathbb{R}^2$.

Proof. Set $\Phi(t) = F\big(s,G(t,\mathbf{x})\big)$. Then, for $\varphi \in C^2(\mathbb{R}^N;\mathbb{R})$,

$$\partial_t \varphi \circ \Phi(t) = \big(\mathcal{L}_{F(s,\cdot)_* W}\varphi\big)\big(G(t,x)\big).$$

If $\varphi_s = \varphi \circ F(s,\,\cdot\,)$, then, because $F(s+h,\,\cdot\,) = F\big(h, F(s,\,\cdot\,)\big)$,

$$\begin{aligned} \partial_s\Big(\big(\mathcal{L}_{F(s,\cdot)_* W}\varphi\big)\big(F(-s,\mathbf{y})\big)\Big) &= \partial_h\Big(\big(\mathcal{L}_{F(h,\cdot)_* W}\varphi_s\big)\big(F(-h,F(-s,\mathbf{y}))\big)\Big)\Big|_{h=0} \\ &= \Big(\big(\mathcal{L}_W \circ \mathcal{L}_V - \mathcal{L}_V \circ \mathcal{L}_W\big)\varphi_s\Big) \circ F(-s,\mathbf{y}) = 0. \end{aligned}$$

Hence $\big(\mathcal{L}_{F(s,\cdot)_* W}\varphi\big)\big(F(-s,\mathbf{y})\big) = \mathcal{L}_W\varphi(\mathbf{y})$, which means that $\big(\mathcal{L}_{F(s,\cdot)_* W}\varphi\big)(\mathbf{y}) = \mathcal{L}_W\varphi\big(F(s,\mathbf{y})\big)$ and therefore that $\partial_t\varphi \circ \Phi(t) = \mathcal{L}_W\varphi\big(\Phi(t)\big)$. Equivalently, $\dot{\Phi}(t) = W\big(\Phi(t)\big)$, and so, since $\Phi(0) = F(s,\mathbf{x})$, $F\big(s,G(t,\mathbf{x})\big) = \Phi(t) = G\big(t,F(s,x)\big)$. □

Theorem 4.5.3. *Let $X(\,\cdot\,,\mathbf{x})$ be the solution to (4.5.4). If the V_k's commute, then $X(t,\mathbf{x})(w) = E\big((t,w(t)),\mathbf{x}\big)$.*

Proof. By (4.5.3),

$$E\big((t,w(t)),\mathbf{x}\big) = \mathbf{x} + \int_0^t \partial_{\xi_0} E\big((\tau,w(\tau),\mathbf{x}\big)\,d\tau + \sum_{k=1}^N \int_0^t \partial_{\xi_k} E\big((\tau,w(\tau)),\mathbf{x}\big) \bullet dw(\tau)_k,$$

and so all that remains is to prove that $\partial_{\xi_k} E(\boldsymbol{\xi},\mathbf{y}) = V_k\big(E(\boldsymbol{\xi},\mathbf{y})\big)$. To this end, apply Lemma 4.5.2 to see that $E\big(\boldsymbol{\xi},E(\boldsymbol{\eta},\mathbf{y})\big) = E\big(\boldsymbol{\eta},E(\boldsymbol{\xi},\mathbf{y})\big)$, and then apply this to check that if $\Phi(t) = E\big(t\boldsymbol{\xi},E(t\boldsymbol{\eta},\mathbf{y})\big)$, then $\dot{\Phi}(t) = V_{\boldsymbol{\xi}+\boldsymbol{\eta}}\big(\Phi(t)\big)$ with $\Phi(0) = \mathbf{y}$. Hence $E(\boldsymbol{\xi}+\boldsymbol{\eta},\mathbf{y}) = E\big(\boldsymbol{\eta},E(\boldsymbol{\xi},\mathbf{y})\big)$. Taking $\boldsymbol{\eta} = t\mathbf{e}_k$, where $(\mathbf{e}_k)_\ell = \delta_{k,\ell}$, and then differentiating with respect to t at $t=0$, arrive at the desired conclusion. □

Besides confirming the connection between integral curves and Stratonovich integral equations, Theorem 4.5.3 shows that, when the V_k's commute, the solution to (4.5.4) is a continuous function of w. In fact, it has as many continuous Fréchet derivatives as a function of w as the V_k's have as functions of $\mathbf{x}$.

When the V_k's don't commute, one has to work harder. Our goal now is to show that if

$$X_n(t,\mathbf{x}) = \mathbf{x} + \int_0^t V_0\big(X_n(\tau,\mathbf{x})\big)\,d\tau + \sum_{k=1}^M \int_0^t V_k\big(X_n(\tau,\mathbf{x})\big)\dot{w}_n(\tau)\,d\tau, \quad (4.5.6)$$

where, for all $m \ge 0$, $w_n(m2^{-n}) = w(m2^{-n})$ and w_n is linear on the intervals $\big[m2^{-n},(m+1)2^{-n}\big]$, then $X_n(\,\cdot\,,\mathbf{x})$ converges in $\mathcal{P}(\mathbb{R}^N)$ to the solution $X(\,\cdot\,,\mathbf{x})$ of (4.5.4). When the V_k's commute, $X_n(t,\mathbf{x}) = E\big((t,w_n(t)),\mathbf{x}\big)$, and so, to prove convergence in the non-commuting case, we will use an approximation scheme that is modeled on the commuting case in the same sense as the Euler approximation scheme was modeled on the case when the coefficients are constant. In particular, we should expect that the difference between $\partial_{\xi_k} E(\boldsymbol{\xi},\mathbf{x})$ and $V_k\big(E(\boldsymbol{\xi},\mathbf{x})\big)$ will be small when $|\boldsymbol{\xi}|$ is small. With this in mind, for $n \ge 0$, determine $\widetilde{X}_n(\,\cdot\,,\mathbf{x})$ by $\widetilde{X}_n(0,\mathbf{x}) = \mathbf{x}$ and

$$\widetilde{X}_n(t,\mathbf{x}) = E\big(\Delta_n(t),\widetilde{X}_n(\lfloor t\rfloor_n,\mathbf{x})\big)$$

where $\Delta_n(0)(w) = (0,\mathbf{0})$ and

$$\Delta_n(t)(w) = \big(t - m2^{-n}, w(t) - w(m2^{-n})\big) \quad \text{for } m2^{-n} < t \le (m+1)2^{-n}.$$

Then, by (4.5.3),

$$\widetilde{X}_n(t,\mathbf{x}) = \mathbf{x} + \int_0^t E_0\big(\Delta_n(\tau), \widetilde{X}_n(\lfloor\tau\rfloor_n, \mathbf{x})\big)\,d\tau \\ + \sum_{k=1}^M \int_0^t E_k\big(\Delta_n(\tau), \widetilde{X}_n(\lfloor\tau\rfloor_n, \mathbf{x})\big) \bullet dw(\tau)_k,$$

where $E_k(\boldsymbol{\xi},\mathbf{y}) := \partial_{\xi_k} E(\boldsymbol{\xi},\mathbf{y})$. After converting the Stratonovich integrals to Itô ones, applying (3.3.2), and using the estimates in (4.5.5), one sees that, for each $p \in [2,\infty)$,

$$\mathbb{E}^{\mathcal{W}}\big[\|\widetilde{X}_n(\,\cdot\,,\mathbf{x})\|_{[0,t]}^p\big] \\ \le 2^{p-1}|\mathbf{x}|^p + C_p(1+t)^{p-1}\int_0^t \mathbb{E}^{\mathcal{W}}\Big[e^{p\nu\Delta_n(\tau)}\Big(1+\big|\widetilde{X}_n(\lfloor\tau\rfloor_n,\mathbf{x})\big|^p\Big)\Big]\,d\tau$$

for some $C_p < \infty$. Further, since $\Delta_n(t)$ is independent of $\widetilde{X}_n(\lfloor\tau\rfloor_n,\mathbf{x})$ and $\sup_{n\ge0}\sup_{t\ge0}\mathbb{E}^{\mathcal{W}}\big[e^{p\nu|\Delta_n(t)|}\big] < \infty$, there is a $C_p < \infty$ such that

$$\mathbb{E}^{\mathcal{W}}\big[\|\widetilde{X}_n(\,\cdot\,,\mathbf{x})\|_{[0,t]}^p\big] \le 2^{p-1}|\mathbf{x}|^p \\ + C_p(1+t)^{p-1}\int_0^t \mathbb{E}^{\mathcal{W}}\big[1+\|\widetilde{X}_n(\,\cdot\,,\mathbf{x})\|_{[0,\tau]}^p\big]\,d\tau,$$

and so, by Lemma 1.2.4,

$$\sup_{n\ge0}\mathbb{E}^{\mathcal{W}}\big[1+\|\widetilde{X}_n(\,\cdot\,,\mathbf{x})\|_{[0,t]}^p\big] \le K_p e^{K_p t}(1+|\mathbf{x}|)^p \tag{4.5.7}$$

for some $K_p < \infty$.

Next let $X(\,\cdot\,,\mathbf{x})$ be the solution to (4.5.4), and set

$$W_k(\boldsymbol{\xi},\mathbf{x}) = E_k(\boldsymbol{\xi},\mathbf{x}) - V_k\big(E(\boldsymbol{\xi},\mathbf{x})\big).$$

Then $\widetilde{X}_n(t,\mathbf{x}) - X(t,\mathbf{x})$ equals

$$\int_0^t \Big(V_0\big(\widetilde{X}_n(\tau,\mathbf{x})\big) - V_0\big(X(\tau,\mathbf{x})\big)\Big)\,d\tau \\ + \sum_{k=1}^M \int_0^t \Big(V_k\big(\widetilde{X}_n(\tau,\mathbf{x})\big) - V_k\big(X(\tau,\mathbf{x})\big)\Big) \bullet dw(\tau)_k + \mathcal{E}_n(t,\mathbf{x}),$$

where

$$\mathcal{E}_n(t,\mathbf{x}) := \int_0^t W_0\big(\Delta_n(\tau), \widetilde{X}_n(\lfloor\tau\rfloor_n,\mathbf{x})\big)\,d\tau \\ + \sum_{k=1}^M \int_0^t W_k\big(\Delta_n(\tau), \widetilde{X}_n(\lfloor\tau\rfloor_n,\mathbf{x})\big) \bullet dw(\tau)_k.$$

After converting to Itô form the integrals with integrands involving the V_k's, one sees that, for $p \in [2, \infty)$,

$$\begin{aligned}\mathbb{E}^{\mathcal{W}}\left[\|\widetilde{X}_n(\,\cdot\,,\mathbf{x}) - X(\,\cdot\,,\mathbf{x})\|^p_{[0,t]}\right] \\ \le C_p(1+t)^{p-1}\int_0^t \mathbb{E}^{\mathcal{W}}\left[\|\widetilde{X}_n(\,\cdot\,,\mathbf{x}) - X(\,\cdot\,,\mathbf{x})\|^p_{[0,\tau]}\right] d\tau \\ + 2^{p-1}\mathbb{E}^{\mathcal{W}}\left[\|\mathcal{E}_n(\,\cdot\,,\mathbf{x})\|^p_{[0,t]}\right]\end{aligned} \tag{4.5.8}$$

for some $C_p < \infty$.

In order to estimate $\mathcal{E}_n(t,\mathbf{x})$, we will need the following lemma.

Lemma 4.5.4. *There is a $C < \infty$ such that*

$$|W_k(\boldsymbol{\xi},\mathbf{x})| \vee |\partial_{\xi_k} W_k(\boldsymbol{\xi},\mathbf{x})| \le C|\boldsymbol{\xi}|e^{2\nu|\boldsymbol{\xi}|}(1+|\mathbf{x}|)$$

for all $0 \le k \le M$.

Proof. By Taylor's theorem,

$$\begin{aligned}E\big(\boldsymbol{\eta}, E(\boldsymbol{\xi},\mathbf{x})\big) &= E(\boldsymbol{\xi},\mathbf{x}) + V_{\boldsymbol{\eta}}\big(E(\boldsymbol{\xi},\mathbf{x})\big) + \tfrac{1}{2}\mathcal{L}_{V_{\boldsymbol{\eta}}}V_{\boldsymbol{\eta}}\big(E(\boldsymbol{\xi},\mathbf{x})\big) + R_0(\boldsymbol{\xi},\boldsymbol{\eta},\mathbf{x}) \\ &= \mathbf{x} + V_{\boldsymbol{\xi}}(\mathbf{x}) + \tfrac{1}{2}\mathcal{L}_{V_{\boldsymbol{\xi}}}V_{\boldsymbol{\xi}}(\mathbf{x}) + V_{\boldsymbol{\eta}}(\mathbf{x}) + \mathcal{L}_{V_{\boldsymbol{\xi}}}V_{\boldsymbol{\eta}}(\mathbf{x}) + \tfrac{1}{2}\mathcal{L}_{V_{\boldsymbol{\eta}}}V_{\boldsymbol{\eta}}(\mathbf{x}) \\ &\quad + R_0(\boldsymbol{\xi},\boldsymbol{\eta},\mathbf{x}) + R_1(\boldsymbol{\xi},\mathbf{x}) + R_2(\boldsymbol{\xi},\boldsymbol{\eta},\mathbf{x}) + R_3(\boldsymbol{\xi},\boldsymbol{\eta},\mathbf{x}),\end{aligned}$$

where

$$\begin{aligned}R_0(\boldsymbol{\xi},\boldsymbol{\eta},\mathbf{x}) &= \int_0^1 (1-t)\Big(\mathcal{L}_{V_{\boldsymbol{\eta}}}V_{\boldsymbol{\eta}}\big(E(t\boldsymbol{\eta},E(\boldsymbol{\xi},\mathbf{x}))\big) - \mathcal{L}_{V_{\boldsymbol{\eta}}}V_{\boldsymbol{\eta}}\big(E(\boldsymbol{\xi},\mathbf{x})\big)\Big)\,dt \\ R_1(\boldsymbol{\xi},\mathbf{x}) &= \int_0^1 (1-t)\Big(\mathcal{L}_{\boldsymbol{\xi}}V_{\boldsymbol{\xi}}\big(E(t\boldsymbol{\xi},\mathbf{x})\big) - \mathcal{L}_{\boldsymbol{\xi}}V_{\boldsymbol{\xi}}(\mathbf{x})\Big)\,dt \\ R_2(\boldsymbol{\xi},\boldsymbol{\eta},\mathbf{x}) &= \int_0^1 \Big(\mathcal{L}_{V_{\boldsymbol{\xi}}}V_{\boldsymbol{\eta}}\big(E(t\boldsymbol{\xi},\mathbf{x})\big) - \mathcal{L}_{V_{\boldsymbol{\xi}}}V_{\boldsymbol{\eta}}(\mathbf{x})\Big)\,dt \\ R_3(\boldsymbol{\xi},\boldsymbol{\eta},\mathbf{x}) &= \frac{\mathcal{L}_{V_{\boldsymbol{\eta}}}V_{\boldsymbol{\eta}}\big(E(\boldsymbol{\xi},\mathbf{x})\big) - \mathcal{L}_{V_{\boldsymbol{\eta}}}V_{\boldsymbol{\eta}}(\mathbf{x})}{2}.\end{aligned}$$

At the same time,

$$E(\boldsymbol{\xi}+\boldsymbol{\eta},\mathbf{x}) = \mathbf{x} + V_{\boldsymbol{\xi}+\boldsymbol{\eta}}(\mathbf{x}) + \tfrac{1}{2}\mathcal{L}_{V_{\boldsymbol{\xi}+\boldsymbol{\eta}}}V_{\boldsymbol{\xi}+\boldsymbol{\eta}}(x) + R_1(\boldsymbol{\xi}+\boldsymbol{\eta},\mathbf{x}).$$

Hence, $E\big(\boldsymbol{\eta},E(\boldsymbol{\xi},\mathbf{x})\big) - E(\boldsymbol{\xi}+\boldsymbol{\eta},\mathbf{x}) = \frac{1}{2}[V_{\boldsymbol{\xi}},V_{\boldsymbol{\eta}}](\mathbf{x}) + R(\boldsymbol{\xi},\boldsymbol{\eta},\mathbf{x})$, where

$$R(\boldsymbol{\xi},\boldsymbol{\eta},\mathbf{x}) = R_0(\boldsymbol{\xi},\boldsymbol{\eta},\mathbf{x}) + R_1(\boldsymbol{\xi},\mathbf{x}) + R_2(\boldsymbol{\xi},\boldsymbol{\eta},\mathbf{x}) + R_3(\boldsymbol{\xi},\boldsymbol{\eta},\mathbf{x}) - R_1(\boldsymbol{\xi}+\boldsymbol{\eta},\mathbf{x}).$$

Therefore $W_k(\boldsymbol{\xi},\mathbf{x}) = \frac{1}{2}[V_{\boldsymbol{\xi}},V_k] + \partial_{\eta_k}R(\boldsymbol{\xi},\mathbf{0},\mathbf{x})$ and

$$\begin{aligned}\partial_{\eta_k} R(\boldsymbol{\xi},\mathbf{0},\mathbf{x}) = &\int_0^1 \Big(\mathcal{L}_{V_{\boldsymbol{\xi}}} V_k\big(E(\boldsymbol{\xi},\mathbf{x})\big) - \mathcal{L}_{V_{\boldsymbol{\xi}}} V_k(\mathbf{x})\Big)\,d\tau\\ &+ \int_0^1 (1-t)\Big(\big(\mathcal{L}_{V_{\boldsymbol{\xi}}} V_k + \mathcal{L}_{V_k} V_{\boldsymbol{\xi}}\big)\big(E(t\boldsymbol{\xi},\mathbf{x})\big)\\ &\qquad - \big(\mathcal{L}_{V_{\boldsymbol{\xi}}} V_k + \mathcal{L}_{V_k} V_{\boldsymbol{\xi}}\big)(\mathbf{x}) + t\mathcal{L}_{V_{\boldsymbol{\xi}}} V_{\boldsymbol{\xi}} \partial_{\xi_k} E(\boldsymbol{\xi},\mathbf{x})\Big)\,dt,\end{aligned}$$

and so the estimates in (4.5.5) now imply the asserted estimate for $W_k(\boldsymbol{\xi},\mathbf{x})$. To prove the one for $\partial_{\xi_k} W_k(\boldsymbol{\xi},\mathbf{x})$, observe that, since $[V_k, V_k] = 0$,

$$[V_{\boldsymbol{\xi}}, V_k] = \sum_{\ell \neq k} \xi_\ell [V_\ell, V_k],$$

and therefore that $\partial_{\xi_k} W_k(\boldsymbol{\xi},\mathbf{x}) = \partial_{\xi_k}\partial_{\eta_k} R(\boldsymbol{\xi},\mathbf{0},\mathbf{x})$. Hence, another application of (4.5.5) give the asserted result. □

Returning to the estimation of $\mathcal{E}_n(t,\mathbf{x})$, set

$$\widetilde{W}(\boldsymbol{\xi},\mathbf{x}) = W_0(\boldsymbol{\xi},\mathbf{x}) + \frac{1}{2}\sum_{k=1}^M \partial_{\xi_k} W_k(\boldsymbol{\xi},\mathbf{x}).$$

Then

$$\begin{aligned}\mathcal{E}_n(t,\mathbf{x}) = &\int_0^t \widetilde{W}\big(\Delta_n(\tau), \widetilde{X}_n(\lfloor\tau\rfloor_n,\mathbf{x})\big)\,d\tau\\ &+ \sum_{k=1}^M \int_0^t W_k\big(\Delta_n(\tau), \widetilde{X}_n(\lfloor\tau\rfloor_n,\mathbf{x})\big)\,dw(\tau)_k,\end{aligned}$$

and therefore

$$\begin{aligned}\mathbb{E}^{\mathcal{W}}\big[\|\mathcal{E}_n(\,\cdot\,,\mathbf{x})\|_{[0,t]}^p\big] \le (2t)^{p-1} &\int_0^t \mathbb{E}^{\mathcal{W}}\big[|\widetilde{W}\big(\Delta_n(\tau), \widetilde{X}_n(\lfloor\tau\rfloor_n,\mathbf{x})\big)|^p\big]\,d\tau\\ &+ 2^{p-1}K_p \sum_{m=1}^M \int_0^t \mathbb{E}^{\mathcal{W}}\big[\big|W_k\big(\Delta_n(\tau), \widetilde{X}_n(\lfloor\tau\rfloor_n,\mathbf{x})\big)\big|^p\big]\,d\tau.\end{aligned}$$

By Lemma 4.5.4 and the independence of $\Delta_n(\tau)$ and $\widetilde{X}_n(\lfloor\tau\rfloor_n,\mathbf{x})$,

$$\mathbb{E}^{\mathcal{W}}\big[|\widetilde{W}\big(\Delta_n(\tau), \widetilde{X}_n(\lfloor\tau\rfloor_n,\mathbf{x})\big)|^p\big]$$

and

$$\sum_{m=1}^M \int_0^t \mathbb{E}^{\mathcal{W}}\big[\big|W_k\big(\Delta_n(\tau), \widetilde{X}_n(\lfloor\tau\rfloor_n,\mathbf{x})\big)\big|^p\big]$$

are dominated by a constant times

$$\mathbb{E}^{\mathcal{W}}\big[|\Delta_n(\tau)|^p e^{2p\nu|\Delta_n(\tau)|}\big]\mathbb{E}^{\mathcal{W}}\big[1+|\widetilde{X}_n(\lfloor\tau\rfloor_n,\mathbf{x})|^p\big].$$

Thus, since $\mathbb{E}^{\mathcal{W}}\big[|\Delta_n(\tau)|^p e^{2p\nu|\Delta_n(\tau)|}\big]$ is dominated by a constant times $2^{-\frac{np}{2}}$, we can now use (4.5.7) to see that there is a $K_p(t)<\infty$ such that

$$\mathbb{E}^{\mathcal{W}}\big[\|\mathcal{E}_n(\,\cdot\,,\mathbf{x})\|_{[0,t]}^p\big]\le 2^{-\frac{np}{2}}K_p(t)(1+|\mathbf{x}|)^p.$$

After combining this with (4.5.8) and applying Lemma 1.2.4, we conclude that

$$\mathbb{E}^{\mathcal{W}}\big[\|\widetilde{X}_n(\,\cdot\,,\mathbf{x})-X(\,\cdot\,,\mathbf{x})\|_{[0,t]}^p\big]^{\frac{1}{p}}\le K_p(t)(1+|\mathbf{x}|)2^{-\frac{n}{2}} \tag{4.5.9}$$

for some $K_p(t)<\infty$. In particular, this means that

$$\|\widetilde{X}_n(\,\cdot\,,\mathbf{x})-X(\,\cdot\,,\mathbf{x})\|_{[0,t]}\longrightarrow 0\quad (\text{a.s.},\mathcal{W})\text{ and in }L^p(\mathcal{W};\mathbb{R}).$$

Although (4.5.9) confirms the idea that solutions to (4.5.4) are limits of integral curves, it is not the connection advertised.

Theorem 4.5.5. *Let $X(\,\cdot\,,\mathbf{x})$ be the solution to* (4.5.4), *and, for each $n\ge 0$, let $X_n(\,\cdot\,,\mathbf{x})$ be the integral curve described in* (4.5.6). *Then, for each $\alpha\in(0,\frac{1}{2})$, $p\in[2,\infty)$, and $t>0$ there exists a $C_{\alpha,p}(t)<\infty$ such that*

$$E^{\mathcal{W}}\big[\|X_n(\,\cdot\,,\mathbf{x})-X(\,\cdot\,,\mathbf{x})\|_{[0,t]}^p\big]^{\frac{1}{p}}\le C_{\alpha,p}(t)(1+|\mathbf{x}|)2^{-\alpha n}.$$

In particular, for each $t>0$, $\|X_n(\,\cdot\,,\mathbf{x})-X(\,\cdot\,,\mathbf{x})\|_{[0,t]}\longrightarrow 0$ both (a.s., $\mathcal{W}$) *and in $L^p(\mathcal{W};\mathbb{R})$.*

Proof. Observe that

$$X_n(t,\mathbf{x})=E\big(2^n(t-m2^{-n})\Delta_n((m+1)2^{-n}),X_n(m2^{-n},\mathbf{x})\big)$$

for $t\in I_{m,n}:=[m2^{-n},(m+1)2^{-n}]$, and so, by induction on $m\ge 0$, $X_n(m2^{-n},\mathbf{x})=\widetilde{X}_n(m2^{-n},\mathbf{x})$ for all $m\ge 0$. At the same time, by (4.5.5), there is a $C<\infty$ such that

$$\begin{aligned}&|X_n(t,\mathbf{x})-X_n(m2^{-n},\mathbf{x})|\vee|\widetilde{X}_n(t,\mathbf{x})-\widetilde{X}_n(m2^{-n},\mathbf{x})|\\&\le C\|\Delta_n(\,\cdot\,)\|_{I_{m,n}}e^{\nu\|\Delta_n(\cdot)\|_{I_{m,n}}}\big(1+|\widetilde{X}_n(m2^{-n},\mathbf{x})|\big)\end{aligned}$$

for $t\in I_{m,n}$, and so

$$\begin{aligned}&\sup_{\tau\in[0,t]}|X_n(\tau,\mathbf{x})-X_n(\lfloor\tau\rfloor_n,\mathbf{x})|\vee|\widetilde{X}_n(\tau,\mathbf{x})-\widetilde{X}_n(\lfloor\tau\rfloor_n,\mathbf{x})|\\&\le 2^{-\alpha n}C\big(1+\|w\|_{\alpha,[0,t]}\big)e^{\nu(1+\|w\|_{[0,t]})}\big(1+\|\widetilde{X}_n(\,\cdot\,,\mathbf{x})\|_{[0,t]}\big),\end{aligned}$$

where

$$\|w\|_{\alpha,[0,t]} := \sup_{0\le\tau_1<\tau_2\le t} \frac{|w(\tau_2)-w(\tau_1)|}{(\tau_2-\tau_1)^\alpha}.$$

Hence, since $X_n(\lfloor\tau\rfloor_n,\mathbf{x}) = \widetilde{X}_n(\lfloor\tau\rfloor_n,\mathbf{x})$,

$$\begin{aligned}
\big|X_n(\tau,\mathbf{x})-X(\tau,\mathbf{x})\big| &\le \big|X_n(\tau,\mathbf{x})-\widetilde{X}_n(\tau,\mathbf{x})\big| + \big|\widetilde{X}_n(\tau,\mathbf{x})-X(\tau,\mathbf{x})\big| \\
&\le \big|X_n(\tau,\mathbf{x})-X_n(\lfloor\tau\rfloor_n,\mathbf{x})\big| + \big|\widetilde{X}_n(\tau,\mathbf{x})-\widetilde{X}_n(\lfloor\tau\rfloor_n,\mathbf{x})\big| \\
&\quad + \big|\widetilde{X}_n(\tau,\mathbf{x})-X(\tau,\mathbf{x})\big| \\
&\le 2^{1-n\alpha}C\big(1+\|w\|_{\alpha,[0,t]}\big)e^{\nu(1+2\|w\|_{[0,t]})}\big(1+\|\widetilde{X}_n(\,\cdot\,,\mathbf{x})\|_{[0,t]}\big) \\
&\quad + \|\widetilde{X}_n(\,\cdot\,,\mathbf{x})-X(\,\cdot\,,\mathbf{x})\|_{[0,t]}.
\end{aligned}$$

By (4.5.9), the L^p-norm of the term in the final line satisfies an estimate of the required form. As for the term in the line above, use (2.3.1), (4.5.7), and Theorem 2.1.2 to see that

$$\mathbb{E}^{\mathcal{W}}\Big[\Big(\big(1+\|w\|_{\alpha,[0,t]}\big)e^{\nu(1+2\|w\|_{[0,t]})}\big(1+\|\widetilde{X}_n(\,\cdot\,,\mathbf{x})\|_{[0,t]}\big)\Big)^p\Big]^{\frac{1}{p}} < \infty$$

for all $t\ge 0$ and $p\in[2,\infty)$. □

Corollary 4.5.6. *Define $S(\mathbf{x})$ to be the closure in $\mathcal{P}(\mathbb{R}^N)$ of the set of continuously differentiable paths $p:[0,\infty)\longrightarrow\mathbb{R}^N$ with $p(0)=\mathbf{x}$ for which there exists a smooth function $\theta\in\mathcal{P}(\mathbb{R}^N)$ such that*

$$\dot p(t) = V_0\big(p(t)\big) + \sum_{k=1}^M \theta(t)_k V_k\big(p(t)\big). \tag{4.5.10}$$

If $\mathbb{P}_{\mathbf{x}}\in\mathbf{M}_1\big(\mathcal{P}(\mathbb{R}^N)\big)$ is the distribution of the solution to (4.5.4), then $\mathbb{P}_{\mathbf{x}}\big(S(\mathbf{x})\big)=1$. In particular, if $M\subseteq\mathbb{R}^N$ is a closed, differentiable manifold and $V_k(\mathbf{y})$ is tangent to M for all $0\le k\le M$ and $\mathbf{y}\in M$, then

$$\mathbf{x}\in M \implies \mathbb{P}_{\mathbf{x}}\big(\psi(t)\in M \quad \textit{for all } t\in[0,\infty)\big)=1.$$

Proof. Since, if $\mathbf{x}\in M$, every $p\in S(\mathbf{x})$ stays on M, the last assertion follows from the first. Next observe that for any piecewise constant $\theta:[0,\infty)\longrightarrow\mathbb{R}^N$, the solution to (4.5.10) is the limit in $\mathcal{P}(\mathbb{R}^N)$ of solutions to (4.5.10) for smooth θ's. Thus, $S(\mathbf{x})$ contains all limits in $\mathcal{P}(\mathbb{R}^N)$ of solutions to (4.5.10) with piecewise constant θ's. In particular, if $X_n(\,\cdot\,,\mathbf{x})(w)$ is the solution to (4.5.6), then $X_n(\,\cdot\,,\mathbf{x})(w)\in S(\mathbf{x})$ for all $n\ge 0$ and w, and so, by Theorem 4.5.5, $\mathbb{P}_{\mathbf{x}}\big(S(\mathbf{x})\big) = \mathcal{W}\big(X(\,\cdot\,,\mathbf{x})\in S(\mathbf{x})\big)=1$. □

The original version of a result like the one in Corollary 4.5.6 goes back to work of E. Wong and M. Zakai. There is a refinement, known as the *support theorem* (cf. [21] or [18]), which states that $S(\mathbf{x})$ is the smallest closed set of $\mathcal{P}(\mathbb{R}^N)$ to which $\mathbb{P}_{\mathbf{x}}$ gives measure 1. That is, not only does $\mathbb{P}_{\mathbf{x}}\big(S(\mathbf{x})\big)=1$ but also, for each $p\in S(\mathbf{x})$, $\epsilon>0$, and $t>0$,

$$\mathbb{P}_{\mathbf{x}}\big(\|\psi(\,\cdot\,) - p(\,\cdot\,)\|_{[0,t]} < \epsilon\big) > 0.$$

In a different direction, one can apply these ideas to show that $X(t,\,\cdot\,)$ is a diffeomorphism of $\mathbb{R}^N$ onto itself. We already know that $X(t,\,\cdot\,)$ is continuously differentiable and, by Theorem 4.5.1, that its Jacobian matrix is non-degenerate. Thus, what remains is to show that it is has a globally defined inverse, and (4.5.9) provides a way to do so. Namely, for each $n \geq 0$ and $w \in \mathcal{W}(\mathbb{R}^M)$, $X_n(1,\,\cdot\,)(w)$ is a diffeomorphism and its inverse is equal to $\breve{X}_n(1,\,\cdot\,)(\breve{w}^1)$ where $\breve{w}^1(\tau) = w(\tau \vee 1) - w\big((1-\tau)^+\big)$ and $\breve{X}_n(\,\cdot\,,\mathbf{x})(w)$ is the solution to

$$\breve{X}_n(t,\mathbf{x})(w) = \mathbf{x} - \int_0^t V_0\big(\breve{X}_n(\tau,\mathbf{x})\big)(w)\,d\tau - \sum_{k=1}^M V_k\big(\breve{X}_n(\tau,\mathbf{x})\big)(w)\dot{w}_n(\tau)_k\,d\tau$$

for $t \in [0,1]$. To see this, simply apply the flow property to the pair $\begin{pmatrix} t \\ X_n(t,\mathbf{x})(w) \end{pmatrix}$. In view of this and the fact that $\breve{w}^1$ has the same distribution under $\mathcal{W}$ as w, one should expect that $X(1,\,\cdot\,)^{-1}(w)$ equals $\breve{X}(1,\,\cdot\,)(\breve{w}^1)$, where

$$\breve{X}(t,\mathbf{x}) = \mathbf{x} - \int_0^t V_0\big(\breve{X}(\tau,\mathbf{x})\big)\,d\tau - \sum_{k=1}^M V_k\big(\breve{X}(\tau,\mathbf{x})\big) \bullet dw(\tau)_k. \tag{4.5.11}$$

Indeed, because $X_n(1,\mathbf{x}) = \widetilde{X}_n(1,\mathbf{x})$, all that we have to check is that, $\mathcal{W}$-almost surely,

$$\widetilde{X}_n(1,\,\cdot\,) \longrightarrow \widetilde{X}(1,\,\cdot\,) \quad \text{and} \quad \breve{X}_n(1,\,\cdot\,) \longrightarrow \breve{X}(1,\,\cdot\,)$$

uniformly on compact subsets of $\mathbb{R}^N$.

The proof of these convergence results is essentially the same as the proof in §3.4 of the corresponding sort of result in the Itô context. That is, one sets $\widetilde{D}_n(t,\mathbf{x}) = \widetilde{X}(t,\mathbf{x}) - \widetilde{X}_n(t,\mathbf{x})$, estimates the moments of $\widetilde{D}_n(t,\mathbf{y}) - \widetilde{D}_n(t,\mathbf{x})$ in terms of $|\mathbf{y}-\mathbf{x}|$, and then applies Kolmogorov's convergence criterion. More precisely, for $p \geq 1$, note that

$$\begin{aligned}|\widetilde{D}_n(t,\mathbf{y}) - \widetilde{D}_n(t,\mathbf{x})|^{2p} \leq \big(&|X(t,\mathbf{y}) - X(t,\mathbf{x})| + |\widetilde{X}_n(t,\mathbf{y}) - \widetilde{X}_n(t,\mathbf{x})|\big)^p \\ &\times \big(|\widetilde{D}_n(t,\mathbf{x})| + |\widetilde{D}_n(t,\mathbf{y})|\big)^p,\end{aligned}$$

and therefore that

$$\begin{aligned}&\mathbb{E}^{\mathcal{W}}\big[|\widetilde{D}_n(t,\mathbf{y}) - \widetilde{D}_n(t,\mathbf{x})|^{2p}\big]^{\frac{1}{2p}} \\ &\quad\leq \Big(\mathbb{E}^{\mathcal{W}}\big[|X(t,\mathbf{y}) - X(t,\mathbf{x})|^{2p}\big]^{\frac{1}{2p}} + \mathbb{E}^{\mathcal{W}}\big[|\widetilde{X}_n(t,\mathbf{y}) - \widetilde{X}_n(t,\mathbf{x})|^{2p}\big]^{\frac{1}{2p}}\Big)^{\frac{1}{2}} \\ &\qquad\times \Big(\mathbb{E}^{\mathcal{W}}\big[|\widetilde{D}_n(t,\mathbf{y})|^{2p}\big]^{\frac{1}{2p}} + \mathbb{E}^{\mathcal{W}}\big[|\widetilde{D}_n(t,\mathbf{x})|^{2p}\big]^{\frac{1}{2p}}\Big)^{\frac{1}{2}}.\end{aligned}$$

By (4.5.9), there is a $C_p(t) < \infty$ such that

$$\mathbb{E}^{\mathcal{W}}\big[|\widetilde{D}_n(t,\mathbf{y})|^{2p}\big]^{\frac{1}{2p}} + \mathbb{E}^{\mathcal{W}}\big[|\widetilde{D}_n(t,\mathbf{x})|^{2p}\big]^{\frac{1}{2p}} \le C_p(t)(1+|\mathbf{x}|+|\mathbf{y}|)2^{-\frac{n}{2}}.$$

As for the first factor on the right, observe that

$$\begin{aligned}X(t,\mathbf{y}) - X(t,\mathbf{x}) = \mathbf{y}-\mathbf{x} &+ \int_0^t \Big(V_0\big(X(\tau,\mathbf{y})\big) - V_0\big(X(\tau,\mathbf{x})\big)\Big)\,d\tau \\ &+ \sum_{k=1}^M \int_0^t \Big(V_k\big(X(\tau,\mathbf{y})\big) - V_0\big(X(\tau,\mathbf{x})\big)\Big) \bullet dw(\tau)_k,\end{aligned}$$

and so, after converting to Itô integrals, applying (3.3.2) and Lemma 1.2.4, one sees that

$$\mathbb{E}^{\mathcal{W}}\big[|X(t,\mathbf{y}) - X(t,\mathbf{x})|^{2p}\big]^{\frac{1}{2p}} \le C_p(t)|\mathbf{y}-\mathbf{x}|$$

for some $C_p(t) < \infty$. To derive the analogous estimate for $\widetilde{X}_n(t,\mathbf{y}) - \widetilde{X}_n(t,\mathbf{x})$, remember that

$$\begin{aligned}&\widetilde{X}_n(t,\mathbf{y}) - \widetilde{X}_n(t,\mathbf{x}) \\ &= \mathbf{y}-\mathbf{x} + \int_0^t \Big(E_0\big(\Delta_n(\tau), \widetilde{X}_n(\lfloor\tau\rfloor_n,\mathbf{y})\big) - E_0\big(\Delta_n(\tau), \widetilde{X}_n(\lfloor\tau\rfloor_n,\mathbf{x})\big)\Big)\,d\tau \\ &+ \sum_{k=1}^M \int_0^t \Big(E_k\big(\Delta_n(\tau), \widetilde{X}_n(\lfloor\tau\rfloor_n,\mathbf{y})\big) - E_k\big(\Delta_n(\tau), \widetilde{X}_n(\lfloor\tau\rfloor_n,\mathbf{x})\big)\Big) \bullet dw(\tau)_k.\end{aligned}$$

Thus, by again converting to Itô integrals, applying (3.3.2), using the estimates in (4.5.5), and, remembering that $\Delta_n(\tau)$ is independent of $X_n(\lfloor\tau\rfloor_n,\mathbf{x})$, one arrives that the same sort of estimate in terms of $|\mathbf{y}-\mathbf{x}|$. After combining these, one has that

$$\mathbb{E}^{\mathcal{W}}\big[|\widetilde{D}_n(t,\mathbf{y}) - \widetilde{D}_n(t,\mathbf{x})|^{2p}\big]^{\frac{1}{2p}} \le C_p(t)(1+|\mathbf{x}|+|\mathbf{y}|)^{\frac{1}{2}}2^{-\frac{n}{4}}|\mathbf{y}-\mathbf{x}|^{\frac{1}{2}}$$

for some $C_p(t) < \infty$. Hence, by taking $p > N$, Kolmogorov's criterion says that there exists an $\alpha > 0$ such that

$$\mathbb{E}^{\mathcal{W}}\big[\|\widetilde{D}_n(t,\,\cdot\,)\|^{2p}_{[-R,R]^N}\big]^{\frac{1}{2p}} \le C(t,R)2^{-\alpha n}$$

for some $C(t,R) < \infty$. Since the same sort of estimate holds for

$$\breve{D}_n(t,\,\cdot\,) := \breve{X}(t,\mathbf{x}) - \breve{X}_n(t,\mathbf{x}),$$

we have now proved that $X(1,\,\cdot\,)^{-1}(w) = \breve{X}(1,\,\cdot\,)(\breve{w}^1)$ $\mathcal{W}$-almost surely.

Of course, for general $t \in [0,\infty)$, it must be true that $X(t,\mathbf{x})^{-1}(w) = \breve{X}(t,\mathbf{x})(\breve{w}^t)$ (a.s., $\mathcal{W}$), where $\breve{w}^t(\tau) = w(\tau \vee t) - w((t-\tau)^+)$, and, as long as $t = m2^{-n}$, for some $(m,n) \in \mathbb{N}^2$, there is no problem about verifying this. However, when $t \notin \{m2^{-n} : (m,n) \in \mathbb{N}^2\}$, one encounters two annoying problems. The first of these is that it is no longer true that $\frac{d}{d\tau}\breve{w}_n^t = -\dot{w}_n$, and the second is that $X_n(t,\,\cdot\,) \neq \widetilde{X}_n(t,\,\cdot\,)$. Perhaps the easiest way to avoid these problems is to modify the approximation scheme. Namely, given $t > 0$, use time increments based on $2^{-n}t$ instead of 2^{-n}. That is, replace w_n by w_n^t, the polygonal approximation of w that equals w at times $m2^{-n}t$ and is linear on intervals $[m2^{-n}t, (m+1)2^{-n}t]$, and define $\Delta_n^t(\tau)(w) = w(\tau) - w(m2^{-n}t)$ for $\tau \in [m2^{-n}t, (m+1)2^{-n}t]$. Without any changes in the argument, one can show the corresponding approximation schemes have the same convergence properties as $\{X_n : n \geq 0\}$ and $\{\widetilde{X}_n : n \geq 0\}$, and the problems described above do not arise.

Theorem 4.5.7. *Let $X(\,\cdot\,,\mathbf{x})$ be the solution to* (4.5.4). *Then, for each $t \geq 0$, $X(t,\,\cdot\,)$ is $\mathcal{W}$-almost surely a diffeomorphism from $\mathbb{R}^N$ onto itself. Furthermore, for each $t > 0$, the distribution of $X(t,\,\cdot\,)^{-1}$ is the same as the distribution of $\breve{X}(t,\,\cdot\,)$, where $\breve{X}(\,\cdot\,,\mathbf{x})$ is the solution to* (4.5.11).

4.6 Exercises

Exercise 4.1. Let $\big(M(t), \mathcal{F}_t, \mathbb{P}\big)$ be a continuous local martingale, and choose stopping times $\zeta_k \nearrow \infty$ accordingly. Show that $\big(M(t), \mathcal{F}_t, \mathbb{P}\big)$ is a martingale if there is a $p \in (1,\infty)$ such that $\sup_{k\geq 1} \mathbb{E}^{\mathbb{P}}\big[|M(t\wedge\zeta_k)|^p\big] < \infty$ for all $t \geq 0$. Next, assume that $M(\,\cdot\,)$ is non-negative, and show that $\big(M(t), \mathcal{F}_t, \mathbb{P}\big)$ is always a supermartingale (i.e., $\mathbb{E}\big[M(t)|\mathcal{F}_s\big] \geq M(s)$ for $0 \leq \mathrm{s} \leq \mathrm{t}$) and that it is a martingale if and only if $\mathbb{E}^{\mathbb{P}}\big[M(t)\big] = \mathbb{E}^{\mathbb{P}}\big[M(0)\big]$ for all $t \geq 0$.

Exercise 4.2. Let $\big(M_1, \mathcal{F}_t, \mathbb{P}\big)$ and $\big(M_2, \mathcal{F}_t, \mathbb{P}\big)$ be a pair of continuous local martingales. Show that $\langle M_1, M_2\rangle = 0$ if $M_1(\,\cdot\,)$ is independent of $M_2(\,\cdot\,)$, and give an example in which $\langle M_1, M_2\rangle = 0$ but $M_1(\,\cdot\,)$ is not independent of $M_2(\,\cdot\,)$.

Next let $\big(X(t), \mathcal{F}_t, \mathbb{P}\big)$ be an $\mathbb{R}^N$-valued continuous semi-martingale, and set $A(t) = \big(\!\big(\langle X_i, X_j\rangle\big)\!\big)_{1\leq i,j\leq N}$. Given $\varphi_1, \varphi_2 \in C^2(\mathbb{R}^N;\mathbb{R})$, show that

$$\big\langle \varphi_1\circ X, \varphi_2\circ X\big\rangle(dt) = \big(\nabla\varphi_1\circ X, A(dt)\nabla\varphi_2\circ X\big)_{\mathbb{R}^N}.$$

Exercise 4.3. The estimate in (4.3.1) is one fourth of Burkholder's inequality. In this exercise, you are to prove another fourth of it. Namely, you are to show that for each $p \in [2,\infty)$ and all $\mathbb{R}$-valued continuous martingales $\big(M(t), \mathcal{F}_t, \mathbb{P}\big)$ with $M(0) = 0$,

$$\mathbb{E}^{\mathbb{P}}\big[\langle\!\langle M\rangle\!\rangle^{\frac{p}{2}}\big] \leq \left(\frac{2^{\frac{1}{2}}p^{\frac{3}{2}}}{p-1}\right)^p \mathbb{E}^{\mathbb{P}}\big[|M(t)|^p\big]. \tag{4.6.1}$$

The missing half of Burkholder's inequality states that estimates like those in (4.3.1) and (4.6.1) hold for all $p \in (1,\infty)$. In fact, together with B. Davis and R. Gundy, Burkholder showed that they hold even when $p \in (0,1]$.

The outline that follows is based on ideas introduced by A. Garcia.

(i) The first step is to show that it suffices to treat the case in which both M and $\langle\langle M\rangle\rangle$ are uniformly bounded.

(ii) Assuming that M and $\langle\langle M\rangle\rangle$ are bounded, set $\xi(t) = \langle\langle M\rangle\rangle(t)^{\frac{p}{4}-\frac{1}{2}}$, and take $M' = I_\xi^M$. After showing that

$$\langle\langle M'\rangle\rangle(t) = \int_0^t \langle\langle M\rangle\rangle^{\frac{p}{2}-1}(\tau)\,\langle\langle M\rangle\rangle(d\tau) = \frac{2}{p}\langle\langle M\rangle\rangle(t)^{\frac{p}{2}},$$

conclude that $\mathbb{E}^{\mathbb{P}}\big[\langle\langle M\rangle\rangle(t)^{\frac{p}{2}}\big] = \frac{p}{2}\mathbb{E}^{\mathbb{P}}\big[M'(t)^2\big]$.

(iii) Continuing part (ii), apply Theorem 4.3.1 to see that

$$M(t)\langle\langle M\rangle\rangle(t)^{\frac{p}{4}-\frac{1}{2}} = M'(t) + \int_0^t M(\tau)\,d\langle\langle M\rangle\rangle(\tau)^{\frac{p}{4}-\frac{1}{2}},$$

and conclude that $\|M'\|_{[0,t]} \le 2\|M\|_{[0,t]}\langle\langle M\rangle\rangle(t)^{\frac{p}{4}-\frac{1}{2}}$. Now combine this with the result in (ii), Hölder's inequality, and Doob's inequality to complete the proof of (4.6.1).

Exercise 4.4. Let $\big(M(t),\mathcal{F}_t,\mathbb{P}\big)$ be an $\mathbb{R}^N$-valued, continuous local martingale on a probability space $(\Omega,\mathcal{F},\mathbb{P})$, and set $A(t) = \big(\langle\langle M_i, M_j\rangle\rangle\big)_{1\le i,j\le N}$. Under the condition that $A(t) = \int_0^t a(\tau)\,d\tau$, it was shown in Theorem 4.4.1 that, by moving to another probability space, $M(\,\cdot\,)$ can be represented as a Brownian stochastic integral. In this exercise you are to show that, aside from a random time change, the same is true even when the condition fails to hold.

(i) Show that the condition on $A(\,\cdot\,)$ in Theorem 4.4.1 holds with an $a \le \mathbf{I}$ if $\mathrm{Trace}\big(A(t)-A(s)\big) \le t-s$ for all $0 \le s < t$.

(ii) Set $\beta(t) = t + \mathrm{Trace}\big(A(t)\big)$, and show that β maps $[0,\infty)$ homeomorphically onto itself. Next, set $\zeta_s = \beta^{-1}(s)$, and show that $\{\zeta_s : s \ge 0\}$ is an increasing family of bounded stopping times.

(iii) Set $\widecheck{M}(t) = M(\zeta_t)$ and $\check{\mathcal{F}}_t = \mathcal{F}_{\zeta_t}$. Using Hunt's stopping time theorem, show that $\big(\widecheck{M}(t),\check{\mathcal{F}}_t,\mathbb{P}\big)$ is a continuous local martingale. Further, show that

$$\big(\langle\langle \widecheck{M}_i, \widecheck{M}_j\rangle\rangle\big)_{1\le i,j\le N} = \int_0^t \check{a}(\tau)\,d\tau,$$

where $\check{a} \le \mathbf{I}$.

(iv) Apply Theorem 4.4.1 to find a Brownian motion $\big(\widetilde{B}(t),\widetilde{\mathcal{F}}_t,\widetilde{\mathbb{P}}\big)$ on a probability space $(\widetilde{\Omega},\widetilde{\mathcal{F}},\widetilde{\mathbb{P}})$ and an measurable map $F:\widetilde{\Omega}\longrightarrow\Omega$ such that $\mathbb{P} = F_*\widetilde{\mathbb{P}}$ and

$$\widetilde{M}(t) = \widetilde{M}(0) + \int_0^t \widetilde{\sigma}(\tau)\, d\widetilde{B}(\tau),$$

where $\widetilde{M} = \breve{M} \circ F$ and $\widetilde{\sigma} = \breve{a}^{\frac{1}{2}} \circ F$.

(v) Note that $M(t) = \breve{M}\big(\beta(t)\big)$, and conclude that

$$(M \circ F)(t) = (M \circ F)(0) + \int_0^{(\beta\circ F)(t)} \breve{a}^{\frac{1}{2}} \circ F(\tau)\, d\widetilde{B}(\tau).$$

Exercise 4.5. It is a little disappointing that, in deriving Itô's formulation of the Stratonovich integral, we mollified both X and Y and not just Y. Thus, it may be comforting to know that one need mollify only Y if $\langle X, Y\rangle(dt) = c(t)dt$, where c is a progressively measurable, locally integrable function. What follows are some steps that lead to this conclusion. The notation here is the same as that used earlier when we introduced Itô's formulation of the Stratonovich integral.

Consider

$$\begin{aligned} D_n &:= \int_0^1 \big(X(\tau) - X_n(\tau)\big)\, dY_n(\tau) \\ &= \sum_{m<2^n} 2^n \left(\int_{I_{m,n}} \Big(\big(X(\tau) - X_{m,n}\big) - 2^n(\tau - m2^{-n})\Delta^X_{m,n} \Big)\, d\tau \right) \Delta^Y_{m,n}, \end{aligned}$$

where $I_{m,n} := [m2^{-n}, (m+1)2^{-n}]$ and, for any $F : [0,\infty) \longrightarrow \mathbb{R}$, $\Delta^F_{m,n} := F_{m+1,n} - F_{m,n}$ with $F_{m,n} := F(m2^{-n})$. Since

$$\lim_{n\to\infty} \int_0^1 X_n(\tau)\, dY_n(\tau) = \int_0^1 X(\tau) \bullet dY(t),$$

it suffices to show that $D_n \longrightarrow 0$.

Obviously

$$4^n \int_{I_{m,n}} (\tau - m2^{-n})\Delta^X_{m,n}\, d\tau = \frac{1}{2}\Delta^X_{m,n},$$

and use integration by parts to see that

$$2^n \int_{I_{m,n}} \big(X(\tau) - X_{m,n}\big)\, d\tau = \Delta^X_{m,n} - 2^n \int_{I_{m,n}} (\tau - m2^{-n})\, dX(\tau).$$

Now set

$$Z_n(t) = \int_0^t \big(\tfrac{1}{2} - 2^n(\tau - \lfloor \tau \rfloor_n)\big)\, dX(\tau),$$

and, proceeding as in the proof of Corollary 4.3.4, write D_n as

$$\sum_{m<2^n} \Delta^{Z_n}_{m,n}\Delta^{Y}_{m,n} = \langle Z_n, Y\rangle(1) + \int_0^1 \big(Z_n(\tau) - Z_n(\lfloor\tau\rfloor_n)\big)\,dY(\tau)$$
$$+ \int_0^1 \big(Y(\tau) - Y(\lfloor\tau\rfloor_n)\big)\,dZ_n(\tau).$$

Show that the stochastic integrals on the right tend to 0 as $n \to \infty$. To handle the term $\langle Z_n, Y\rangle(1)$, first assume that $t \rightsquigarrow c(t)$ is continuous, and observe that

$$\langle Z_n, Y\rangle(1) = \frac{1}{2}\int_0^1 c(\tau)\,d\tau - 2^n \int_0^1 (\tau - \lfloor\tau\rfloor_n)c(\tau)\,d\tau$$

and that

$$\begin{aligned}
2^n &\int_0^1 (\tau - \lfloor\tau\rfloor_n)c(\tau)\,d\tau \\
&= \frac{1}{2}2^{-n}\sum_{m<2^n} c(m2^{-n}) + 2^n \int_0^1 (\tau - \lfloor\tau\rfloor_n)\big(c(\tau) - c(\lfloor\tau\rfloor_n)\big)\,d\tau \\
&\longrightarrow \frac{1}{2}\int_0^1 c(\tau)\,d\tau.
\end{aligned}$$

Hence $D_n \longrightarrow 0$ when c is continuous. Finally, when c isn't continuous, choose continuous c_k's so that

$$\lim_{k\to\infty}\int_0^1 |c_k(\tau) - c(\tau)|\,d\tau = 0,$$

and use the fact that

$$2^n \int_0^1 (\tau - \lfloor\tau\rfloor_n)|c_k(\tau) - c(\tau)|\,d\tau \le \int_0^1 |c_k(\tau) - c(\tau)|\,d\tau.$$

Chapter 5
Addenda

In this concluding chapter we will take up five topics that are somewhat tangential but nonetheless closely related to those in the preceding chapters.

5.1 The martingale problem

Given an operator L of the form in (1.2.5), in Chapter 1 we constructed solutions to Kolmogorov's forward equation (1.2.6), and in Chapter 2 we constructed Markov processes whose transition probability functions satisfy (1.2.3). However, only when the coefficients of L are constant did we prove that these quantities are uniquely determined by L. For instance, the first step in Itô's procedure is to write the diffusion coefficient matrix a as $\sigma\sigma^\top$, and there are many ways in which that can be done. On the other hand, we never showed that the distribution of the solution to (3.0.2) does not depend on the choice of σ or, more generally, on whatever construction procedure one adopts.

In this section we will begin by addressing the preceding uniqueness question. We will then discuss some of the benefits of answering it.

5.1.1 *Formulation*

In order to address the uniqueness question raised above, it is necessary to first formulate the question precisely. Thus, let L be given by (1.2.5), and assume that the coefficients a and b are locally bounded, Borel measurable functions. Given $\mathbf{x} \in \mathbb{R}^N$, we will say that $\mathbb{P}_{\mathbf{x}}$ solves the **martingale problem** for L starting at $\mathbf{x}$ if $\mathbb{P}_{\mathbf{x}}$ is a Borel measure on $\mathcal{P}(\mathbb{R}^N)$, $\mathbb{P}_{\mathbf{x}}\big(\psi(0) = \mathbf{x}\big) = 1$, and

D. W. Stroock, *Elements of Stochastic Calculus and Analysis*,
CRM Short Courses, https://doi.org/10.1007/978-3-319-77038-3_5

$$\left(\varphi(\psi(t)) - \int_0^t L\varphi(\tau)\,d\tau, \mathcal{B}_t, \mathbb{P}_{\mathbf{x}}\right)$$

is a martingale for every $\varphi \in C_c^\infty(\mathbb{R}^N;\mathbb{C})$.

If for each $\mathbf{x} \in \mathbb{R}^N$ there exists precisely one solution, then we will say that the martingale problem for L is **well-posed**.

5.1.2 *Some elementary properties*

Let $\mathbb{P}_{\mathbf{x}}$ be a solution to the martingale problem for L starting at $\mathbf{x}$, and set

$$V(t) = \mathbf{x} + \int_0^t b(\psi(t))\,d\tau \quad \text{and} \quad M(t) = \psi(t) - V(t). \tag{5.1.1}$$

Next, choose $\eta \in C_c^\infty\big(\mathbb{R}^N;[0,1]\big)$ so that $\eta = 1$ on $\overline{B(\mathbf{0};1)}$ and $\eta = 0$ off $B(\mathbf{0},2)$, and define $\eta_R(\mathbf{y}) = \eta(R^{-1}\mathbf{y})$ for $R > 0$. Then, for any $R > 0$

$$\Big(\eta_R\big(\psi(t)\big)\psi(t) \\ - \int_0^t \Big((\eta_R b)\big(\psi(\tau)\big) + L\eta_R\big(\psi(\tau)\big)\psi(\tau) + \big(a\nabla\eta_R\big)\big(\psi(\tau)\big)\Big)\,d\tau, \mathcal{B}_t, \mathbb{P}_{\mathbf{x}}\Big)$$

is an $\mathbb{R}^N$-valued martingale. Therefore, if $\zeta_R = \inf\{t \geq 0 : |\psi(t)| \geq R\}$, then $\big(M(t\wedge\zeta_R), \mathcal{B}_t, \mathbb{P}_{\mathbf{x}}\big)$ is an $\mathbb{R}^N$-valued local martingale, and so $\big(\psi(t), \mathcal{B}_t, \mathbb{P}_{\mathbf{x}}\big)$ is an $\mathbb{R}^N$-valued semi-martingale. Hence, by Corollary 4.3.3, for any $\varphi \in C^{1,2}(\mathbb{R}\times\mathbb{R}^N;\mathbb{C})$,

$$\left(\varphi(t,\psi(t)) - \int_0^t (\partial_\tau + L)\varphi(\tau,\psi(\tau))\,d\tau, \mathcal{B}_t, \mathbb{P}_{\mathbf{x}}\right)$$

is a local martingale. In particular, this shows that $\langle \psi_i, \psi_j\rangle(dt) = a_{ij}\big(\psi(t)\big)\,dt$.

Theorem 5.1.1. *Assume that*

$$\Lambda := \sup_{\mathbf{y}\in\mathbb{R}^N} \frac{\operatorname{Trace}\big(a(\mathbf{y})\big) + 2\big(\mathbf{y}, b(\mathbf{y})\big)^+_{\mathbb{R}^N}}{1+|\mathbf{y}|^2} < \infty, \tag{5.1.2}$$

and let $\mathbb{P}_{\mathbf{x}}$ be a solution to the martingale problem for L starting at $\mathbf{x}$. Then, for each $p \in [2,\infty)$ and $T > 0$ there is a $C_p(T) < \infty$ such that

$$\mathbb{E}^{\mathbb{P}_{\mathbf{x}}}\big[\|\psi(\,\cdot\,)\|^p_{[0,T]}\big] \leq C_p(T)(1+|\mathbf{x}|^p).$$

Hence, if $\varphi \in C^{1,2}\big((0,T)\times\mathbb{R}^N;\mathbb{C}\big) \cap C\big([0,T]\times\mathbb{R}^N;\mathbb{C}\big)$ and

$$\sup_{(t,\mathbf{y})\in[0,T]\times\mathbb{R}^N} \frac{|\varphi(t,\mathbf{y})| + |(\partial_t + L)\varphi(t,\mathbf{y})|}{1+|\mathbf{y}|^k} < \infty$$

for some $k \in \mathbb{N}$, then

$$\left(\varphi(t\wedge T, \psi(t\wedge T)) - \int_0^{t\wedge T} (\partial_\tau + L)\varphi(\tau,\psi(\tau))\,d\tau, \mathcal{B}_t, \mathbb{P}_{\mathbf{x}}\right) \tag{5.1.3}$$

is a martingale.

Proof. Take $M(t)$ as in (5.1.1). Then $\big(M(t), \mathcal{B}_t, \mathbb{P}_{\mathbf{x}}\big)$ is a continuous local $\mathbb{R}^N$-valued martingale and, by Theorem 4.3.1,

$$\begin{aligned}
&\big(1+|\psi(t)|^2\big)^{\frac12}\\
&= \big(1+|\mathbf{x}|^2\big)^{\frac12} + \int_0^t \big(1+|\psi(t)|^2\big)^{-\frac12}\big(\psi(\tau), dM(\tau)\big)_{\mathbb{R}^N}\\
&\quad + \frac12\int_0^t \left(\frac{\mathrm{Trace}\big(a(\psi(\tau))\big) + 2\big(\psi(\tau), b(\psi(\tau))\big)_{\mathbb{R}^N}}{(1+|\psi(\tau)|^2)^{\frac12}} - \frac{\big(\psi(\tau), a(\psi(\tau))\psi(\tau)\big)_{\mathbb{R}^N}}{(1+|\psi(\tau)|^2)^{\frac32}}\right) d\tau\\
&\le \big(1+|\mathbf{x}|^2\big)^{\frac12} + \left|\int_0^t \big(1+|\psi(\tau)|^2\big)^{-\frac12}\big(\psi(\tau), dM(\tau)\big)_{\mathbb{R}^N}\right|\\
&\qquad + \frac12\int_0^t \frac{\mathrm{Trace}\big(a(\psi(\tau))\big) + 2\big(\psi(\tau), b(\psi(\tau))\big)^+_{\mathbb{R}^N}}{(1+|\psi(\tau)|^2)^{\frac12}}\,d\tau.
\end{aligned}$$

Hence, if $\zeta_R = \inf\{t \ge 0 : |\psi(t)| \ge R\}$, then, by (4.3.1),

$$\begin{aligned}
&3^{1-p}\mathbb{E}^{\mathbb{P}_{\mathbf{x}}}\Big[\big(1+\|\psi\|^2_{[0,t\wedge\zeta_R]}\big)^{\frac p2}\Big]\\
&\le \big(1+|\mathbf{x}|^2\big)^{\frac p2}\\
&\quad + K_p t^{\frac p2 -1}\int_0^t \mathbb{E}^{\mathbb{P}_{\mathbf{x}}}\left[\left(\frac{\big(\psi(\tau\wedge\zeta_R), a(\psi(\tau\wedge\zeta_R))\psi(\tau\wedge\zeta_R)\big)_{\mathbb{R}^N}}{1+|\psi(\tau\wedge\zeta_R)|^2}\right)^{\frac p2}\right] d\tau\\
&\quad + t^{p-1}\int_0^t \mathbb{E}^{\mathbb{P}_{\mathbf{x}}}\left(\frac{\mathrm{Trace}\big(a(\psi(\tau\wedge\zeta_R))\big) + 2\big(\psi(\tau\wedge\zeta_R), b(\psi(\tau\wedge\zeta_R))\big)^+_{\mathbb{R}^N}}{(1+|\psi(\tau\wedge\zeta_R)|^2)^{\frac12}}\right)^p d\tau\\
&\le \big(1+|\mathbf{x}|^2\big)^{\frac p2} + K_p\big((\Lambda t)^{\frac p2} + (\Lambda t)^p\big)\int_0^t \mathbb{E}^{\mathbb{P}_{\mathbf{x}}}\big[\big(1+|\psi(\tau\wedge\zeta_R)|^2\big)^{\frac p2}\big].
\end{aligned}$$

Starting from this and applying Lemma 1.2.4, one gets the estimate in the first assertion after $R \to \infty$.

Given the first assertion, one sees that if $\varphi \in C^{1,2}(\mathbb{R}\times\mathbb{R}^N;\mathbb{C})$ and

$$(*)\qquad \sup_{(t,\mathbf{y})\in[0,\infty)\times\mathbb{R}^N} \frac{|\varphi(t,\mathbf{y})| + |(\partial_t + L)\varphi(t,\mathbf{y})|}{1+|\mathbf{y}|^k} < \infty,$$

then the local martingale

$$\left(\varphi\big(t,\psi(t)\big) - \int_0^t (\partial_\tau + L)\varphi\big(\tau,\psi(\tau)\big)\,d\tau, \mathcal{B}_t, \mathbb{P}_{\mathbf{x}}\right)$$

is a martingale. Finally, let $\varphi \in C^{1,2}\big((0,T) \times \mathbb{R}^N;\mathbb{C}\big) \cap C\big([0,T] \times \mathbb{R}^N;\mathbb{C}\big)$ satisfying (5.1.3) be given, and, for $\theta \in \big(0, \frac{T}{2}\big)$, choose $\varphi_\theta \in C^{1,2}\big(\mathbb{R} \times \mathbb{R}^N;\mathbb{C}\big)$ satisfying $(*)$ so that $\varphi_\theta = \varphi$ on $[\theta, T-\theta] \times \mathbb{R}^N$. Then the preceding applied to φ_θ shows that

$$\mathbb{E}^{\mathbb{P}_{\mathbf{x}}}\Big[\varphi\big(t,\psi(t)\big) - \varphi\big(s,\psi(s)\big)\,\Big|\,\mathcal{B}_s\Big] = \mathbb{E}^{\mathbb{P}_{\mathbf{x}}}\left[\int_s^t (\partial_\tau + L)\varphi\big(\tau,\psi(\tau)\big)\,d\tau\,\middle|\,\mathcal{B}_s\right]$$

for $\theta \le s < t \le T-\theta$, and so the required conclusion follows after $\theta \searrow 0$. □

5.1.3 *Some uniqueness criteria*

As was pointed out in §2.2.1, the property in (2.2.6) is a pathspace version of Kolmogorov's forward equation and, as such, will be shared by any Markovian measure on $\mathcal{P}(\mathbb{R}^N)$ associated via (1.2.3) with the operator L in (1.2.5). In particular, as we showed, when σ and b are uniformly Lipschitz continuous, the distribution of the solution to (3.0.2) will satisfy (2.2.6), and therefore the distribution of $X(\,\cdot\,,\mathbf{x})$ will solve the martingale problem for L starting at $\mathbf{x}$. More generally, if a and b are continuous and satisfy (5.1.2), then one can use a pathspace analog of the argument in §1.2.1 to show that there always exists a solution to the associated martingale problem starting at each $\mathbf{x} \in \mathbb{R}^N$. Thus finding solutions poses no problem. On the other hand, uniqueness does. This is not surprising since the easier it is to find solutions to a problem, the harder it should be to prove that there is only one of them. Nonetheless, there are some general uniqueness results. For instance, if $N \in \{1,2\}$, the martingale problem is well-posed for any bounded, measurable a and b as long as $a \ge \epsilon\mathbf{I}$ for some $\epsilon > 0$, and, when $N \ge 3$, it is well-posed if a and b are bounded, a is continuous, b is measurable, and $a \ge \epsilon\mathbf{I}$ for some $\epsilon > 0$. However, when a isn't continuous, N. Nadirashvili in [12] showed that uniqueness will fail in general when $N \ge 3$ even though $a \ge \epsilon\mathbf{I}$. The situation when a can degenerate is much less satisfactory and requires more regularity of the coefficients.

Here I will present two approaches to proving uniqueness. The first is based on existence results for Kolmogorov's backward and related equations. The second approach is more probabilistic and relies on Theorem 4.4.1.

There is a general principle in functional analysis about the relationship between uniqueness of solutions to a problem and existence of solutions to a dual problem. The equations (1.2.6) and (1.2.4) are an example of such a duality. To see this, suppose that $t \rightsquigarrow \mu_t$ satisfies (1.2.6) and that $(t,\mathbf{y}) \rightsquigarrow$

$u(t,\mathbf{y})$ is a solution to (1.2.4). Then

$$\frac{d}{dt}\langle u(T-t,\,\cdot\,),\mu_t\rangle = 0 \quad \text{for}\, t\in[0,T],$$

and so $\langle u(T,\,\cdot\,),\mu_0\rangle = \langle\varphi,\mu_T\rangle$. Therefore, if one knows μ_0 and one can solve (1.2.4) for a sufficiently rich class of φ's, then one knows μ_T. Since the martingale problem is a pathspace version of (1.2.6), one should suspect that uniqueness of solutions to it also follows from existence of solutions to (1.2.4), and that is what we are about to show.

Theorem 5.1.2. *Assume that for each $\varphi\in C_c^\infty(\mathbb{R}^N;\mathbb{R})$ there is a solution $u_\varphi\in C^{1,2}\big((0,\infty)\times\mathbb{R}^N;\mathbb{R}\big)\cap C_b\big([0,\infty)\times\mathbb{R}^N;\mathbb{R}\big)$ to*

$$\partial_t u_\varphi = Lu_\varphi \;\text{ in } [0,\infty)\times\mathbb{R}^N \quad\text{and}\quad \lim_{t\searrow 0} u_\varphi(t,\,\cdot\,)=\varphi. \tag{5.1.4}$$

Then, for each $\mathbf{x}\in\mathbb{R}^N$ there is at most one solution to the martingale problem for L starting at $\mathbf{x}$.

Proof. Take $v(t,\,\cdot\,) = u_\varphi(T-t,\,\cdot\,)$, and apply Theorem 5.1.1 to conclude that

$$\Big(u_\varphi(T-t\wedge T,\psi(t\wedge T)),\mathcal{F}_t,\mathbb{P}_\mathbf{x}\Big)$$

is a bounded local martingale and is therefore a martingale for any solution $\mathbb{P}_\mathbf{x}$ to the martingale problem for L starting at $\mathbf{x}$. In particular,

$$\mathbb{E}^{\mathbb{P}_\mathbf{x}}\big[\varphi(\psi(T))\,\big|\,\mathcal{B}_t\big] = u_\varphi\big(T-t,\psi(t)\big)\ (\text{a.s.},\mathbb{P}_\mathbf{x})\quad\text{for } t\in[0,T].$$

Now suppose that $\mathbb{P}'_\mathbf{x}$ is a second solution. Then,

$$\mathbb{E}^{\mathbb{P}_\mathbf{x}}\big[\varphi(\psi(T))\big] = u_\varphi(T,\mathbf{x}) = \mathbb{E}^{\mathbb{P}'_\mathbf{x}}\big[\varphi(\psi(T))\big].$$

Next, assume that

$$(*)\qquad \mathbb{E}^{\mathbb{P}_\mathbf{x}}\big[\varphi_1(\psi(t_1))\cdots\varphi_n(\psi(t_n))\big] = \mathbb{E}^{\mathbb{P}'_\mathbf{x}}\big[\varphi_1(\psi(t_1))\cdots\varphi_n(\psi(t_n))\big]$$

for all $0\le t_1<\cdots<t_n$ and $\varphi_1,\dots,\varphi_n\in C_c^\infty(\mathbb{R}^N;\mathbb{R})$ and therefore for all bounded, Borel measurable φ_j's. Then if $0\le t_1<\cdots<t_{n+1}$ and $\varphi_1,\dots,\varphi_{n+1}\in C_c^\infty(\mathbb{R}^N;\mathbb{R})$,

$$\begin{aligned}
&\mathbb{E}^{\mathbb{P}_\mathbf{x}}\big[\varphi_1(\psi(t_1))\cdots\varphi_{n+1}(\psi(t_{n+1}))\big]\\
&= \mathbb{E}^{\mathbb{P}_\mathbf{x}}\big[\varphi_1(\psi(t_1))\cdots\varphi_n(\psi(t_n))u_{\varphi_{n+1}}(t_{n+1}-t_n,\psi(t_n))\big]\\
&= \mathbb{E}^{\mathbb{P}'_\mathbf{x}}\big[\varphi_1(\psi(t_1))\cdots\varphi_n(\psi(t_n))u_{\varphi_{n+1}}(t_{n+1}-t_n,\psi(t_n))\big]\\
&= \mathbb{E}^{\mathbb{P}'_\mathbf{x}}\big[\varphi_1(\psi(t_1))\cdots\varphi_{n+1}(\psi(t_{n+1}))\big].
\end{aligned}$$

Hence, by induction, $(*)$ holds for all $n\ge 1$, and so $\mathbb{P}_\mathbf{x}=\mathbb{P}'_\mathbf{x}$. □

A closely related uniqueness criterion is based on the equation $\lambda u - Lu = \varphi$ for $\lambda > 0$. Although solutions to this equation can be generated by taking the Laplace transform of solutions to (5.1.4), there are technical reasons why it is often easier to construct its solutions directly.

Theorem 5.1.3. *Assume that for each $\varphi \in C_c^\infty(\mathbb{R}^N;\mathbb{R})$ and $\lambda > 0$ there is a bounded $u_{\lambda,\varphi} \in C^2(\mathbb{R}^N;\mathbb{R})$ that satisfies*

$$\lambda u_{\lambda,\varphi} - Lu_{\lambda,\varphi} = \varphi. \tag{5.1.5}$$

Then, for each $\mathbf{x} \in \mathbb{R}^N$, there is at most one solution to the martingale problem for L starting at $\mathbf{x}$.

Proof. By taking $v(t,\,\cdot\,) = e^{-\lambda t}u_{\lambda,\varphi}(\,\cdot\,)$, one sees that

$$\left(e^{-\lambda t}u_\varphi\big(\psi(t)\big) + \int_0^t e^{-\lambda\tau}\varphi\big(\psi(\tau)\big)\,d\tau, \mathcal{B}_t, \mathbb{P}_{\mathbf{x}}\right)$$

is a martingale for any solution $\mathbb{P}_{\mathbf{x}}$ to the martingale problem for L starting at $\mathbf{x}$. Hence, for $0 \le t < T$,

$$\mathbb{E}^{\mathbb{P}_{\mathbf{x}}}\left[\int_t^T e^{-\lambda\tau}\varphi\big(\psi(\tau)\big)\,d\tau \,\middle|\, \mathcal{B}_t\right] = e^{-\lambda t}u_{\lambda,\varphi}\big(\psi(t)\big) - e^{-\lambda T}\mathbb{E}^{\mathbb{P}_{\mathbf{x}}}\left[u_{\lambda,\varphi}\big(\psi(T)\big) \,\middle|\, \mathcal{B}_t\right],$$

which, when $T \to \infty$, leads to

$$\int_0^\infty e^{-\lambda\tau}\mathbb{E}^{\mathbb{P}_{\mathbf{x}}}\left[\varphi\big(\psi(t+\tau)\big)\,d\tau \,\middle|\, \mathcal{B}_t\right]d\tau = u_{\lambda,\varphi}\big(\psi(t)\big)$$

$\mathbb{P}_{\mathbf{x}}$-almost surely. Now suppose that $\mathbb{P}_{\mathbf{x}}$ and $\mathbb{P}'_{\mathbf{x}}$ are two solutions, and let $\Gamma \in \mathcal{B}_t$. Then, from the preceding, we know that

$$\int_0^\infty e^{-\lambda\tau}\mathbb{E}^{\mathbb{P}_{\mathbf{x}}}\left[\varphi\big(\psi(t+\tau)\big),\Gamma\right]d\tau = \int_0^\infty e^{-\lambda\tau}\mathbb{E}^{\mathbb{P}'_{\mathbf{x}}}\left[\varphi\big(\psi(t+\tau)\big),\Gamma\right]d\tau$$

for all $\lambda > 0$, and therefore, by the uniqueness of the Laplace transform,

$$\mathbb{E}^{\mathbb{P}_{\mathbf{x}}}\left[\varphi\big(\psi(t+\tau)\big),\Gamma\right] = \mathbb{E}^{\mathbb{P}'_{\mathbf{x}}}\left[\varphi\big(\psi(t+\tau)\big),\Gamma\right].$$

Starting from this and arguing as in the proof of Theorem 5.1.2, one concludes that $\mathbb{P}_{\mathbf{x}} = \mathbb{P}'_{\mathbf{x}}$. □

To apply Theorem 5.1.2 or 5.1.3, one needs to know that either (5.1.4) or (5.1.5) has solutions. In §3.4.2 we showed that if $a = \sigma\sigma^\top$ and both σ and b have three bounded derivatives, then (5.1.4) has a solution for every $\varphi \in C_b^2(\mathbb{R}^N;\mathbb{R})$, and therefore the associated martingale problem is well-posed. However, there are two reasons why this is a rather weak result. One is that, as we will show in Theorem 5.1.4, the martingale problem is well-posed if $\sigma = a^{\frac{1}{2}}$ and b are uniformly Lipschitz continuous. The second reason is that, when $a \geq \epsilon\mathbf{I}$ for some $\epsilon > 0$ and both a and b are Hölder continuous, then standard theorems (cf. [4] and [5]) from the theory of partial differential equations guarantee that both (5.1.4) and (5.1.5) have solutions.

The basic structure of the proof in [22] that the martingale problem is well-posed for any bounded a which is continuous and satisfies $a \geq \epsilon\mathbf{I}$ is the same as that of Theorem 5.1.2. What makes it more difficult is that when a is only continuous, the solutions to (5.1.4) are not classical solutions. Instead, their second derivatives exist only in the sense of Sobolev, and so, before they can be used to prove uniqueness for a martingale problem, one has to show that all solutions to that martingale problem satisfy *a priori* estimates that justify the use of these non-classical solutions to (5.1.4).

In that it does not involve partial differential equations and is purely probabilistic, the following represents a quite different approach to proving uniqueness.

Theorem 5.1.4. *If $\sigma = a^{\frac{1}{2}}$ and b are uniformly Lipschitz continuous, then the martingale problem for L is well-posed.*

Proof. Write $\psi = M + V$ as in (5.1.1). Because $\big(M(t), \mathcal{B}_t, \mathbb{P}_{\mathbf{x}}\big)$ is a continuous local $\mathbb{R}^N$-valued martingale, Theorem 4.4.1 says that there is a Brownian motion $\big(\widetilde{B}(t), \widetilde{\mathcal{F}}_t, \widetilde{\mathbb{P}}\big)$ on a probability space $(\widetilde{\Omega}, \widetilde{\mathcal{F}}, \widetilde{\mathbb{P}})$ and a measurable map $F\colon \widetilde{\Omega} \longrightarrow \mathcal{P}(\mathbb{R}^N)$ such that $\mathbb{P}_{\mathbf{x}} = F_*\widetilde{\mathbb{P}}$ and

$$(\psi \circ F)(t) = \mathbf{x} + \int_0^t \sigma\big((\psi \circ F)(\tau)\big)\, d\widetilde{B}(\tau) + \int_0^t b\big((\psi \circ F)(\tau)\big)\, d\tau,$$

where $\sigma = a^{\frac{1}{2}}$. Define $\widetilde{X}_n(0, \mathbf{x}) = \mathbf{x}$ and

$$\begin{aligned}\widetilde{X}_n(t, \mathbf{x}) =& \widetilde{X}_n(\lfloor t \rfloor_n, \mathbf{x}) + \sigma\big(\widetilde{X}_n(\lfloor t \rfloor, \mathbf{x})\big(\widetilde{B}(t) - \widetilde{B}(\lfloor t \rfloor_n)\big) \\ &+ b\big(\widetilde{X}_n(\lfloor t \rfloor_n, \mathbf{x}\big)(t - \lfloor t \rfloor_n)\end{aligned}$$

for $t > 0$. Then, for each $n \geq 0$, $\widetilde{X}_n(\,\cdot\,, \mathbf{x})$ has the same distribution under $\widetilde{\mathbb{P}}$ as the $X_n(\,\cdot\,, \mathbf{x})$ in (2.2.1) has under $\mathcal{W}$. Furthermore, $\widetilde{\mathbb{P}}$-almost surely, $\widetilde{X}_n(\,\cdot\,, \mathbf{x})$ converges to $(\psi \circ F)(\,\cdot\,)$ uniformly on compacts, and so $(\psi \circ F)(\,\cdot\,)$ has the same distribution under $\widetilde{\mathbb{P}}$ as $X(\,\cdot\,, \mathbf{x})$ has under $\mathcal{W}$. Since $\mathbb{P}_{\mathbf{x}}$ is the distribution of $(\psi \circ F)(\,\cdot\,)$ under $\widetilde{\mathbb{P}}$, this proves that $\mathbb{P}_{\mathbf{x}}$ is uniquely determined. □

5.2 Exponential semi-martingales

This section describes two applications of exponential semi-martingales as integrating factors.

5.2.1 *The Feynman–Kac formula*

Let $\mathbb{P}_{\mathbf{x}}$ be a solution starting at $\mathbf{x}$ to the martingale problem for an L whose coefficients satisfy (5.1.2), and write $\psi = M + V$ as in (5.1.1). Also, let $H\colon \mathbb{R}^N \longrightarrow \mathbb{C}$ be a locally bounded, Borel measurable function, and set

$$E_H(t) = \exp\left(\int_0^t H(\psi(\tau))\,d\tau\right).$$

Then, for any $\varphi \in C^{1,2}\big((0,T)\times\mathbb{R}^N;\mathbb{C}\big)\cap C\big([0,T]\times\mathbb{R}^N;\mathbb{C}\big)$

$$\begin{aligned}
&E_H(t\wedge T)\varphi(t\wedge T,\psi(t\wedge T))\\
&\quad = \psi(0,\mathbf{x}) + \int_0^{t\wedge T} E_H(\tau)\Big(\nabla\varphi(\tau,\psi(\tau)),dM(\tau)\Big)_{\mathbb{R}^N}\\
&\qquad + \int_0^{t\wedge T} E(\tau)\Big(H(\tau)\varphi(\tau,\psi(\tau)) + (\partial_\tau + L)\varphi(\tau,\psi(\tau))\Big)\,d\tau,
\end{aligned}$$

and so

$$\left(\begin{aligned}&E(t\wedge T)\varphi(t\wedge T,\psi(t\wedge T))\\ &\quad -\int_0^{t\wedge T} E(\tau)\Big(H(\tau)\varphi(\tau,\psi(\tau)) + (\partial_\tau + L)\varphi(\tau,\psi(\tau))\Big)\,d\tau, \mathcal{B}_t, \mathbb{P}_{\mathbf{x}}\end{aligned}\right)$$

is a local martingale. Now assume that the real part of H is bounded above. If $\varphi \in C^{1,2}\big((0,T)\times\mathbb{R}^N;\mathbb{C}\big)\cap C\big([0,T]\times\mathbb{R}^N;\mathbb{C}\big)$ and

$$\sup_{(t,\mathbf{y})\in[0,T]\times\mathbb{R}^N} \frac{|\varphi(t,\mathbf{y})| + |H(\mathbf{y})\varphi(t,\mathbf{y}) + (\partial_t + L)\varphi(t,\mathbf{y})|}{1+|\mathbf{y}|^k} < \infty$$

for some $k \in \mathbb{N}$, then one can proceed as in the proof of Theorem 5.1.1 to check that this local martingale is a martingale.

The preceding considerations lead to a version of a famous formula which is known as the **Feynman–Kac formula**. Feynman introduced the formula to give a pathspace representation of solutions to Schrödinger's equation. Unfortunately, although his formula transformed the way physicists think about quantum mechanics, from a mathematical standpoint, even today his formula is problematic. On the other hand, M. Kac realized that Feynman's formula could be put on a firm mathematical foundation if, instead of using it for solutions to Schrödinger's equations, one applied it to solutions of parabolic equations.

Theorem 5.2.1. *Assume that a and b satisfy* (5.1.2), *and let $\mathbb{P}_{\mathbf{x}}$ be a solution to the martingale problem for L starting at $\mathbf{x}$. If $u \in C^{1,2}\big((0,T)\times\mathbb{R}^N;\mathbb{C}\big)\cap C\big([0,T]\times\mathbb{R}^N;\,\mathbb{C}\big)$ satisfies*

$$\sup_{(t,\mathbf{y})\in[0,T]\times\mathbb{R}^N} \frac{|u(t,\mathbf{y})|}{1+|\mathbf{y}|^k} < \infty$$

for some $k \in \mathbb{N}$ *and is a solution to* $\partial_t u = Lu + Hu$ *for some* $H \in C(\mathbb{R}^N; \mathbb{C})$ *whose real part is bounded above, then*

$$u(T,\mathbf{x}) = \mathbb{E}^{\mathbb{P}_\mathbf{x}}\big[E_H(T)u\big(0,\psi(T)\big)\big]. \tag{5.2.1}$$

Proof. In view of the preceding, we know that

$$\big(E_H(t\wedge T)u\big((T-t)^+,\psi(t\wedge T)\big),\mathcal{F}_t,\mathbb{P}_\mathbf{x}\big)$$

is a martingale, and therefore the expected value of $E_H(T)u\big(0,\psi(T)\big)$ is the same as that of $E_H(0)u\big(T,\psi(0)\big)$. □

Kac made many imaginative applications of Theorem 5.2.1, one of which is the following. Consider the equation

$$\partial_t u = \tfrac{1}{2}\Big(\partial_x^2 u - x^2 u\Big) \text{ on } (0,\infty)\times\mathbb{R} \quad \text{with } \lim_{t\searrow 0} u(t,\,\cdot\,) = \varphi. \tag{5.2.2}$$

This is one of the few heat equations whose fundamental solution has an explicit expression. Observe that if v is a solution to the Ornstein–Uhlenbeck equation

$$\partial_t v(t,x) = \tfrac{1}{2}\partial_x^2 v(t,x) - x\partial_x v(t,x) \quad \text{with } \lim_{t\searrow 0} v(t,\mathbf{x}) = e^{\frac{x^2}{2}}\varphi(x),$$

then $u(t,x) = e^{-\frac{t+x^2}{2}}v(t,x)$ solves (5.2.2). Hence (cf. Exercise 1.1), if $\varphi \in C(\mathbb{R};\mathbb{R})$ has at most exponential growth and

$$\begin{aligned} h(t,x,y) &= e^{-\frac{t+x^2}{2}} g\left(\frac{1-e^{-2t}}{2}, y - e^{-t}x\right) e^{\frac{y^2}{2}} \\ &= \frac{1}{\sqrt{2\pi\sinh t}} \exp\left(-\frac{x^2}{2\tanh t} + \frac{xy}{\sinh t} - \frac{y^2}{2\tanh t}\right), \end{aligned} \tag{5.2.3}$$

then

$$u_\varphi(t,\mathbf{x}) = \int_\mathbb{R} h(t,x,y)\varphi(y)\,dy,$$

is a solution to 5.2.2. In particular, when $\varphi = \mathbf{1}$, (5.2.1) says that

$$\mathbb{E}^{\mathcal{W}}\left[\exp\left(-\frac{1}{2}\int_0^t \big(x+w(\tau)\big)^2\,d\tau\right)\right] = \int_\mathbb{R} h(t,x,y)\,dy = \frac{1}{\sqrt{\cosh t}} e^{-\frac{\tanh t}{2}x^2}.$$

Further, because, for each $\alpha > 0$, $t \rightsquigarrow w(\alpha t)$ has the same distribution under $\mathcal{W}$ as $t \rightsquigarrow \alpha^{\frac{1}{2}}w(t)$, one knows that

$$\exp\left(-\frac{\alpha}{2}\int_0^t \left(x+w(\tau)\right)^2 d\tau\right) \quad \text{and} \quad \exp\left(-\frac{1}{2}\int_0^{\alpha^{\frac{1}{2}}t} \left(\alpha^{\frac{1}{4}}x+w(\tau)\right)^2 d\tau\right)$$

have the same distribution. Hence,

$$\begin{aligned} &\mathbb{E}^{\mathcal{W}}\left[\exp\left(-\frac{\alpha}{2}\int_0^t \left(x+w(\tau)\right)^2 d\tau\right)\right] \\ &\qquad = \left(\cosh(\alpha^{\frac{1}{2}}t)\right)^{-\frac{1}{2}} \exp\left(-\frac{\alpha^{\frac{1}{2}}\tanh(\alpha^{\frac{1}{2}}t)}{2}x^2\right). \end{aligned} \tag{5.2.4}$$

This computation can be used to give further quantitative evidence of the fuzziness of Brownian paths. Namely, take $t = 1$ and integrate both sides of (5.2.4) with respect to x to obtain

$$\mathbb{E}^{\mathcal{W}}\left[\exp\left(-\frac{\alpha}{2}\left(\int_0^1 w(\tau)^2\,d\tau - \left(\int_0^1 w(\tau)\,d\tau\right)^2\right)\right)\right] = \sqrt{\frac{\alpha^{\frac{1}{2}}}{\sinh\alpha^{\frac{1}{2}}}},$$

and then, again using Brownian scaling,

$$\mathbb{E}^{\mathcal{W}}\left[\exp\left(-\frac{\alpha}{2}\operatorname{var}_{[0,t]}\left(w(\,\cdot\,)\right)\right)\right] = \sqrt{\frac{(\alpha t)^{\frac{1}{2}}}{\sinh(\alpha t)^{\frac{1}{2}}}}$$

$$\text{where } \operatorname{var}_{[0,t]}\left(w(\,\cdot\,)\right) = \frac{1}{t}\int_0^t w(\tau)^2\,d\tau - \left(\frac{1}{t}\int_0^t w(\tau)\,d\tau\right)^2$$

is the variance of the path $w \restriction [0,t]$. From this it follows that

$$\mathcal{W}\left(\operatorname{var}_{[0,t]}\left(w(\,\cdot\,)\right) \le \frac{1}{r}\right) \le e^{\frac{\alpha}{2r}-\frac{\sqrt{\alpha t}}{2}}\sqrt{\frac{2\sqrt{\alpha t}}{1-e^{-2\sqrt{\alpha t}}}}$$

for all $\alpha > 0$, and so, by taking $\alpha = \frac{r^2 t}{4}$, we have that

$$\mathcal{W}\left(\operatorname{var}_{[0,t]}\left(w(\,\cdot\,)\right) \le \frac{1}{r}\right) \le e^{-\frac{rt}{4}}\sqrt{\frac{rt}{1-e^{-rt}}}.$$

In other words, the probability that a Brownian path has small variance is exponentially small.

5.2.2 *The Cameron–Martin–Girsanov formula*

Given an $\mathbb{R}$-valued local martingale $\left(M(t), \mathcal{F}_t, \mathbb{P}\right)$ with $M(0) = 0$, an application of Theorem 4.3.1 shows that

$$E_M(t) = 1 + \int_0^t E_M(\tau)\, dM(\tau) \quad \text{where } E_M(t) := \exp\bigl(M(t) - \tfrac{1}{2}\langle\langle M\rangle\rangle(t)\bigr).$$

In particular, $(E_M(t), \mathcal{F}_t, \mathbb{P})$ is a non-negative, continuous local martingale and $\langle E_M, M'\rangle(dt) = E_M(t)\langle M, M'\rangle(dt)$ for all $M' \in M_{\text{loc}}(\mathbb{P}; \mathbb{R})$. Hence, by Exercise 4.1, $(E_M(t), \mathcal{F}_t, \mathbb{P})$ is always a supermartingale, and it is a martingale if and only if $\mathbb{E}^{\mathbb{P}}\bigl[E_M(t)\bigr] = 1$ for all $t \ge 0$.

The problem of determining when $\mathbb{E}^{\mathbb{P}}\bigl[E_M(t)\bigr] = 1$ is a challenging one, and the following result of A. Novikov gives a frequently cited criterion for when it is.

Theorem 5.2.2. *If*

$$\mathbb{E}^{\mathbb{P}}\bigl[e^{\frac{1}{2}\langle\langle M\rangle\rangle(t)}\bigr] < \infty \quad \textit{for all } t \ge 0, \tag{5.2.5}$$

then $(E_M(t), \mathcal{F}_t, \mathbb{P})$ *is a martingale.*

Proof. Because $\mathbb{E}^{\mathbb{P}}\bigl[\langle\langle M\rangle\rangle(t)\bigr] < \infty$ for all $t \ge 0$, we know that $(M(t), \mathcal{F}_t, \mathbb{P})$ is a martingale and therefore that $(e^{\alpha M(t)}, \mathcal{F}_t, \mathbb{P})$ is a submartingale for all $\alpha \in \mathbb{R}$.

Define $\zeta_k = \inf\{t \ge 0 : |M(t)| \vee \langle\langle M\rangle\rangle(t) \ge k\}$. Because $\mathbb{E}^{\mathbb{P}}\bigl[E_M(t\wedge\zeta_k)\bigr] = 1$, (5.2.5) implies that, for all $k \ge 1$ and $t \ge 0$,

$$\mathbb{E}^{\mathbb{P}}\bigl[e^{\frac{1}{2}M(t\wedge\zeta_k)}\bigr] = \mathbb{E}^{\mathbb{P}}\bigl[E_M(t\wedge\zeta_k)^{\frac{1}{2}} e^{\frac{1}{4}\langle\langle M\rangle\rangle(t\wedge\zeta_k)}\bigr] \le \mathbb{E}^{\mathbb{P}}\bigl[e^{\frac{1}{2}\langle\langle M\rangle\rangle(t)}\bigr]^{\frac{1}{2}},$$

and therefore that

$$\mathbb{E}^{\mathbb{P}}\bigl[e^{\frac{1}{2}M(t)}\bigr] \le \mathbb{E}^{\mathbb{P}}\bigl[e^{\frac{1}{2}\langle\langle M\rangle\rangle(t)}\bigr]^{\frac{1}{2}}. \tag{5.2.6}$$

Next, for $\lambda \in (0,1)$, determine $p_\lambda \in (1,\infty)$ by the equation $2\lambda p_\lambda(p_\lambda - \lambda) = 1 - \lambda^2$, and for $p \in [1, p_\lambda]$, set $\alpha(\lambda, p) = \frac{\lambda p(p-\lambda)}{1-\lambda^2}$. Then

$$E_{\lambda M}(t)^{p^2} = \bigl(e^{\alpha(\lambda,p)M(t)}\bigr)^{1-\lambda^2} E_{pM}(t)^{\lambda^2},$$

and so, by Hölder's inequality, the submartingale property for $e^{\alpha(\lambda,p)M(t)}$, and Doob's stopping time theorem,

$$(*) \qquad \mathbb{E}^{\mathbb{P}}\bigl[E_{\lambda M}(t\wedge\zeta)^{p^2}\bigr] \le \mathbb{E}^{\mathbb{P}}\bigl[e^{\alpha(\lambda,p)M(t)}\bigr]^{1-\lambda^2} \mathbb{E}^{\mathbb{P}}\bigl[E_{pM}(t\wedge\zeta)\bigr]^{\lambda^2}$$

for any stopping time ζ.

Since $\mathbb{E}^{\mathbb{P}}\bigl[E_{pM}(t\wedge\zeta_k)\bigr] = 1$, by taking $p = p_\lambda$, we see that

$$\mathbb{E}^{\mathbb{P}}\bigl[E_{\lambda M}(t\wedge\zeta_k)^{p_\lambda^2}\bigr] \le \mathbb{E}^{\mathbb{P}}\bigl[e^{\frac{1}{2}M(t)}\bigr]^{1-\lambda^2} \le \mathbb{E}^{\mathbb{P}}\bigl[e^{\frac{1}{2}\langle\langle M\rangle\rangle(t)}\bigr]^{\frac{1-\lambda^2}{2}},$$

and so, by Exercise 4.1, $(E_{\lambda M}(t), \mathcal{F}_t, \mathbb{P})$ is martingale for every $\lambda \in (0,1)$. Finally, take $p = 1$ and $\zeta = t$. Then, by $(*)$, Jensen's inequality, and (5.2.6)

$$1 = \mathbb{E}^{\mathbb{P}}\big[E_{\lambda M}(t)\big] \le \mathbb{E}^{\mathbb{P}}\big[e^{\frac{\lambda}{1+\lambda}M(t)}\big]^{1-\lambda^2}\mathbb{E}^{\mathbb{P}}\big[E_M(t)\big]^{\lambda^2}$$
$$\le \mathbb{E}^{\mathbb{P}}\big[e^{\frac{1}{2}M(t)}\big]^{2\lambda(1-\lambda)}\mathbb{E}^{\mathbb{P}}\big[E_M(t)\big]^{\lambda^2} \le \mathbb{E}^{\mathbb{P}}\big[e^{\frac{1}{2}\langle\langle M\rangle\rangle(t)}\big]^{\lambda(1-\lambda)}\mathbb{E}^{\mathbb{P}}\big[E_M(t)\big]^{\lambda^2},$$

which, after $\lambda \nearrow 1$, leads to $\mathbb{E}^{\mathbb{P}}\big[E_M(t)\big] = 1$. □

Novikov's criterion is useful, but it is far from definitive. For example, if $\big(B(t), \mathcal{F}_t, \mathbb{P}\big)$ is an $\mathbb{R}$-valued Brownian motion and $M(t) = \int_0^t B(\tau)\,dB(\tau)$, then $\langle\langle M\rangle\rangle(t) = \int_0^t B(\tau)^2\,d\tau$. Since $M(t) = \frac{B(t)^2-t}{2}$ and $\mathbb{E}^{\mathbb{P}}\big[e^{\frac{1}{2}B(t)^2}\big] = \infty$ for $t \ge 2$, (5.2.6) implies that M does not satisfy (5.2.5). Nonetheless, as we are about to show, $\big(E_M(t), \mathcal{F}_t, \mathbb{P}\big)$ is a martingale. To prove that it is, choose a sequence $\{\varphi_k : k \ge 1\} \subseteq C_{\mathrm{b}}\big(\mathbb{R}; [0,\infty)\big)$ so that $\varphi_k(y) \nearrow e^{\frac{y^2}{2}}$ for all $y \in \mathbb{R}$, and let $h(t,x,y)$ be the kernel in (5.2.3). Then, by (5.2.1),

$$\mathbb{E}^{\mathbb{P}}\big[E_M(t)\big] = \lim_{k\to\infty} e^{-\frac{t}{2}}\mathbb{E}^{\mathbb{P}}\big[e^{-\frac{1}{2}\langle\langle M\rangle\rangle(t)}\varphi_k\big(B(t)\big)\big]$$
$$= \lim_{k\to\infty} e^{-\frac{t}{2}}\int_{\mathbb{R}} h(t,0,y)\varphi_k(y)\,dy = e^{-t}\int_{\mathbb{R}} g\left(\frac{1-e^{-2t}}{2}, y\right)e^{y^2}\,dy = 1.$$

My major interest in these considerations is their application to the martingale problem. In order to explain these applications, we will need the following fact about Borel probability measures on $\mathcal{P}(\mathbb{R}^N)$.

Lemma 5.2.3. *For each $k \ge 1$, let $\mathbb{P}_k$ be a Borel probability measure on $\mathcal{P}(\mathbb{R}^N)$, and assume that $\mathbb{P}_{k+1} \restriction \mathcal{B}_k = \mathbb{P}_k \restriction \mathcal{B}_k$. Then there exists a unique Borel probability measure $\mathbb{P}$ on $\mathcal{P}(\mathbb{R}^N)$ such that $\mathbb{P} \restriction \mathcal{B}_k = \mathbb{P}_k \restriction \mathcal{B}_k$ for all $k \ge 1$. In particular, if $\mathbb{P} \in \mathbf{M}_1\big(\mathcal{P}(\mathbb{R}^N)\big)$ and $\big(E(t), \mathcal{B}_t, \mathbb{P}\big)$ is a non-negative martingale with expectation value* 1, *then there exists a unique $\widetilde{\mathbb{P}} \in \mathbf{M}_1\big(\mathcal{P}(\mathbb{R}^N)\big)$ such that*

$$\widetilde{\mathbb{P}}(\Gamma) = \mathbb{E}^{\mathbb{P}}\big[E(t), \Gamma\big] \quad \textit{for all } t \ge 0 \textit{ and } \Gamma \in \mathcal{B}_t.$$

Proof. This is a relatively straight forward application of the fact (cf. Theorem 9.1.9 in [20]) that a family of Borel probability measures on a Polish space (i.e., a complete, separable metric space) E is relatively compact in the topology of weak convergence if and only if it is tight in the sense that for each $\epsilon > 0$ there is a compact subset $K \subseteq E$ such that $K\complement$ is assigned measure less than ϵ by all members of the family. To apply this result here, one also has to know that $K \subseteq C\big([0,\infty); \mathbb{R}^N\big)$ is compact if and only if $\{\psi \restriction [0,k] : \psi \in K\}$ is compact in $C\big([0,k]; \mathbb{R}^N\big)$ for each $k \ge 1$.

The uniqueness assertion is trivial. To prove the existence, let $\epsilon > 0$ be given, and for each $k \ge 1$ choose a compact subset K_k of $C\big([0,k]; \mathbb{R}^N\big)$ such that

$$\max_{1\le j\le k} \mathbb{P}_j\big(\{\psi : \psi \restriction [0,k] \notin K_k\}\big) \le 2^{-k}\epsilon,$$

in which case

$$\sup_{j\geq 1}\mathbb{P}_j\big(\{\psi:\psi\restriction[0,k]\notin K_k\}\big)\leq 2^{-k}\epsilon.$$

Finally, set

$$K=\{\psi:\psi\restriction[0,k]\in K_k \text{ for all } k\geq 1\}.$$

Then K is a compact subset of $C\big([0,\infty);\mathbb{R}^N\big)$, and

$$\mathbb{P}_j(K\complement)\leq\sum_{k=1}^{\infty}\mathbb{P}_j\big(\{\psi:\psi\restriction[0,k]\notin K_k\}\big)\leq\epsilon \quad \text{for all } j\in\mathbb{Z}^+.$$

Hence, the sequence $\{\mathbb{P}_k : k\geq 1\}$ is relatively compact. Furthermore, any limit $\mathbb{P}$ will have the property that $\mathbb{P}\restriction\mathcal{B}_k=\mathbb{P}_k\restriction\mathcal{B}_k$ for all $k\geq 1$. Therefore the sequence converges and its limit has the required property.

To prove the final assertion, simply define $\mathbb{P}_k(\Gamma)=\mathbb{E}^{\mathbb{P}}\big[E(k),\Gamma\big]$ for $k\geq 1$ and $\Gamma\in\mathcal{B}_E$. □

Theorem 5.2.4. *Let $\mathbb{P}_{\mathbf{x}}$ be a solution to the martingale problem for L starting at $\mathbf{x}$, and let $\beta\colon\mathbb{R}^N\longrightarrow\mathbb{R}^N$ be a Borel measurable function for which $(\beta,a\beta)_{\mathbb{R}^N}$ is bounded. Write $\psi=M+V$ as in (5.1.1), and define*

$$M_\beta(t)=\int_0^t\big(\beta(\psi(\tau)),dM(\tau)\big)_{\mathbb{R}^N} \quad \textit{and} \quad E_\beta(t)=e^{M_\beta(t)-\frac{1}{2}\langle\langle M_\beta\rangle\rangle(t)}.$$

Then $\big(E_\beta(t),\mathcal{B}_t,\mathbb{P}_{\mathbf{x}}\big)$ is a non-negative martingale with expectation value 1. Finally, if $\mathbb{P}^\beta_{\mathbf{x}}\in\mathbf{M}_1\big(\mathcal{P}(\mathbb{R}^N)\big)$ is determined by

$$\mathbb{P}^\beta_{\mathbf{x}}(\Gamma)=\mathbb{E}^{\mathbb{P}_{\mathbf{x}}}\big[E_\beta(t),\Gamma\big] \quad \textit{for } t\geq 0 \textit{ and } \Gamma\in\mathcal{B}_t$$

and L^β is given by

$$L^\beta\varphi=L\varphi+\big(a\beta,\nabla\varphi\big)_{\mathbb{R}^N},$$

then $\mathbb{P}^\beta_{\mathbf{x}}$ solves the martingale problem for L^β starting at $\mathbf{x}$.

Proof. Because

$$\langle\langle M_\beta\rangle\rangle(t)=\int_0^t\big(\beta(\psi(\tau)),a(\psi(\tau))\beta(\psi(\tau))\big)_{\mathbb{R}^N}\,d\tau$$

is bounded, $\big(E_\beta(t),\mathcal{B}_t,\mathbb{P}_{\mathbf{x}}\big)$ is a non-negative martingale with expectation value 1. Hence, by Lemma 5.2.3, $\mathbb{P}^\beta_{\mathbf{x}}$ exists.

Turning to the second part, it is clear that $\mathbb{P}^\beta_{\mathbf{x}}\big(\psi(0)=\mathbf{x}\big)=1$. Next let $\varphi\in C^\infty_{\mathrm{c}}(\mathbb{R}^N;\mathbb{R})$ be given. Because

$$\langle E_\beta,\varphi\circ\psi\rangle(dt)=E_\beta(t)\Big(\beta\big(\psi(t)\big),a\big(\psi(t)\big)\nabla\varphi\big(\psi(t)\big)\Big)_{\mathbb{R}^N}dt,$$

Theorem 4.3.1 says that

$$\begin{aligned}
&E_\beta(t)\varphi(\psi(t)) - \varphi(\mathbf{x}) \\
&\quad = \int_0^t E_\beta(\tau)\Big(\varphi(\psi(\tau))\beta(\psi(\tau)) + a\nabla\varphi(\psi(\tau)), dM(\tau)\Big)_{\mathbb{R}^N} \\
&\qquad + \int_0^t E_\beta(\tau)\Big(L\varphi(\psi(\tau)) + \big(\beta(\psi(\tau)), a\nabla\varphi(\psi(\tau))\big)_{\mathbb{R}^N}\Big)\, d\tau.
\end{aligned}$$

Hence,

$$\left(E_\beta(t)\varphi(\psi(t)) - \int_0^t E_\beta(\tau)L^\beta\varphi(\psi(\tau))\, d\tau, \mathcal{B}_t, \mathbb{P}_{\mathbf{x}}\right)$$

is a local martingale. In fact, because $\mathbb{E}^{\mathbb{P}_{\mathbf{x}}}\big[E_\beta(t)^2\big] < \infty$ for all $t \geq 0$, it is a martingale. Finally, observe that if $0 \leq s < t$ and $\Gamma \in \mathcal{B}_s$, then

$$\begin{aligned}
&\mathbb{E}^{\mathbb{P}^\beta_{\mathbf{x}}}\left[\varphi(\psi(t)) - \varphi(\psi(s)) - \int_s^t L^\beta\varphi(\psi(\tau))\, d\tau, \Gamma\right] \\
&= \mathbb{E}^{\mathbb{P}_{\mathbf{x}}}\left[E_\beta(t)\left(\varphi(\psi(t)) - \varphi(\psi(s)) - \int_s^t L^\beta\varphi(\psi(\tau))\, d\tau\right), \Gamma\right] \\
&= \mathbb{E}^{\mathbb{P}_{\mathbf{x}}}\left[E_\beta(t)\varphi(\psi(t)) - E_\beta(s)\varphi(\psi(s)) - \int_s^t E_\beta(\tau)L^\beta\varphi(\psi(\tau))\, d\tau, \Gamma\right] = 0,
\end{aligned}$$

and therefore

$$\left(\varphi(\psi(t)) - \int_0^t L^\beta\varphi(\psi(\tau))\, d\tau, \mathcal{B}_t, \mathbb{P}^\beta_{\mathbf{x}}\right)$$

is a martingale. □

Because Cameron and Martin gave the first examples of the result in Theorem 5.2.4 and V. Girsanov gave a formulation (cf. Exercise 5.2) that plays an important role in R. Merton's interpretation of the Black–Scholes model, the relationship between $\mathbb{P}^\beta_{\mathbf{x}}$ and $\mathbb{P}_{\mathbf{x}}$ is sometimes called the **Cameron–Martin–Girsanov formula**

Corollary 5.2.5. *Assume that the martingale problem for L is well posed. Then for any Borel measurable function* $\beta\colon \mathbb{R}^N \longrightarrow \mathbb{R}^N$ *for which* $(\beta, a\beta)_{\mathbb{R}^N}$ *is bounded, the martingale problem for the operator* L^β *in Theorem* 5.2.4 *is well posed, and the solution* $\mathbb{P}^\beta_{\mathbf{x}}$ *to the martingale problem for* L^β *starting at* $\mathbf{x}$ *is related to the solution* $\mathbb{P}_{\mathbf{x}}$ *to the martingale problem for L starting at* $\mathbf{x}$ *by the prescription in Theorem* 5.2.4.

Proof. The existence of $\mathbb{P}^\beta_{\mathbf{x}}$ is proved in Theorem 5.2.4. To prove the uniqueness, let $\widetilde{\mathbb{P}}^\beta_{\mathbf{x}}$ be a solution for L^β, write

$$\psi = M^\beta + V^\beta \quad \text{where } V^\beta(t) = \mathbf{x} + \int_0^t (b + a\beta)(\psi(\tau))\, d\tau,$$

and set $\widetilde{E}_{-\beta}(t)$ equal to

$$\exp\left(-\int_0^t \left(\beta(\psi(\tau)), dM^\beta(\tau)\right)_{\mathbb{R}^N} - \frac{1}{2}\int_0^t \Big(\beta(\psi(\tau)), a\beta(\psi(\tau))\Big)_{\mathbb{R}^N} d\tau\right).$$

Just as before, $\big(\widetilde{E}_{-\beta}(t), \mathcal{B}_t, \widetilde{P}^\beta_{\mathbf{x}}\big)$ is a non-negative martingale with expectation value 1, and therefore there is a $\widetilde{\mathbb{P}}_{\mathbf{x}}$ such that $d\widetilde{\mathbb{P}}_{\mathbf{x}} \restriction \mathcal{B}_t = \widetilde{E}_{-\beta}(t)\, d\widetilde{\mathbb{P}}^\beta_{\mathbf{x}} \restriction \mathcal{B}_t$ for all $t \geq 0$, and this $\widetilde{\mathbb{P}}_{\mathbf{x}}$ will be a solution to the martingale problem for L starting at $\mathbf{x}$. Thus, $\widetilde{\mathbb{P}}_{\mathbf{x}} = \mathbb{P}_{\mathbf{x}}$. But this means that $d\widetilde{\mathbb{P}}^\beta_{\mathbf{x}} \restriction \mathcal{B}_t = E_{-\beta}(t)^{-1}\, d\mathbb{P}_{\mathbf{x}} \restriction \mathcal{B}_t$ for all $t \geq 0$, and therefore, since $\widetilde{E}_{-\beta}(t)^{-1} = E_\beta(t)$, $\mathbb{P}^\beta_{\mathbf{x}}$ is uniquely determined and $d\mathbb{P}^\beta_{\mathbf{x}} \restriction \mathcal{B}_t = E_\beta(t)\, d\mathbb{P}_{\mathbf{x}} \restriction \mathcal{B}_t$ for all $t \geq 0$. □

The preceding considerations can be used to shed light on the problem of determining when $\big(E_M(t), \mathcal{F}_t, \mathbb{P}\big)$ is a martingale. Namely, suppose that $b\colon \mathbb{R} \longrightarrow \mathbb{R}$ is a locally Lipschitz continuous function, and consider the integral equation

$$X(t,x) = x + w(t) + \int_0^t b\big(X(\tau,x)\big)\, d\tau,$$

where $w(\,\cdot\,)$ is an $\mathbb{R}$-valued Wiener path. In general, the solution to this equation will have a finite lifetime because $X(\,\cdot\,,x)$ may become infinite in a finite length of time. To handle this possibility, choose $\eta \in C^\infty_{\mathrm{c}}\big(\mathbb{R},[0,1]\big)$ so that $\eta = 1$ on $[-1,1]$, and set $b_k(\mathbf{y}) = \eta(k^{-1}\mathbf{y})b(\mathbf{y})$ for $k \geq 1$. Since b_k is globally Lipschitz, for each $k \geq 1$ there is a unique $X_k(\,\cdot\,,x)$ that satisfies

$$X_k(t,x) = x + w(t) + \int_0^t b_k\big(X_k(\tau,x)\big)\, d\tau.$$

Now take $\zeta_k = \inf\{t \geq 0 : |X_k(t,x)| \geq k\}$. Then, since $b_{k+1} = b_k$ on $[-k,k]$,

$$|X_{k+1}(t\wedge\zeta_k,x) - X_k(t\wedge\zeta_k)| \leq \|b_k\|_{\mathrm{Lip}} \int_0^t |X_{k+1}(\tau\wedge\zeta_k,x) - X_k(\tau\wedge\zeta_k)|\, d\tau,$$

and so, by Lemma 1.2.4, $X_{k+1}(\,\cdot\,,x) \restriction [0,\zeta_k) = X_k(\,\cdot\,,x) \restriction [0,\zeta_k)$. In particular, this means that $\zeta_k \leq \zeta_{k+1}$ and therefore that the **explosion time** $\mathfrak{e} := \lim_{k\to\infty} \zeta_k$ exists in $[0,\infty]$. Next, set

$$E(t) = \exp\left(\int_0^t b\big(w(\tau)\big)dw(\tau) - \frac{1}{2}\int_0^t b\big(w(\tau)\big)^2 d\tau\right).$$

Then, because

$$E(t) = \exp\left(\int_0^t b_k\big(w(\tau)\big)dw(\tau) - \frac{1}{2}\int_0^t b_k\big(w(\tau)\big)^2 d\tau\right) \quad \text{if } \zeta_k > t,$$

one can apply Corollary 5.2.5 to see that

$$\mathcal{W}(\zeta_k > t) = \mathbb{E}^{\mathcal{W}}\big[E(t), \zeta^0_k > t\big] \quad \text{where } \zeta^0_k = \inf\{t \geq 0 : |x + w(t)| \geq k\}.$$

Since $\zeta_k \nearrow \mathfrak{e}$ and $\zeta_k^0 \nearrow \infty$, we now see that

$$\mathcal{W}(\mathfrak{e} > t) = \mathbb{E}^{\mathcal{W}}\big[E(t)\big].$$

In other words, *the expected value of* $E(t)$ *is the probability that* $X(\,\cdot\,, x)$ *has not exploded by time* t. In the example following Theorem 5.2.2, $b(x) = x$ and so $X(t,x) = e^t x + \int_0^t e^{t-\tau}\,dw(\tau)$, which never explodes and explains why $E_M(t)$ is a martingale if $M(t) = \int_0^t w(\tau)\,dw(\tau)$.

Although the preceding example looks very special, using clever tricks one can show that it is always possible to interpret $\mathbb{E}^{\mathbb{P}}\big[E_M(t)\big]$ as the probability that some process has not exploded by time t, and this interpretation leads to **Kazamaki's criterion** which says that $\big(E_M(t), \mathcal{F}_t, \mathbb{P}\big)$ is a martingale if, for each $T > 0$,

$$\sup_{t\in[0,T]} \mathbb{E}^{\mathbb{P}}\big[e^{\frac{1}{2}M(t)}\big] < \infty.$$

(For more on this topic, see [15].) Since $\mathbb{E}^{\mathbb{P}}\big[e^{\frac{1}{2}M(t)}\big] \le \mathbb{E}^{\mathbb{P}}\big[e^{\frac{1}{2}\langle\langle M\rangle\rangle(t)}\big]^{\frac{1}{2}}$, Kazamaki's criterion is sharper than Novikov's. Nonetheless, it fails to predict the right answer when applied to the preceding example.

5.3 Brownian motion on a submanifold

As was pointed out in Corollary 4.5.6, if L is given by (4.5.1) and the restriction to a submanifold M of each vector field V_k is tangent to M, then the solution $\mathbb{P}_{\mathbf{x}}$ to the martingale problem for L starting at a point $\mathbf{x} \in M$ will give $\mathcal{P}(M) := C\big([0,\infty); M\big)$ measure 1. In this section, we will show how, by taking advantage of this observation, one can construct Brownian motion on an embedded submanifold of $\mathbb{R}^N$.

5.3.1 *A little background*

Let M be an m-dimensional, connected Riemannian manifold with Riemann metric g. Recall that the gradient grad f of a continuously differentiable f on M is the vector field[1] on M with the property that its g-inner product with any other vector field Y is equal to the action of Y on f. Thus, if (U, Φ) is a coordinate chart (i.e., U is an open subset of M and Φ is a diffeomorphism from U into $\mathbb{R}^m$) and g^{Φ} is the associated matrix representation of g, then

[1]Remember that in differential geometry a vector field is identified with its associated directional derivative operator.

$$\operatorname{grad} f = \sum_{i,i'1}^{m} (g^{\Phi})^{-1}_{ii'} (\partial^{\Phi}_{i'} f) \partial^{\Phi}_{i},$$

where

$$\partial^{\Phi}_{i} f = (\partial_i f \circ \Phi^{-1}) \circ \Phi$$

and ∂_i denotes differentiation with respect to the ith coordinate. Next, remember that the divergence $\operatorname{div}(Y)$ of a vector field Y on M is the function on M such that

$$\int_M f \operatorname{div}(Y)\, d\lambda_M = -\int_M Y f\, d\lambda_M \quad \text{for all } f \in C^1_c(M;\mathbb{R}),$$

where λ_M, the Riemannian measure on M determined by g, is given by

$$\int_U \varphi\, d\lambda_M = \int_{\Phi(U)} \varphi \circ \Phi^{-1}(\boldsymbol{\xi}) \sqrt{g^{\Phi} \circ \Phi^{-1}(\boldsymbol{\xi})}\, d\boldsymbol{\xi}.$$

If f has compact support in U and $Y^{\Phi}_i = Y\Phi_i$, integration by parts shows that

$$\begin{aligned}\int_U Y f\, d\lambda_M &= \sum_{i=1}^{m} \int_{\Phi(U)} Y^{\Phi}_i \circ \Phi^{-1}(\boldsymbol{\xi}) \partial_i (f \circ \Phi^{-1})(\boldsymbol{\xi}) \big(\sqrt{\det(g^{\Phi})}\big) \circ \Phi^{-1}(\boldsymbol{\xi})\, d\boldsymbol{\xi} \\ &= -\sum_{i=1}^{m} \int_{\Phi(U)} f \circ \Phi^{-1}(\boldsymbol{\xi}) \partial_i \Big(Y^{\Phi}_i \big(\sqrt{\det(g^{\Phi})}\big)\Big) \circ \Phi^{-1}(\boldsymbol{\xi})\, d\boldsymbol{\xi},\end{aligned}$$

and so

$$\operatorname{div}(Y) = \frac{1}{\sqrt{\det(g^{\Phi})}} \sum_{i=1}^{m} \partial^{\Phi}_i \Big(\sqrt{\det(g^{\Phi})} Y^{\Phi}_i\Big) \text{ on } U.$$

Equivalently, if $Y^{*\lambda_M}$ denotes the formal adjoint of Y as an operator on $L^2(\lambda_M;\mathbb{R})$, then $\operatorname{div}(Y) = -(Y^{*\lambda_M} + Y)$. Finally, the Laplacian Δ_M is defined on $C^2(M;\mathbb{R})$ by

$$\Delta_M f = \operatorname{div}\big(\operatorname{grad} f\big).$$

Since

$$\int_M \big(\operatorname{grad} f_1, \operatorname{grad} f_2\big)_M d\lambda_M = \int_M (\operatorname{grad} f_1) f_2\, d\lambda_M = -\int_M \big(\operatorname{div}(\operatorname{grad} f_1)\big) f_2\, d\lambda_M,$$

one sees that

$$\big(\Delta_M f_1, f_2\big)_{L^2(M;\lambda_M)} = -\int_M \big(\operatorname{grad} f_1, \operatorname{grad} f_2\big)_M d\lambda_M.$$

Thus Δ_M is a non-positive definite and symmetric as an operator on $L^2(M;\lambda_M)$. Further, using the preceding coordinate expressions for grad and div, one finds that

$$\Delta_M f = \frac{1}{\sqrt{\det(g^\Phi)}} \sum_{i,i'=1}^{m} \partial_i^\Phi \left(\sqrt{\det(g^\Phi)} (g^\Phi)_{ii'}^{-1} \partial_{i'}^\Phi f \right). \tag{5.3.1}$$

5.3.2 *Brownian motion*

A Brownian motion on M is an M-valued stochastic process whose distribution solves the martingale problem for $\frac{1}{2}\Delta_M$. Because the expression for Δ_M in (5.3.1) is only local and depends on the particular coordinate chart with which one is working, one cannot use it to write down a stochastic integral that the Brownian paths on M will satisfy globally. Thus, if one is going to use (5.3.1), the best that one can do is write a stochastic integral equation that the paths will satisfy locally. As a consequence, one ends up constructing a collection of locally defined paths that have to be patched together. In his address (cf. [8]) at the 1962 meeting of the International Mathematics Congress, Itô showed how to carry out such a patching procedure, but, although it is a technical *tour de force*, his construction is wanting from both an æsthetic and geometric standpoint. In the 1970's, J. Eells, D. Ellworthy, and Malliavin realized that, by moving to the orthonormal frame bundle, one can give a much more pleasing construction, and their approach (cf. [17]) has become the standard one. However, a good many preparations have to be made before one can adopt their procedure, and so we will restrict our attention here to a situation that requires fewer preparations and has considerable intuitive appeal. Namely, from now on, M will be an m-dimensional, compact, embedded submanifold of $\mathbb{R}^N$ with the Riemannian structure that it inherits from $\mathbb{R}^N$. Thus, if $\mathbf{x} \in M$, then the tangent space $T_\mathbf{x}M$ at $\mathbf{x}$ is to be thought of as an m-dimensional subspace of $\mathbb{R}^N$ and the Riemannian inner product of its elements is their Euclidean inner product as elements of $\mathbb{R}^N$.

The goal in this section is to show that a Brownian motion on M can be constructed by "projecting" the increments of a Brownian motion on $\mathbb{R}^N$ onto M.

Lemma 5.3.1. *Given an orthonormal basis* $\mathfrak{e} = (\mathbf{e}_1, \dots, \mathbf{e}_N)$, *define* $y_i^\mathfrak{e} = (\mathbf{y}, \mathbf{e}_i)_{\mathbb{R}^N}$ *for* $1 \le i \le N$. *Then, for each* $\mathbf{x} \in M$, *there exists an* $r > 0$, *an orthonormal basis* $\mathfrak{e}$, *and smooth functions*[2]

$$F_j \colon \left\{ (y_1^\mathfrak{e}, \dots, y_m^\mathfrak{e}) : \mathbf{y} \in B_{\mathbb{R}^N}(\mathbf{x}, r) \right\} \longrightarrow \mathbb{R}, \quad m+1 \le j \le N,$$

such that $F_j(y_1, \dots, y_m) - y_j$ *vanishes to first order at* $(x_1^\mathfrak{e}, \dots, x_m^\mathfrak{e})$ *and*

$$M \cap B_{\mathbb{R}^N}(\mathbf{x}, r) = \left\{ \mathbf{y} \in B_{\mathbb{R}^N}(\mathbf{x}, r) : y_j^\mathfrak{e} = F_j\left(y_1^\mathfrak{e}, \dots, y_m^\mathfrak{e}\right) \text{ for } m+1 \le j \le N \right\}.$$

[2]Here I use a subscript on B to indicate in which space the ball lies.

Hence, if $\Phi(\mathbf{y}) = \big(y_1^{\mathfrak{e}}, \dots, y_m^{\mathfrak{e}}\big)$ *for* $\mathbf{y} \in M \cap B(\mathbf{x}, r)$, *then* $\big(M \cap B_{\mathbb{R}^N}(\mathbf{x}, r), \Phi\big)$ *is a coordinate chart and*

$$\partial_i^{\Phi} = \partial_{\mathbf{e}_i} + \sum_{j=m+1}^{N} (\partial_{\mathbf{e}_i} F_j)\partial_{\mathbf{e}_j}.$$

In particular,

$$\sup_{\mathbf{y} \in M \cap B(\mathbf{x},r) \setminus \{\mathbf{x}\}} \frac{\|g^{\Phi}(\mathbf{y}) - \mathbf{I}_{\mathbb{R}^m}\|_{\text{H.S.}}}{|\mathbf{y} - \mathbf{x}|^2} < \infty.$$

Proof. The initial assertion is a standard result in elementary differential geometry. To prove it, choose an orthonormal basis $\mathfrak{e} = (\mathbf{e}_1, \dots, \mathbf{e}_N)$ for $\mathbb{R}^N$ such that $(\mathbf{e}_1, \dots, \mathbf{e}_m)$ is a basis for $T_{\mathbf{x}}M$. Next, choose a coordinate chart (U, Ψ) with the properties that $\mathbf{x} \in U$ and $\partial_i^{\Psi} f(\mathbf{x}) = \partial_{\mathbf{e}_i} f(\mathbf{x})$ for $1 \le i \le m$, and define $\Phi(\mathbf{y}) = (y_1^{\mathfrak{e}}, \dots, y_m^{\mathfrak{e}})$ for $\mathbf{y} \in \mathbb{R}^N$. Then

$$\partial_i(\Phi \circ \Psi^{-1})_{i'}\big(\Psi(\mathbf{x})\big) = (\partial_{\mathbf{e}_i} y_{i'}^{\mathfrak{e}})(\mathbf{x}) = \delta_{i,i'},$$

and so $\Phi \circ \Psi^{-1}$ is a diffeomorphism in a neighborhood W of $\Psi(\mathbf{x})$. Thus, there exists an $r > 0$ such that $M \cap B_{\mathbb{R}^N}(\mathbf{x}, r) \subseteq \Psi^{-1}(W)$, and the function $\mathbf{F} := \Psi^{-1} \circ (\Phi \circ \Psi^{-1})^{-1}$ is smooth on $B_{\mathbb{R}^m}\big((x_1^{\mathfrak{e}}, \dots, x_m^{\mathfrak{e}}), r\big)$. Since $\Phi \circ \mathbf{F}$ is the identity map on $B_{\mathbb{R}^m}\big((x_1^{\mathfrak{e}}, \dots, x_m^{\mathfrak{e}}), r\big)$, it follows that, for any $\mathbf{y} \in B_{\mathbb{R}^N}(\mathbf{x}, r)$, $\mathbf{y} \in M$ if and only if $y_j^{\mathfrak{e}} = \big(\mathbf{F}(y_1^{\mathfrak{e}}, \dots, y_m^{\mathfrak{e}}), \mathbf{e}_j\big)_{\mathbb{R}^N}$ for all $1 \le j \le N$, and so

$$\Phi^{-1} = \sum_{i=1}^{m} y_i^{\mathfrak{e}} \mathbf{e}_i + \sum_{j=m+1}^{N} F_j \mathbf{e}_j \text{ on } B_{\mathbb{R}^m}(\mathbf{x}, r) \quad \text{when } F_j := (\mathbf{F}, \mathbf{e}_j)_{\mathbb{R}^N}.$$

Hence, $\big(M \cap B_{\mathbb{R}^N}(\mathbf{x}, r), \Phi\big)$ is a coordinate chart, and

$$\partial_i^{\Phi} = \partial_{\mathbf{e}_i} + \sum_{j=m+1}^{N} (\partial_{\mathbf{e}_i} F_j)\partial_{\mathbf{e}_j} \quad \text{for } 1 \le i \le m.$$

Finally, since ∂_i^{Φ} is tangent to M and $T_{\mathbf{x}}M = \text{span}\big(\{\mathbf{e}_1, \dots, \mathbf{e}_m\}\big)$,

$$(\partial_{\mathbf{e}_i} F_j)(x_1^{\mathfrak{e}}, \dots, x_m^{\mathfrak{e}}) = 0 \quad \text{for all } 1 \le i \le m \text{ and } m+1 \le j \le N.$$

Given the preceding, it is clear that

$$g^{\Phi}(\mathbf{y})_{ii'} = \delta_{i,i'} + \sum_{j=m+1}^{M} (\partial_{\mathbf{e}_i} F_j)(\partial_{\mathbf{e}_{i'}} F_j)$$

and therefore that $\|g^{\Phi}(\mathbf{y}) - \mathbf{I}_{\mathbb{R}^m}\|_{\text{H.S.}} \le C|\mathbf{y} - \mathbf{x}|^2$ for some $C < \infty$. □

Referring to Lemma 5.3.1 and using (5.3.1), one sees that

$$\Delta_M f(\mathbf{x}) = \sum_{i=1}^{m} (\partial_i^{\Phi})^2 f(\mathbf{x}). \tag{5.3.2}$$

Although (5.3.2) is simpler looking than (5.3.1), it holds at only one point. For that reason, we will now see how to replace it by a similar expression that holds at all points in M.

Lemma 5.3.2. *There exists a smooth map*

$$\mathbf{y} \in \mathbb{R}^N \longmapsto \Pi^M(\mathbf{y}) \in \mathrm{Hom}(\mathbb{R}^N; \mathbb{R}^N)$$

such that $\|\Pi^M(\mathbf{y})\|_{\mathrm{op}} \le 1$ *for all* $\mathbf{y} \in \mathbb{R}^N$ *and* $\Pi^M(\mathbf{y})$ *is orthogonal projection onto* $T_{\mathbf{y}}M$ *for* $\mathbf{y} \in M$.

Proof. First observe that it suffices to construct Π^M in a ball $B_{\mathbb{R}^N}(\mathbf{x}, r)$ for each $\mathbf{x} \in M$. Indeed, once that is done, one can use a standard partition of unity argument to construct a global choice of Π^M. Thus let $\mathbf{x} \in M$ be given, choose $r > 0$, Φ, and $F_{m+1}, \dots, F_N$ as in Lemma 5.3.1, and set

$$\mathbf{W}_i = \mathbf{e}_i + \sum_{j=m+1}^{N} (\partial_{\mathbf{e}_i} F_j)\mathbf{e}_j. \tag{5.3.3}$$

Then $\partial_i^{\Phi} = \mathcal{L}_{\mathbf{W}_i}$, and so, for each $\mathbf{y} \in M \cap B_{\mathbb{R}^N}(\mathbf{x}, r)$, $\big(\mathbf{W}_1(\mathbf{y}), \dots, \mathbf{W}_m(\mathbf{y})\big)$ is a basis for $T_{\mathbf{y}}M$. Now apply the Gram–Schmidt orthogonalization procedure to produce from the $\mathbf{W}_i$'s smooth vector fields $\mathbf{E}_1, \dots, \mathbf{E}_m$ on $M \cap B_{\mathbb{R}^N}(\mathbf{x}, r)$ such that, for each $\mathbf{y} \in M \cap B_{\mathbb{R}^N}(\mathbf{x}, r)$, $\big(\mathbf{E}_1(\mathbf{y}), \dots, \mathbf{E}_m(\mathbf{y})\big)$ is an orthonormal basis for $T_{\mathbf{y}}M$. Finally, for $\mathbf{y} \in B_{\mathbb{R}^N}(\mathbf{x}, r)$, define

$$\Pi^M(\mathbf{y})\boldsymbol{\xi} = \sum_{i=1}^{m} \Big(\boldsymbol{\xi}, \mathbf{E}_i\big(\Phi^{-1}(y_1^{\mathfrak{e}}, \dots, y_m^{\mathfrak{e}})\big)\Big)_{\mathbb{R}^N} \mathbf{E}_i\big(\Phi^{-1}(y_1^{\mathfrak{e}}, \dots, y_m^{\mathfrak{e}})\big)$$

for $\mathbf{y} \in B_{\mathbb{R}^N}(\mathbf{x}, r)$ and $\boldsymbol{\xi} \in \mathbb{R}^N$. □

Given an orthonormal basis $\mathfrak{e} = (\mathbf{e}_1, \dots, \mathbf{e}_N)$, consider the operator

$$L = \sum_{j=1}^{N} (D_j^{\mathfrak{e},M})^2 \quad \text{where } D_j^{\mathfrak{e},M} := \mathcal{L}_{\Pi^M \mathbf{e}_j}, \tag{5.3.4}$$

and observe that L is independent of the choice of $\mathfrak{e}$. We will now show that $Lf = \Delta_M f$ on M. To this end, let $\mathbf{x} \in M$, and take $\mathfrak{e}, r > 0, \Phi$, and $F_{m+1}, \dots, F_N$ as in Lemma 5.3.1. Observe that

$$Lf(\mathbf{x}) = \frac{1}{2}\sum_{i=1}^{m} (D_i^{\mathfrak{e},M})^2 f(\mathbf{x}) = \frac{1}{2}\sum_{m=1}^{m} \partial_{\mathbf{e}_i} D_i^{\mathfrak{e},M} f(\mathbf{x})$$

and that $D_i^{\mathfrak{e},M} f = \big(\mathbf{e}_i, \operatorname{grad} f\big)_{\mathbb{R}^N}$, and therefore

$$Lf(\mathbf{x}) = \frac{1}{2}\sum_{i=1}^{m} \partial_{\mathbf{e}_i}\Big(\big(\mathbf{e}_i, \operatorname{grad} f\big)_{\mathbb{R}^N}\Big)(\mathbf{x}).$$

Finally, if $\mathbf{W}_1, \dots, \mathbf{W}_m$ are the vector fields in (5.3.3),

$$\big(\mathbf{e}_i, \operatorname{grad} f\big)_{\mathbb{R}^N} = \big(\mathbf{W}_i, \operatorname{grad} f\big)_{\mathbb{R}^N} - \sum_{j=m+1}^{N} \partial_{\mathbf{e}_i} F_j \big(\mathbf{e}_j, \operatorname{grad} f\big)_{\mathbb{R}^N},$$

and so, since $\partial_{\mathbf{e}_i} F_j$ and $\big(\mathbf{e}_j, \operatorname{grad} f\big)_{\mathbb{R}^N}$ both vanish at $\mathbf{x}$ for $m+1 \le j \le N$,

$$\partial_{\mathbf{e}_i}\big(\mathbf{e}_i, \operatorname{grad} f\big)(\mathbf{x}) = \partial_{\mathbf{e}_i}\big(\mathbf{W}_i, \operatorname{grad} f\big)(\mathbf{x}) = \partial_{\mathbf{e}_i}\partial_i^{\Phi} f(\mathbf{x}) = (\partial_i^{\Phi})^2 f(\mathbf{x}).$$

Hence, by (5.3.2), $Lf(\mathbf{x}) = \Delta_M f(\mathbf{x})$.

Theorem 5.3.3. *Let $\mathcal{W}$ be Wiener measure on $W(\mathbb{R}^N)$, and define Π^M as in Lemma* 5.3.2. *Then, for each $\mathbf{x} \in M$ and orthonormal basis $(\mathbf{e}_1, \dots, \mathbf{e}_N)$ for $\mathbb{R}^N$, the distribution of the solution $X(\,\cdot\,, \mathbf{x})$ to the Stratonovich integral equation*

$$X(t, \mathbf{x}) = \mathbf{x} + \sum_{j=1}^{N} \int_0^t \Pi^M\big(X(\tau, \mathbf{x})\big)\mathbf{e}_j \bullet dw(\tau)_j$$

is a Brownian motion on M starting at $\mathbf{x}$.

Proof. There is hardly anything to do. By Corollary 4.5.6, we know that, for any $\mathbf{x} \in M$, $X(\,\cdot\,, \mathbf{x}) \in \mathcal{P}(M)$ (a.s., $\mathcal{W}$) and its distribution solves the martingale problem for $\frac{1}{2}\sum_{j=1}^{N}(D_j^{\mathfrak{e},M})^2$ starting at $\mathbf{x}$. Hence, since, $\frac{1}{2}\Delta_M f = \frac{1}{2}\sum_{j=1}^{N}(D_j^{\mathfrak{e},M})^2 f$ on M, there is nothing more to do. □

Besides providing an elegant construction of Brownian motion on M, in conjunction with Theorem 4.2.6, Theorem 5.3.3 has intuitive appeal. Namely, if w_n is the polygonal approximation of the $\mathbb{R}^N$-valued Wiener path w used in (4.5.6) and

$$X_n(t, \mathbf{x}) = \mathbf{x} + \int_0^t \Pi^M\big(X_n(\tau, \mathbf{x})\big)\dot{w}_n(\tau)\, d\tau,$$

then $X_n(\,\cdot\,, \mathbf{x})(w)$ is obtained by integrating the projection of $\dot{w}_n(\,\cdot\,)$ onto $T_{X_n(\cdot,\mathbf{x})}M$, and so, as the $\mathcal{W}$-almost sure limit of the $X_n(\,\cdot\,, \mathbf{x})$'s, for $\mathcal{W}$-almost every w, $X(\,\cdot\,, \mathbf{x})(w)$ can be thought of as the path whose "tangent" is the projection of "$\dot{w}(\,\cdot\,)$" onto $T_{X(\cdot,\mathbf{x})}M$. It is in this sense that *the Brownian motion on M can be obtained by projecting the Brownian increments on $\mathbb{R}^N$ onto M.*

Another interesting fact about this construction comes from writing the operator L in (5.3.4) in the form in (1.2.5) instead of Hörmander form. When one does so, one finds that a on M equals the matrix representation of Π^M and that b on M equals

$$\mathcal{N} := \sum_{i=1}^{N} D_i^{\mathfrak{e},M} \Pi^M \mathbf{e}_i$$

for any choice of orthonormal basis $\mathfrak{e}$. The vector field $\mathcal{N}$ is familiar to differential geometers, who call it the **mean curvature normal**. The reason why they call it a normal is that $\mathcal{N}(\mathbf{x}) \perp T_{\mathbf{x}}M$ for all $\mathbf{x} \in M$. To see this, for a given $\mathbf{x} \in M$, choose $\mathfrak{e}$ as in Lemma 5.3.1. Then

$$\mathcal{N}(\mathbf{x}) = \sum_{i=1}^{m} \partial_{\mathbf{e}_i} \Pi^M \mathbf{e}_i(\mathbf{x}).$$

Next, note that, for any $1 \le i' \le m$, at $\mathbf{x}$

$$\begin{aligned} \left(\partial_{\mathbf{e}_i} \Pi^M \mathbf{e}_i, \mathbf{e}_{i'}\right)_{\mathbb{R}^N} &= \partial_{\mathbf{e}_i}\left(\Pi^M \mathbf{e}_i, \mathbf{e}_{i'}\right)_{\mathbb{R}^N} = \partial_{\mathbf{e}_i}\left(\Pi^M \mathbf{e}_i - \mathbf{e}_i, \mathbf{e}_{i'}\right)_{\mathbb{R}^N} \\ &= \partial_{\mathbf{e}_i}\left(\Pi^M \mathbf{e}_i - \mathbf{e}_i, \mathbf{e}_{i'} - \Pi^M \mathbf{e}_{i'}\right)_{\mathbb{R}^N} = 0, \end{aligned}$$

since $\left|\left(\Pi^M(\mathbf{y})\mathbf{e}_i - \mathbf{e}_i, \mathbf{e}_{i'} - \Pi^M(\mathbf{y})\mathbf{e}_{i'}\right)_{\mathbb{R}^N}\right| \le C|\mathbf{y} - \mathbf{x}|^2$ for some $C < \infty$. To understand its relationship to curvature requires more sophistication. Be that as it may, $\mathcal{N}$ vanishes if and only if M is what geometers call a minimal surface, and so Brownian motion on M will be an $\mathbb{R}^N$-valued martingale if and only if M is a minimal surface. For more information on this topic, see [17].

5.4 The Kalman–Bucy filter[3]

Let $\left(B(t), \mathcal{F}_t, \mathbb{P}\right)$ be an $\mathbb{R}^2$-valued Brownian motion on a probability space $(\Omega, \mathcal{F}, \mathbb{P})$, and let X_0 be an $\mathbb{R}$-valued Gaussian random variable that is independent of $\sigma\left(\{B(t) : t \ge 0\}\right)$. Given bounded, Borel measurable functions $\boldsymbol{\alpha}\colon [0,\infty) \longrightarrow \mathbb{R}^2$ and $\boldsymbol{\beta}\colon [0,\infty) \longrightarrow \mathbb{R}^2$, consider the system of equations

$$\begin{aligned} X_1(t) &= X_0 + \int_0^t \alpha_1(\tau)\, dB_1(\tau) + \int_0^t \beta_1(\tau) X_1(\tau)\, d\tau \\ X_2(t) &= \int_0^t \alpha_2(\tau)\, dB_2(\tau) + \int_0^t \beta_2(\tau) X_1(\tau)\, d\tau. \end{aligned} \tag{5.4.1}$$

These equations determine $X_1(\,\cdot\,)$ and $X_2(\,\cdot\,)$ up to a $\mathbb{P}$-null set. Indeed, using the same argument as we used in the discussion about the Ornstein–Uhlenbeck process in §3.1, one sees that, for any $0 \le s \le t$,

$$X_1(t) = e^{\int_s^t \beta_1(\tau)\, d\tau}\left(X_1(s) + \int_s^t e^{-\int_s^\tau \beta_1(\sigma)\, d\sigma} \alpha_1(\tau)\, dB_1(\tau)\right), \tag{5.4.2}$$

[3]This section is adapted from §6.2 in [14].

and, since $X_1(0) = X_0$, it follows from this that $X_1(\,\cdot\,)$ and $X_2(\,\cdot\,)$ exist and are $\mathbb{P}$-almost surely uniquely determined by (5.4.1).

Think of $X_1(t)$ as some sort of noisy signal and of $X_2(t)$ as a randomly corrupted transcription of $X_1(t)$, and assume that one can observe only the corrupted version. Then the corresponding filtering problem is that of predicting the value of $X_1(t)$ on the basis of ones observations of $X_2(\,\cdot\,)$ during the time interval $[0,t]$. To be more precise, let $\mathcal{F}_t^{(2)}$ denote the $\mathbb{P}$-completion of $\sigma\big(\{X_2(\tau) : \tau \in [0,t]\}\big)$. Then the filtering problem is to understand the random variable

$$\widehat{X}_1(t) := \mathbb{E}^{\mathbb{P}}\big[X_1(t) \mid \mathcal{F}_t^{(2)}\big],$$

which is the $\mathcal{F}_t^{(2)}$-measurable element of $L^2(\mathbb{P};\mathbb{R})$ whose $L^2(\mathbb{P};\mathbb{R})$-distance from $X_1(t)$ is minimal. In a more general setting, a satisfactory solution to the analogous problem is difficult, if not impossible, but, as Kalman and Bucy showed, the one here admits a remarkably nice solution as long as there exists a $\delta > 0$ such that

$$|\alpha_2(t)| \ge \delta \quad \text{for all } t \ge 0. \tag{5.4.3}$$

The basic reason why this filtering problem has a good solution is that all the quantities involved belong to the closed Gaussian family $\mathfrak{G}$ which is the closure in $L^2(\mathbb{P};\mathbb{R})$ of the linear subspace spanned by

$$\{\mathbf{1}, X_0\} \cup \big\{(\boldsymbol{\xi}, B(t))_{\mathbb{R}^2} : t \ge 0 \text{ and } \boldsymbol{\xi} \in \mathbb{R}^2\big\}.$$

Indeed, as (5.4.2) makes clear, $X_1(t) \in \mathfrak{G}$ for all $t \ge 0$, and therefore $X_2(t) \in \mathfrak{G}$ for all $t \ge 0$. Further, from (5.4.2), we know that, for $0 \le s < t$,

$$\mathbb{E}^{\mathbb{P}}\big[X_1(t)^2\big] = e^{2\int_s^t \beta_1(\tau)\,d\tau}\left(\mathbb{E}^{\mathbb{P}}[X_1(s)^2] + \int_s^t e^{-2\int_0^\tau \beta_1(\sigma)\,d\sigma}\alpha_1(\tau)^2\,d\tau\right). \tag{5.4.4}$$

In addition, because $\sigma\big(\{X_1(\tau) : \tau \ge 0\}\big)$ and $\sigma\big(\{B_2(\tau) : t \ge 0\}\big)$ are independent and α_2 satisfies (5.4.3), part (i) of Exercise 3.7 says that for each $t > 0$ there exists an $\epsilon_t > 0$ such that

$$\epsilon_t \|f\|^2_{L^2([0,t];\mathbb{R})} \le \mathbb{E}^{\mathbb{P}}\left[\left(\int_0^t f(\tau)\,dX_2(\tau)\right)^2\right] \le \epsilon_t^{-1}\|f\|^2_{L^2([0,t];\mathbb{R})}$$

for $f \in L^2([0,t];\mathbb{R})$, and part (iii) of that exercise says that there exist a $c(t) \in \mathbb{R}$ and a $g(t,\,\cdot\,) \in L^2\big([0,\infty);\mathbb{R}\big)$ such that $g(t,\,\cdot\,)$ vanishes on (t,∞) and

$$\widehat{X}_1(t) = c(t) + \int_0^t g(t,\tau)\,dX_2(\tau). \tag{5.4.5}$$

In particular, this means that $\widehat{X}_1(t) \in \mathfrak{G}$.

The challenge now is to find expressions for $c(t)$ and $g(t,\tau)$, and the following lemma is a first step in that direction.

Lemma 5.4.1. *For each $T > 0$ there is a $K(T) < \infty$ such that*

$$\mathbb{E}^{\mathbb{P}}\big[|\widehat{X}_1(t) - \widehat{X}_1(s)|^2\big]^{\frac{1}{2}} \vee \|g(t,\,\cdot\,) - g(s,\,\cdot\,)\|_{L^2([0,\infty);\mathbb{R})} \le K(T)(t-s)^{\frac{1}{2}}$$

for $0 \le s \le t \le T$. In particular, there is a continuous version of $t \rightsquigarrow \widehat{X}_1(t)$. Finally,

$$c(t) = e^{\int_0^t \beta_1(\tau)\,d\tau}\mathbb{E}^{\mathbb{P}}[X_0]\left(1 - \int_0^t e^{-\int_\tau^t \beta_1(\sigma)\,d\sigma} g(t,\tau)\beta_2(\tau)\,d\tau\right),$$

and so $c(\,\cdot\,)$ is continuous.

Proof. Set $L_t = \overline{\operatorname{span}\big(\{X_2(\tau) : \tau \in [0,t]\}\big)}^{L^2([0,t];\mathbb{R})}$. By part (iii) of Exercise 3.7, we know that $\widehat{X}_1(t) = \Pi_t X_t(t)$, where Π_t denotes orthogonal projection onto $\{\mathbf{1}\} \oplus L_t$. Thus,

$$\mathbb{E}^{\mathbb{P}}\big[|\widehat{X}_1(t) - \widehat{X}_1(s)|^2\big]^{\frac{1}{2}} \le \mathbb{E}^{\mathbb{P}}\big[|\Pi_t(X_1(t) - X_1(s))|^2\big]^{\frac{1}{2}} + \mathbb{E}^{\mathbb{P}}\big[|\Pi_t X_1(s) - \widehat{X}_1(s)|^2\big]^{\frac{1}{2}}.$$

Because $\mathbb{E}^{\mathbb{P}}\big[|\Pi_t(X_1(t) - X_1(s))|^2\big]^{\frac{1}{2}} \le \mathbb{E}^{\mathbb{P}}\big[|X_1(t) - X_1(s)|^2\big]^{\frac{1}{2}}$, and, by (5.4.2), there is a $C(T) < \infty$ such that

$$(*) \qquad \mathbb{E}^{\mathbb{P}}\big[|X_1(t) - X_1(s)|^2\big]^{\frac{1}{2}} \le C(T)(t-s)^{\frac{1}{2}},$$

the first term on the right poses no problem. To handle the second term, it suffices to show that there is a $C(T) < \infty$ such that

$$\mathbb{E}^{\mathbb{P}}\big[(X_1(s) - \widehat{X}_1(s))Y\big] \le C(T)(t-s)^{\frac{1}{2}}\|Y\|_{L^2(\mathbb{P};\mathbb{R})}$$

for all $0 \le s < t \le T$ and $Y \in \mathbf{1} \oplus L_t$, and, by part (iii) of Exercise 3.7, every such Y equals $a + \int_0^t f(\tau)\,dX_2(\tau)$ for some $a \in \mathbb{R}$ and $f \in L^2([0,t];\mathbb{R})$. Hence, since $X_1(s) - \widehat{X}_1(s) \perp \mathbf{1} \oplus L_s$ and $X_1(s) - \widehat{X}_1(s)$ is $\mathcal{F}_s$-measurable,

$$\begin{aligned}
\mathbb{E}^{\mathbb{P}}\big[(X_1(s) - \widehat{X}_1(s))Y\big] &= \mathbb{E}^{\mathbb{P}}\left[(X_1(s) - \widehat{X}_1(s))\left(\int_s^t f(\tau)\,dX_2(\tau)\right)\right] \\
&= \mathbb{E}^{\mathbb{P}}\left[(X_1(s) - \widehat{X}_1(s))\left(\int_s^t f(\tau)\beta_2(\tau)X_1(\tau)\,d\tau\right)\right] \\
&\le \|X_1(s)\|_{L^2(\mathbb{P};\mathbb{R})}\mathbb{E}^{\mathbb{P}}\left[\left(\int_s^t f(\tau)\beta_2(\tau)X_1(\tau)\,d\tau\right)^2\right]^{\frac{1}{2}} \\
&\le \|X_1(s)\|_{L^2(\mathbb{P};\mathbb{R})}\|f\|_{L^2([0,t];\mathbb{R})}\left(\int_s^t \beta_2(\tau)^2\mathbb{E}^{\mathbb{P}}[X_1(\tau)^2]\,d\tau\right)^{\frac{1}{2}},
\end{aligned}$$

which, together with (5.4.4), makes it is clear that there exists an $A(T) < \infty$ such that

$$\mathbb{E}^{\mathbb{P}}\big[(X_1(s) - \widehat{X}_1(s))Y\big] \le A(T)(t-s)^{\frac{1}{2}}\|f\|_{L^2([0,t];\mathbb{R})}.$$

Since, by (5.4.3),

$$\mathbb{E}^{\mathbb{P}}[Y^2] = \mathbb{E}^{P}\left[\left(a + \int_0^t f(\tau)\beta_2(\tau)X_1(\tau)\,d\tau\right)^2\right] + \int_0^t f(\tau)^2\alpha_2(\tau)^2\,d\tau \ge \delta^2\|f\|^2_{L^2([0,t];\mathbb{R})},$$

we now know that there is a $C(T) < \infty$ for which $(*)$ holds and therefore that $\mathbb{E}^{\mathbb{P}}\big[|\widehat{X}_1(t) - \widehat{X}_1(s)|^2\big]^{\frac{1}{2}}$ satisfies the asserted estimate. Similarly,

$$\mathbb{E}^{\mathbb{P}}\big[|\widehat{X}_1(t) - \widehat{X}_1(s)|^2\big] = \big(c(t) - c(s)\big)^2 + \mathbb{E}^{\mathbb{P}}\left[\left(\int_0^t \big(g(t,\tau) - g(s,\tau)\big)\,dX_2(\tau)\right)^2\right] \ge \epsilon_t\big\|g(t,\,\cdot\,) - g(s,\,\cdot\,)\big\|^2_{L^2([0,t];\mathbb{R})},$$

and so

$$\|g(t,\,\cdot\,) - g(s,\,\cdot\,)\|_{L^2([0,\infty);\mathbb{R})} \le \epsilon_t^{-1}\|\widehat{X}_1(t) - \widehat{X}_1(s)\|_{L^2(\mathbb{P};\mathbb{R})}.$$

We have therefore proved the required estimates.

To check the asserted expression for $c(t)$, simply take the expected value of the expression for $\widehat{X}_1(t)$, and use the fact that

$$\mathbb{E}^{\mathbb{P}}[\widehat{X}_1(t)] = \mathbb{E}^{\mathbb{P}}[X_1(t)] = e^{\int_0^t \beta_1(\tau)\,d\tau}\mathbb{E}^{\mathbb{P}}[X_0].$$

Finally, to see that $\widehat{X}_1(\,\cdot\,)$ admits a continuous version, set

$$\overline{X}_1(t) = \widehat{X}_1(t) - \mathbb{E}^{\mathbb{P}}[\widehat{X}_1(t)] = \widehat{X}_1(t) - e^{\int_0^t \beta_1(\tau)\,d\tau}\mathbb{E}^{\mathbb{P}}[X_0].$$

Clearly, $\widehat{X}_1(\,\cdot\,)$ admits a continuous version if and only if $\overline{X}_1(\,\cdot\,)$ does. Furthermore, $\overline{X}_1(t)$ is a centered Gaussian random variable, and

$$\|\overline{X}_1(t) - \overline{X}_1(s)\|^2_{L^2(\mathbb{P};\mathbb{R})} = \mathrm{var}\big(\widehat{X}_1(t) - \widehat{X}_1(s)\big) \le \|\widehat{X}_1(t) - \widehat{X}_1(s)\|^2_{L^2(\mathbb{P};\mathbb{R})}.$$

Hence, for any $p \in [1,\infty)$,

$$\|\overline{X}_1(t) - \overline{X}_1(s)\|^p_{L^p(\mathbb{P};\mathbb{R})} \le A_p\|\widehat{X}(t) - \widehat{X}(s)\|^p_{L^2(\mathbb{P};\mathbb{R})} \le A_pK(T)^p(t-s)^{\frac{p}{2}},$$

where $A_p = \int_{\mathbb{R}} |y|^p\,\gamma_{0,1}(dy)$, and so the existence of a continuous version follows from Theorem 2.1.2. □

Lemma 5.4.1 means we can assume $\widehat{X}_1(\,\cdot\,)$ is continuous, and therefore, since $\widehat{X}_1(t)$ is $\mathcal{F}_t^{(2)}$-measurable for each $t \ge 0$, that it is progressively measurable with respect to $\{\mathcal{F}_t^{(2)} : t \ge 0\}$.

Define

$$\widetilde{X}_2(t) = X_2(t) - \int_0^t \beta_2(\tau)\widehat{X}_1(\tau)\,d\tau = \int_0^t \alpha_2(\tau)\,dB_2(\tau) + \int_0^t \beta_2(\tau)\widetilde{X}_1(\tau)\,d\tau,$$

where $\widetilde{X}_1(t) := X_1(t) - \widehat{X}_1(t)$.

Lemma 5.4.2. $\big(\widetilde{X}_2(t), \mathcal{F}_t^{(2)}, \mathbb{P}\big)$ *is a continuous martingale, and* $\langle\!\langle \widetilde{X}_2 \rangle\!\rangle(t) = \int_0^t \alpha_2(\tau)^2\,d\tau$. *Furthermore, for each* $t \geq 0$, $\mathcal{F}_t^{(2)}$ *equals the* $\mathbb{P}$*-completion* $\overline{\sigma\big(\{\widetilde{X}_2(\tau) : \tau \in [0,t]\}\big)}^{\mathbb{P}}$ *of* $\sigma\big(\{\widetilde{X}_2(\tau) : \tau \in [0,t]\}\big)$.

Proof. Suppose that $\Gamma \in \mathcal{F}_s^{(2)}$. Then, since $\mathcal{F}_s^{(2)} \subseteq \mathcal{F}_s$ and

$$\mathbb{E}^{\mathbb{P}}\big[\widetilde{X}_1(\tau) \mid \mathcal{F}_s^{(2)}\big] = 0 \text{ for } \tau \in [s,\infty),$$

$$\mathbb{E}^{\mathbb{P}}\big[\widetilde{X}_2(t) - \widetilde{X}_2(s), \Gamma\big] = \int_s^t \beta_2(\tau)\mathbb{E}^{\mathbb{P}}\big[\widetilde{X}_1(\tau), \Gamma\big]\,d\tau = 0$$

for $t \geq s$. Hence $\big(\widetilde{X}_2(t), \mathcal{F}_t^{(2)}, \mathbb{P}\big)$ is a continuous martingale. In addition, $\big(\widetilde{X}_2(t), \mathcal{F}_t, \mathbb{P}\big)$ is a semimartingale, and, as such,

$$d\widetilde{X}_2(t)^2 = 2\widetilde{X}_2(t)\,d\widetilde{X}_2(t) + \alpha_2(t)^2\,dt.$$

Hence

$$\begin{aligned}\mathbb{E}^{\mathbb{P}}\big[\widetilde{X}_2(t)^2 - \widetilde{X}_2(s)^2, \Gamma\big] &= 2\mathbb{E}^{\mathbb{P}}\left[\int_s^t \widetilde{X}_2(\tau)\,d\widetilde{X}_2(\tau), \Gamma\right] + \left(\int_s^t \alpha_2(\tau)^2\,d\tau\right)\mathbb{P}(\Gamma)\\ &= \left(\int_s^t \alpha_2(\tau)^2\,d\tau\right)\mathbb{P}(\Gamma),\end{aligned}$$

and therefore $\left(\widetilde{X}_2(t)^2 - \int_0^t \alpha_2(\tau)^2\,d\tau, \mathcal{F}_t^{(2)}, \mathbb{P}\right)$ is a martingale.

Turning to the equality $\mathcal{F}_t^{(2)} = \overline{\sigma\big(\{\widetilde{X}_2(\tau) : \tau \in [0,t]\}\big)}^{\mathbb{P}}$, first observe that the inclusion $\mathcal{F}_t^{(2)} \supseteq \overline{\sigma\big(\{\widetilde{X}_2(\tau) : \tau \in [0,t]\}\big)}^{\mathbb{P}}$ is obvious. To prove the opposite inclusion, write

$$\int_0^t h(\tau)\,d\widetilde{X}_2(\tau) = \int_0^t h(\tau)\,dX_2(\tau) - \int_0^t h(\tau)\beta_2(\tau)\widehat{X}_1(\tau)\,d\tau \text{ for } h \in L^2\big([0,t];\mathbb{R}\big),$$

and observe that, by (5.4.5), for any $h \in L^2([0,t];\mathbb{R})$,

$$\begin{aligned}\int_0^t h(\tau)\beta_2(\tau)\widehat{X}_1(\tau)\,d\tau &= \int_0^t h(\tau)\beta_2(\tau)c(\tau)\,d\tau\\ &\quad + \int_0^t h(\tau)\beta_2(\tau)\left(\int_0^t g(\tau,\sigma)\,dX_2(\sigma)\right)d\tau.\end{aligned}$$

Using the fact that $\tau \in [0,t] \longmapsto g(\tau,\,\cdot\,) \in L^2\big([0,t];\mathbb{R}\big)$ is a continuous map, one can easily check that

$$\int_0^t h(\tau)\beta_2(\tau)\left(\int_0^t g(\tau,\sigma)\,dX_2(\sigma)\right)d\tau = \int_0^t \left(\int_0^t h(\tau)\beta_2(\tau)g(\tau,\sigma)\,d\tau\right)dX_2(\sigma).$$

Hence

$$\int_0^t h(\tau)\beta_2(\tau)c(\tau)\,d\tau + \int_0^t h(\tau)\,d\widetilde{X}_2(\tau) = \int_0^t \big(h(\sigma) - Kh(\sigma)\big)\,dX_2(\sigma),$$

where K is the operator on $L^2\big([0,t];\mathbb{R}\big)$ given by

$$Kh(\sigma) = \int_0^t h(\tau)\beta_2(\tau)g(\tau,\sigma)\,d\tau.$$

We now apply Exercise 5.5 to show that for any there is an $h_t \in L^2\big([0,t];\mathbb{R}\big)$ such that $h_t - Kh_t = \mathbf{1}$. Indeed, since

$$\iint\limits_{[0,t]^2} \beta_2(\tau)^2 g(\tau,\sigma)^2\,d\sigma d\tau < \infty,$$

all that we have to check is that $\mathrm{Null}(\mathbf{I} - K^*) = \{0\}$. To this end, suppose that $h = K^*h$. Then, since $g(\tau,\sigma) = 0$ if $\sigma > \tau$,

$$h(\tau) = \beta_2(\tau)\int_0^\tau g(\tau,\sigma)h(\sigma)\,d\sigma,$$

and so

$$|h(\tau)|^2 \le C\int_0^\tau |h(\sigma)|^2\,d\sigma \quad \text{where } C = \sup_{\tau\in[0,t]} |\beta_2(\tau)|^2\|g(\tau,\,\cdot\,)\|^2_{L^2([0,t];\mathbb{R})} < \infty.$$

Starting from this and proceeding by induction on $n \ge 1$, one sees that $|h(\tau)|^2 \le \frac{C^n\tau^n}{n!}\|h\|^2_{L^2([0,t];\mathbb{R})}$, which means that $h = 0$. Thus, by Exercise 5.5, $\mathrm{Range}(\mathbf{I} - K) = L^2([0,t];\mathbb{R})$, and so we can find an $h_t \in L^2\big([0,t];\mathbb{R}\big)$ such that $h_t - Kh_t = \mathbf{1}$ and therefore

$$X_2(t) = \int_0^t h_t(\tau)\beta_2(\tau)c(\tau)\,d\tau + \int_0^t h_t(\tau)\,d\widetilde{X}_2(\tau).$$

Since this means that $X_2(t)$ is $\overline{\sigma\big(\{\widetilde{X}_2(\tau) : \tau \in [0,t]\}\big)}^{\mathbb{P}}$-measurable, there is nothing more to do. □

Set

$$\widetilde{B}_2(t) = \int_0^t \frac{1}{\alpha_2(\tau)}\,d\widetilde{X}_2(\tau).$$

By Lemma 5.4.2 and Corollary 4.3.2, we know that $\big(\widetilde{B}_2(t), \mathcal{F}_t^{(2)}, \mathbb{P}\big)$ is a Brownian motion, that $\mathcal{F}_t^{(2)} = \overline{\sigma\big(\{\widetilde{B}_2(\tau) : t \in [0,t]\}\big)}^{\mathbb{P}}$ and therefore that

$$\widehat{X}_1(t) = \mathbb{E}^{\mathbb{P}}\big[X_1(t) \mid \overline{\sigma\big(\{\widetilde{B}_2(\tau) : \tau \in [0,t]\}\big)}^{\mathbb{P}}\big].$$

Hence, by part (iii) of Exercise 3.7, there exists an $a(t) \in \mathbb{R}$ and an element $f(t,\,\cdot\,)$ of $L^2([0,t];\mathbb{R})$ such that

$$\widehat{X}_1(t) = a(t) + \int_0^t f(t,\tau)\, d\widetilde{B}_2(\tau).$$

Obviously, $a(t) = \mathbb{E}^{\mathbb{P}}\big[\widehat{X}_1(t)\big] = \mathbb{E}^{\mathbb{P}}[X_1(t)]$, and so $a(t) = e^{\int_0^t \beta_1(\tau)\,d\tau}\mathbb{E}^{\mathbb{P}}[X_0]$. In addition,

$$\begin{aligned}\int_0^s f(t,\tau)\,d\tau &= \mathbb{E}^{\mathbb{P}}\big[\widetilde{B}_2(s)\widehat{X}_1(t)\big] = \mathbb{E}^{\mathbb{P}}\big[\widetilde{B}_2(s)X_1(t)\big] \\ &= \int_0^s \frac{\beta_2(\tau)\mathbb{E}^{\mathbb{P}}[\widetilde{X}_1(\tau)X_1(t)]}{\alpha_2(\tau)}\,d\tau = \int_0^s e^{\int_\tau^t \beta_1(\sigma)\,d\sigma}\frac{\beta_2(\tau)\mathbb{E}^{\mathbb{P}}[\widetilde{X}_1(\tau)X_1(\tau)]}{\alpha_2(\tau)}\,d\tau,\end{aligned}$$

where the last equality is an application of (5.4.2). Thus,

$$\begin{gathered} f(t,s) = e^{\int_s^t \beta_1(\sigma)\,d\sigma}\rho_2(s)D(s) \\ \text{where } \rho_2(s) = \frac{\beta_2(s)}{\alpha_2(s)} \text{ and } D(s) = \mathbb{E}^{\mathbb{P}}\big[\widetilde{X}_1(s)X_1(s)\big], \end{gathered}$$

and so

$$\begin{aligned}\widehat{X}_1(t) = e^{\int_0^t \beta_1(\sigma)\,d\sigma}\Bigg(&\mathbb{E}^{\mathbb{P}}[X_0] \\ &+ \int_0^t e^{-\int_0^\tau \beta_1(\sigma)\,d\sigma}\rho_2(\tau)D(\tau)\,d\widetilde{B}_2(\tau)\Bigg).\end{aligned} \tag{5.4.6}$$

This means that $\widehat{X}_1(t)$ is a Gaussian random variable with mean $e^{\int_0^t \beta_1(\sigma)\,d\sigma} \times \mathbb{E}^{\mathbb{P}}[X_0]$ and variance

$$\int_0^t e^{2\int_\tau^t \beta_1(\sigma)\,d\sigma}\rho_2(\tau)^2 D(\tau)^2\,d\tau.$$

Moreover,

$$\begin{aligned}\widehat{X}_1(t) = e^{\int_0^t \beta_1(\sigma)\,d\sigma}\Bigg(\mathbb{E}^{\mathbb{P}}[X_0] &+ \int_0^t e^{-\int_0^\tau \beta_1(\sigma)\,d\sigma}\frac{\rho_2(\tau)D(\tau)}{\alpha_2(\tau)}\,dX_2(\tau) \\ &- \int_0^t e^{-\int_0^\tau \beta_1(\sigma)\,d\sigma}\rho_2(\tau)^2\widehat{X}_1(\tau)D(\tau)\,d\tau\Bigg).\end{aligned}$$

Hence

$$d\widehat{X}_1(t) = \big(\beta_1(t) - \rho_2(\tau)^2 D(\tau)\big)\widehat{X}_1(t)\,dt + \frac{\rho_2(\tau)D(\tau)}{\alpha_2(\tau)}\,dX_2(t),$$

and so

$$\begin{aligned}\widehat{X}_1(t) =& e^{\int_0^t (\beta_1(\sigma)-\rho_2(\sigma)^2 D(\sigma))\,d\sigma}\mathbb{E}^{\mathbb{P}}[X_0] \\ &+ \int_0^t e^{\int_\tau^t (\beta_1(\sigma)-\rho_2(\sigma)^2 D(\sigma))\,d\sigma}\frac{\rho_2(\tau)D(\tau)}{\alpha_2(\tau)}\,dX_2(\tau).\end{aligned} \tag{5.4.7}$$

Equivalently, referring to (5.4.5),

$$c(t) = e^{\int_0^t (\beta_1(\sigma)-\rho_2(\sigma)^2 D(\sigma))\,d\sigma}\mathbb{E}^{\mathbb{P}}[X_0]$$
$$\text{and}\quad g(t,\tau) = e^{\int_\tau^t (\beta_1(\sigma)-\rho_2(\sigma)^2 D(\sigma))\,d\sigma}\frac{\rho_2(\tau)D(\tau)}{\alpha_2(\tau)}.$$

What remains is to understand $D(s)$. To this end, first observe that

$$D(s) = \mathbb{E}^{\mathbb{P}}\big[X_1(s)^2\big] - \mathbb{E}^{\mathbb{P}}\big[X_1(s)\widehat{X}_1(s)\big] = \mathbb{E}^{\mathbb{P}}\big[X_1(s)^2\big] - \mathbb{E}^{\mathbb{P}}\big[\widehat{X}_1(s)^2\big]$$

and

$$\begin{aligned}\mathbb{E}^{\mathbb{P}}\big[\widetilde{X}_1(s)^2\big] &= \mathbb{E}^{\mathbb{P}}\big[X_1(s)^2\big] - 2\mathbb{E}^{\mathbb{P}}\big[X_1(s)\widehat{X}_1(s)\big] + \mathbb{E}^{\mathbb{P}}\big[\widehat{X}(s)^2\big] \\ &= \mathbb{E}^{\mathbb{P}}\big[X_1(s)^2\big] - \mathbb{E}^{\mathbb{P}}\big[\widehat{X}_1(s)^2\big].\end{aligned}$$

Hence

$$D(s) = \mathbb{E}^{\mathbb{P}}\big[X_1(s)^2\big] - \mathbb{E}^{\mathbb{P}}\big[\widehat{X}_1(s)^2\big] = \mathbb{E}^{\mathbb{P}}\big[\widetilde{X}_1(s)^2\big]$$

is the square of the distance between $X_1(t)$ and $\widehat{X}_1(t)$ in $L^2(\mathbb{P};\mathbb{R})$. Next note that, from (5.4.4), we know that

$$\mathbb{E}^{\mathbb{P}}\big[X_1(s)^2\big] = e^{2\int_0^s \beta_1(\sigma)\,ds}\left(\mathbb{E}^{\mathbb{P}}[X_0^2] + \int_0^s e^{-2\int_0^\tau \beta_1(\sigma)\,d\sigma}\alpha_1(\tau)^2\,d\tau\right),$$

and therefore

$$\partial_s\mathbb{E}^{\mathbb{P}}\big[X_1(s)^2\big] = 2\beta_1(s)\mathbb{E}^{\mathbb{P}}\big[X_1(s)^2\big] + \alpha_1(s)^2.$$

At the same time, by (5.4.6), we know that

$$\mathbb{E}^{\mathbb{P}}\big[\widehat{X}_1(s)^2\big] = e^{2\int_0^s \beta_1(\sigma)\,d\sigma}\left(\mathbb{E}^{\mathbb{P}}[X_0]^2 + \int_0^s e^{-2\int_0^\tau \beta_1(\sigma)\,d\sigma}\rho_2(\tau)^2 D(\tau)^2\,d\tau\right),$$

and so

$$\partial_s\mathbb{E}^{\mathbb{P}}\big[\widehat{X}_1(s)^2\big] = 2\beta_1(s)\mathbb{E}^{\mathbb{P}}\big[\widehat{X}_1(s)^2\big] + \rho_2(s)^2 D(s)^2.$$

After combining these, we see that $D(\,\cdot\,)$ is the solution to

$$\partial_s D(s) = 2\beta_1(s)D(s) - \rho_2(s)^2 D(s)^2 + \alpha_1(s)^2 \quad \text{with } D(0) = \operatorname{var}(X_0). \tag{5.4.8}$$

The elementary theory of ordinary differential equations guarantees that (5.4.8) admits a unique solution. Moreover, (5.4.8) is a *Ricardi equation*, a class of equations that have been well studied and about which a good deal is known (cf. Exercise 5.6), and, because $\mathbb{E}^{\mathbb{P}}\big[\big(X_1(s)-\widehat{X}_1(s)\big)^2\big] \le \operatorname{var}\big(X_1(s)\big)$, our considerations show that

$$\begin{aligned} 0 \le D(s) &\le \operatorname{var}\big(X_1(s)\big) \\ &= e^{2\int_0^s \beta_1(\tau)\,d\tau}\operatorname{var}(X_0) + \int_0^s e^{-2\int_\tau^s \beta_1(\sigma)\,d\sigma}\alpha_1(\tau)^2\,d\tau. \end{aligned} \tag{5.4.9}$$

In terms of the filtering problem for which this is a model, our results have the following interpretation. One observes the process $X_2(\,\cdot\,)$ during the time interval $[0,t]$ and then passes ones observations through the *filter* represented by (5.4.7) to obtain the predicted value $\widehat{X}_1(t)$ of $X_1(t)$. Although in general the $dX_2(t)$-integral in (5.4.7) is a stochastic integral and therefore its pathwise meaning is somewhat ambiguous, when α_2 and β_2 have locally bounded variation it is a Riemann–Stieltjes integral and therefore well-defined path by path.

Proving the natural multi-dimensional analogs of these results requires no new ideas. The only major difficulty is that when $\beta_1(\,\cdot\,)$ and $\beta_2(\,\cdot\,)$ are matrix-valued, in general one will no longer have a closed form solution to the equations

$$\dot{E}_1(t) = \beta_1(t)E_1(t) \quad \text{and} \quad \dot{E}_2(t) = \beta_2(t)E_2(t),$$

and so the use of these solutions as integrating factors is more complicated.

5.5 A soupçon of Malliavin calculus

We have developed a good deal of machinery for studying solutions to Kolmogorov's forward equation, but as yet none of it addresses the regularity of those solutions as a function of the forward variable. In light of the many analytic results that are known about such equations, this is a somewhat embarrassing situation. For example, analysts have shown (cf. [4]) that if a and b are Hölder continuous and a is uniformly elliptic, in the sense that $a \ge \epsilon\mathbf{I}$ for some $\epsilon > 0$, then solutions to the forward equation admit a Hölder continuous density with respect to Lebesgue measure. However, it is not obvious what role ellipticity might play in the theory that we have developed, and, to the best of my knowledge, it was P. Malliavin who first proposed a technique for exploiting ellipticity to derive regularity results using Itô's stochastic integral equations and (3.1.6). Unfortunately, to develop Malliavin's ideas in full requires the introduction of a great many technical minutia (cf. [13]), and so what follows is far from anything approaching a complete exposition of his theory. Instead, the goal here is to explain his basic idea via a few examples.

Stated briefly, given a solution $X(t, \mathbf{x})$ to Itô's equation, Malliavin's idea was the use the integration by parts formula in (3.1.6) to derive, under appropriate ellipticity assumptions, estimates of the form

$$\left|\mathbb{E}^{\mathcal{W}}\big[(\partial_j\varphi)\circ X(t,\mathbf{x})\big]\right| \le C_p(t,\mathbf{x})\mathbb{E}^{\mathcal{W}}\big[|\varphi\circ X(t,\mathbf{x})|^q\big]^{\frac{1}{q}}, \quad \varphi\in C_c^\infty(\mathbb{R}^N;\mathbb{R}),$$

for some $1 < q < \frac{N}{N-1}$. Once one has such an estimate, the following elementary version of a Sobolev embedding theorem shows that the distribution of $X(t, \mathbf{x})$ admits a bounded, continuous density.

Lemma 5.5.1. *Let μ be a Borel probability measure on $\mathbb{R}^N$ and $p \in (N, \infty)$. If there is a $K < \infty$ such that, for all $\varphi \in C_c^\infty(\mathbb{R}^N; \mathbb{R})$,*

$$\max_{1\le j\le N}\left|\langle \partial_{x_j}\varphi, \mu\rangle\right| \le K\|\varphi\|_{L^{p'}(\mu;\mathbb{R})},$$

where $p' = \frac{p}{p-1}$ is the Hölder conjugate of p, then there is an $f \in C_b\big(\mathbb{R}^N; [0,\infty)\big)$ such that $\mu(dy) = f(y)\,dy$. In fact, there is a $C < \infty$, depending only on N and p, such that $\|f\|_u \le CK^N$.

Proof. Begin by observing that the hypothesis says that the map

$$\varphi \rightsquigarrow \langle \partial_{x_j}\varphi, \mu\rangle$$

determines a continuous linear functional on $L^{p'}(\mu; \mathbb{R})$ with norm less than or equal to K. Therefore there exists a $\psi_j \in L^p(\mu; \mathbb{R})$ such that

$$\|\psi_j\|_{L^p(\mu;\mathbb{R})} \le K \quad\text{and}\quad \langle \partial_{x_j}\varphi, \mu\rangle = \int_{\mathbb{R}^N} \varphi\psi_j\,d\mu.$$

In the language of Schwartz distribution theory, this is the statement that $\partial_{x_j}\mu = -\psi_j\mu$.

Set $g_t(\mathbf{y}) = (4\pi t)^{-\frac{N}{2}}e^{-\frac{|\mathbf{y}|^2}{4t}}$. Then

$$\partial_t(g_t*\mu) = \Delta(g_t*\mu) = -\sum_{j=1}^N(\partial_{y_j}g_t)*(\psi_j\mu).$$

Let $\lambda > 0$ be given. After multiplying the preceding equality by $e^{-\lambda t}$ and integrating over (δ, ∞), one obtains

$$e^{-\lambda\delta}g_\delta*\mu = \lambda\int_\delta^\infty e^{-\lambda t}g_t*\mu\,dt + \sum_{j=1}^N\int_\delta^\infty e^{-\lambda t}(\partial_{y_j}g_t)*(\psi_j\mu)\,dt.$$

Hence, if $r_\lambda = \int_0^\infty e^{-\lambda t} g_t\,dt$, then both r_λ and $|\nabla r_\lambda|$ are integrable, and, as $\delta \searrow 0$, the right hand side of the preceding converges to the function

$$f = \lambda r_\lambda * \mu + \sum_{j=1}^{N} (\partial_{y_j} r_\lambda) * (\psi_j \mu).$$

Thus μ, as a tempered distribution, is equal to f. Further, $\partial_{y_j} f = -\psi_j f$ in the sense of distributions, and, since μ is a probability measure, we can take f to be a non-negative function with integral 1. Now set $f_\delta = g_\delta * f$. For any $\epsilon > 0$, $\partial_{y_j}(f_\delta + \epsilon)^{\frac{1}{p}} = -\frac{1}{p}(f_\delta + \epsilon)^{\frac{1}{p-1}} g_\delta * (\psi_j f)$, and so

$$\partial_{y_j}(f + \epsilon)^{\frac{1}{p}} = -\psi_j \frac{1}{p}(f+\epsilon)^{\frac{1}{p}} \frac{f}{f+\epsilon}.$$

Therefore, if $\varphi = f^{\frac{1}{p}}$, then $\|\varphi\|_{L^p(\mathbb{R}^N;\mathbb{R})} = 1$, $\partial_{y_j}\varphi = -\frac{1}{p}\psi_j\varphi$, and $\|\psi_j\varphi\|_{L^p(\mathbb{R}^N;\mathbb{R})} = \|\psi_j\|_{L^p(\mu;\mathbb{R})} \le K$. In particular,

$$(*) \qquad \varphi = \lambda r_\lambda * \varphi + \frac{1}{p}\sum_{j=1}^{N} (\partial_{y_j} r_\lambda) * (\psi_j \varphi).$$

To complete the proof, we must show that r_λ and $|\nabla r_\lambda|$ are in $L^{p'}(\mathbb{R}^N;\mathbb{R})$ for $p > N$, and, since

$$\begin{aligned} \|r_\lambda\|_{L^{p'}(\mathbb{R}^N;\mathbb{R})} &= \lambda^{\frac{N}{2p}-1}\|r_1\|_{L^{p'}(\mathbb{R}^N;\mathbb{R})} \\ \text{and} \quad \|\nabla r_\lambda\|_{L^{p'}(\mathbb{R}^N;\mathbb{R}^N)} &= \lambda^{\frac{N}{2p}-\frac{1}{2}}\|\nabla r_1\|_{L^{p'}(\mathbb{R}^N;\mathbb{R}^N)}, \end{aligned}$$

it suffices to treat the case when $\lambda = 1$. To this end, observe that, by the continuous form of Minkowski's inequality,

$$\begin{aligned} \|r_1\|_{L^{p'}(\mathbb{R}^N;\mathbb{R})} &\le \int_0^\infty e^{-t}\|g_t\|_{L^{p'}(\mathbb{R}^N;\mathbb{R})}\,dt \\ \text{and} \quad \|\nabla r_1\|_{L^{p'}(\mathbb{R}^N;\mathbb{R}^N)} &\le \int_0^\infty e^{-t}\|\nabla g_t\|_{L^{p'}(\mathbb{R}^N;\mathbb{R}^N)}\,dt. \end{aligned}$$

But

$$\begin{aligned} \|g_t\|_{L^{p'}(\mathbb{R}^N;\mathbb{R})} &= t^{-\frac{N}{2p}}\|g_1\|_{L^{p'}(\mathbb{R}^N;\mathbb{R})} \\ \text{and} \quad \|\nabla g_t\|_{L^{p'}(\mathbb{R}^N;\mathbb{R}^N)} &= t^{-\frac{N}{2p}-\frac{1}{2}}\|\nabla g_1\|_{L^{p'}(\mathbb{R}^N;\mathbb{R}^N)}, \end{aligned}$$

and so both the preceding integrals converge when $p > N$. Finally, knowing that r_λ and $|\nabla r_\lambda|$ are in $L^{p'}(\mathbb{R}^N;\mathbb{R})$, observing that $\|\varphi\|_{L^p(\mathbb{R}^N;\mathbb{R})} = 1$ and $\|\psi_j\varphi\|_{L^p(\mathbb{R}^N;\mathbb{R})} = \|\psi_j\|_{L^p(\mu;\mathbb{R})}$, it follows from $(*)$ that φ is a continuous function and that there a $C(p,N) < \infty$ such that

$$\|\varphi\|_{\mathrm{u}} \le C(p,N)\big(\lambda^{\frac{N}{2p}} + \lambda^{\frac{N}{2p}-\frac{1}{2}}K\big)$$

for all $\lambda > 0$. By taking $\lambda = K^2$, one arrives at $\|\varphi\|_{\mathrm{u}} \le 2C(p,N)K^{\frac{n}{p}}$ and thence at $\|f\|_{\mathrm{u}} \le 2^p C(p,N)^p K^N$. □

In addition to Lemma 5.5.1, it will be convenient to have the following criterion for applying (3.1.6). Remember that $T_h : \mathbb{W}(\mathbb{R}^N) \longrightarrow \mathbb{W}(\mathbb{R}^N)$ is the translation map $T_h w = w + h$.

Lemma 5.5.2. *Let $\Phi \in L^p(\mathcal{W};\mathbb{R})$ for some $p > 1$ and $h \in H^1(\mathbb{R}^N)$. Assume that there is $\mathcal{W}$-null set Λ such that $\xi \rightsquigarrow \Phi \circ T_{\xi h}(w)$ is continuously differentiable on $(-1,1)$ when $w \notin \Lambda$, and set*

$$D_h\Phi(w) = \begin{cases} \partial_\xi\big(\Phi \circ T_{\xi h}(w)\big)\big|_{\xi=0} & \text{if } w \notin \Lambda \\ 0 & \text{if } w \in \Lambda. \end{cases}$$

Then $\partial_\xi\big(\Phi \circ T_{\xi h}(w)\big) = (D_h\Phi) \circ T_{\xi h}(w)$ for $\xi \in (-1,1)$ and $w \notin \Lambda$. Furthermore, if $D_h\Phi \in L^p(\mathcal{W};\mathbb{R})$, then, for all $q \in [1,p)$,

$$\lim_{\xi \to 0} \frac{\Phi \circ T_{\xi h} - \Phi}{\xi} = D_h\Phi \text{ in } L^q(\mathcal{W};\mathbb{R}).$$

Proof. That $\partial_\xi\big(\Phi \circ T_{\xi h}(w)\big) = (D_h\Phi) \circ T_{\xi h}(w)$ for $\xi \in (-1,1)$ and $w \notin \Lambda$ is obvious. Thus

$$\frac{\Phi \circ T_{\xi h}(w) - \Phi(w)}{\xi} - D_h\Phi(w) = \int_0^1 \big((D_h\Phi) \circ T_{\theta\xi h}(w) - D_h\Phi(w)\big)\,d\theta$$

for $\xi \in (-1,1)$ and $w \notin \Lambda$, and so

$$\lim_{\xi \to 0} \frac{\Phi \circ T_{\xi h} - \Phi}{\xi} = D_h\Phi \quad (\text{a.s.}, \mathcal{W}).$$

Finally, if $q \in [1,p)$ and $r = \frac{p+q}{2}$, then, by (3.1.4),

$$\begin{aligned} &\left\|\int_0^1 \Big((D_h\Phi) \circ T_{\theta\xi h}(w) - D_h\Phi(w)\Big)\,d\theta\right\|_{L^r(\mathcal{W};\mathbb{R})} \\ &\le \int_0^1 \|D_h\Phi \circ T_{\theta\xi h}\|_{L^r(\mathcal{W};\mathbb{R})}\,d\theta + \|D_h\Phi\|_{L^r(\mathcal{W};\mathbb{R})} \\ &\le \Big(e^{\frac{4\|h\|^2_{H^1(\mathbb{R}^N)}}{p-q}} + 1\Big)\|D_h\Phi\|_{L^r(\mathcal{W};\mathbb{R})} \end{aligned}$$

for $\xi \in (-1,1)$. Hence the convergence is taking place in $L^q(\mathcal{W};\mathbb{R})$ as well. □

5.5.1 *Ellipticity in one dimension*

Although it is ridiculous to do so, we will begin by using Malliavin's ideas to show that the distribution of a one dimensional Wiener path at time $t > 0$ admits a continuous density. Set $X(w) = w(t)$. Given $h \in H^1(\mathbb{R})$, consider the derivative $D_h X$ of X in the direction h, and observe the $D_h X(w) = h(t)$ for all w. Thus, if $\varphi \in C^1_\mathrm{b}(\mathbb{R};\mathbb{R})$, then $D_h(\varphi \circ X) = h(t)\varphi' \circ X$. Now take $h(\tau) = \tau \wedge t$, and apply (3.1.5) to conclude that

$$\mathbb{E}^{\mathcal{W}}\big[\varphi' \circ X\big] = t^{-1}\mathbb{E}^{\mathcal{W}}\big[w(t)\varphi \circ X\big].$$

Equivalently, if μ is the distribution of X, then

$$\int_{\mathbb{R}} \varphi'(y)\,\mu(dy) = t^{-1}\int_{\mathbb{R}} y\varphi(y)\,\mu(dy).$$

Since $\int y^2\,\mu(dy) = \mathbb{E}^{\mathcal{W}}\big[w(t)^2\big] = t$, Lemma 5.5.1 with $N = 1$ and $\psi_t(y) = \frac{y}{t}$ implies that μ has a continuous density that is bounded by $t^{-\frac{1}{2}}$.

Trivial as the preceding is, the basic reasoning there can be applied in non-trivial situations. For instance, consider the solution $X(\,\cdot\,,x)$ to

$$X(t,x) = x + \int_0^t \sigma\big(X(\tau,x)\big)\,dw(\tau) + \int_0^t b\big(X(\tau,x)\big)\,d\tau,$$

where $\sigma, b \in C^3_\mathrm{b}(\mathbb{R};\mathbb{R})$ and there is an $\epsilon > 0$ such that $\sigma \geq \epsilon$. Assuming that $D_h X(t,x)$ exists, one knows that $D_h\big(\varphi \circ X(t,x)\big)$ exists and is equal to $\varphi' \circ X(t,x) D_h X(t,x)$. Hence, if h can be chosen so $D_h X(t,x)$ is strictly positive and its reciprocal is sufficiently integrable, (3.1.6) would say that

$$\mathbb{E}^{\mathcal{W}}\big[\varphi' \circ X(t,x)\big] = \mathbb{E}^{\mathcal{W}}\left[\left(\frac{I(\dot h)}{D_h X(t,x)} - D_h \frac{1}{D_h X(t,x)}\right)\varphi \circ X(t,x)\right],$$

at which point, in order to apply Lemma 5.5.1, we would only need to show that the factor multiplying $\varphi \circ X(t,x)$ is sufficiently integrable.

Given $h \in H^1(\mathbb{R})$ and $\xi \in \mathbb{R}$, remember that $(T_{\xi h})_*\mathcal{W}$ and $\mathcal{W}$ are mutually absolutely continuous, and thereby conclude that $w \rightsquigarrow X(t,x) \circ T_{\xi h}(w)$ is well defined up to a $\mathcal{W}$-null set. Under the assumption that h is smooth, we will now show that, $\mathcal{W}$-almost surely, $\xi \in (-1,1) \longmapsto X(t,x) \circ T_{\xi h}$ can be chosen so that it is a twice continuously function. To this end, choose $\eta \in C^\infty_\mathrm{b}(\mathbb{R};\mathbb{R})$ so the $\eta(\xi) = \xi$ for $\xi \in [-2,2]$, define

$$\Sigma(\mathbf{y}) = \begin{pmatrix} 0 \\ 0 \\ \sigma(y_3) \end{pmatrix}, \quad \text{and} \quad B_h(\mathbf{y}) = \begin{pmatrix} 1 \\ 0 \\ b(y_3) + \eta(y_2)\sigma(y_3)\dot h(y_1) \end{pmatrix}$$

for $\mathbf{y} \in \mathbb{R}^3$, and let $Y_h(\,\cdot\,,\mathbf{y})$ be the solution to

$$Y_h(t,\mathbf{y}) = \mathbf{y} + \int_0^t \Sigma\big(Y_h(\tau,\mathbf{y})\big)\,dw(\tau) + \int_0^t B_h\big(Y_h(\tau,\mathbf{y})\big)\,d\tau.$$

Then, for $\xi \in (-1,1)$,

$$Y_h\big(t,(0,\xi,x)\big) = \begin{pmatrix} t \\ \xi \\ X(t,x)\circ T_{\xi h}\end{pmatrix}.$$

In particular, by the results in §3.4.1, $\mathbf{y} \rightsquigarrow Y_h(t,\mathbf{y})$ can be chosen so that it has two continuous derivatives, and so $X(t,x)\circ T_{\xi h}$ can be chosen so that it is twice continuously differentiable with respect to $\xi \in (-1,1)$, which means that Lemma 5.5.2 applies. In fact, because $D_hX(t,x) = \partial_{y_2}Y_h\big(t,(0,0,x)\big)_3$ and $D_h^2X(t,x) = \partial_{y_2}^2Y_h\big(t,(0,0,x)\big)_3$,

$$\begin{aligned} D_hX(t,x) &= \int_0^t \sigma'\big(X(\tau,x)\big)D_hX(\tau,x)\,dw(\tau) + \int_0^t b'\big(X(\tau,x)\big)D_hX(\tau,x)\,d\tau \\ &\quad + \int_0^t \sigma\big(X(\tau,x)\big)\dot h(\tau)\,d\tau, \\ D_h^2X(t,x) &= \int_0^t \sigma'\big(X(\tau,x)\big)D_h^2X(\tau,x)\,dw(\tau) + \int_0^t b'\big(X(\tau,x)\big)D_h^2X(\tau,x)\,d\tau \\ &\quad + \int_0^t \sigma''\big(X(\tau,x)\big)\big(D_hX(\tau,x)\big)^2\,dw(\tau) \\ &\quad + \int_0^t \Big(b''\big(X(\tau,x)\big)\big(D_hX(\tau,x)\big)^2 \\ &\qquad\qquad + 2\sigma'\big(X(\tau,x)\big)D_hX(\tau,x)\dot h(\tau)\Big)\,d\tau. \end{aligned}$$

Now set

$$\begin{aligned} J(t,x) = \exp\Bigg(&\int_0^t \sigma'\big(X(\tau,x)\big)\,dw(\tau) \\ &+ \int_0^t \Big(b'\big(X(\tau,x)\big) - \tfrac12\sigma'\big(X(\tau,x)\big)^2\Big)\,d\tau\Bigg). \end{aligned}$$

Using Itô's formula and the first of the preceding equations, one finds that

$$\frac{D_hX(t,x)}{J(t,x)} = \int_0^t \frac{\sigma\big(X(\tau,x)\big)}{J(\tau,x)}\dot h(\tau)\,d\tau,$$

and therefore that

$$
\begin{aligned}
D_hX(t,x) &= J(t,x)\int_0^t \frac{\sigma\big(X(\tau,x)\big)}{J(\tau,x)}\dot{h}(\tau)\,d\tau \\
&= \int_0^t \sigma\big(X(\tau,x)\big)\exp\bigg(\int_\tau^t \sigma'\big(X(s,x)\big)\,dw(s) \qquad (5.5.1)\\
&\qquad + \int_\tau^t \big(b' - \tfrac{1}{2}(\sigma')^2\big)\big(X(s,x)\big)\,ds\bigg)\dot{h}(\tau)\,d\tau.
\end{aligned}
$$

Thus, for each $p \in [1,\infty)$ there is a $C_p(h) < \infty$ such that

$$
\mathbb{E}^{\mathbb{P}}\big[|D_hX(t,x)|^p\big]^{\frac{1}{p}} \le C_p(h)t \quad \text{for } (t,x) \in [0,1]\times\mathbb{R}^N. \qquad (5.5.2)
$$

Finally, by using the preceding equations, one can show that

$$
\begin{aligned}
&\frac{D_h^2X(t,x)}{J(t,\mathbf{x})} \\
&= \int_0^t \frac{\sigma''\big(X(\tau,\mathbf{x})D_hX(\tau,x)^2}{J(\tau,x)}\,dw(\tau) \\
&+ \int_0^t \frac{(b''-\sigma'\sigma'')\big(X(\tau,x)\big)\big(D_hX(\tau,x)\big)^2 + 2\sigma'\big(X(\tau,x)\big)D_h(\tau,x)\dot{h}(\tau)}{J(\tau,x)}\,d\tau.
\end{aligned}
$$

After combining this with (5.5.2), one knows that, for each $p \in [2,\infty)$, there is a $C_p(h) < \infty$

$$
\mathbb{E}^{\mathcal{W}}\big[|D_h^2X(t,x)|^p\big]^{\frac{1}{p}} \le C_p(h)t^2 \quad \text{for } (t,x) \in [0,1]\times\mathbb{R}. \qquad (5.5.3)
$$

Remember that $\sigma \ge \epsilon$ for some $\epsilon > 0$, and choose a smooth $h \in H^1(\mathbb{R})$ so that $h(t) = t$ for $t \in [0,1]$. Then, by (5.5.1) and Jensen's inequality

$$
\begin{aligned}
\frac{1}{D_hX(t,x)} \le \frac{1}{\epsilon t^2}\int_0^t \exp\bigg(&-\int_\tau^t \sigma'\big(X(s,x)\big)\,dw(s) \\
&+ \int_\tau^t \big(\tfrac{1}{2}(\sigma')^2 - b'\big)\big(X(s,x)\big)\,ds\bigg)\,d\tau
\end{aligned}
$$

for any $(t,\mathbf{x}) \in (0,1]\times\mathbb{R}^N$. Hence, for each $p \in [2,\infty)$, one can find a $C_p < \infty$ such that

$$
\mathbb{E}^{\mathcal{W}}\big[|D_hX(t,x)|^{-p}\big]^{\frac{1}{p}} \le C_pt^{-1} \quad \text{for } t \in (0,1]. \qquad (5.5.4)
$$

In particular, if $\Psi(t,x) := \frac{1}{D_hX(t,x)}$, then $D_h\Psi(t,x) = -\frac{D_h^2X(t,x)}{D_hX(t,x)^2}$, and so the preceding combined with (5.5.3) imply that there is a $C_p < \infty$ such that

$$
\mathbb{E}^{\mathcal{W}}\big[|D_h\Psi(t,x)|^p\big]^{\frac{1}{p}} \le C_p \quad \text{for } (t,x) \in (0,1]\times\mathbb{R}. \qquad (5.5.5)
$$

Given $\varphi \in C_c^\infty(\mathbb{R};\mathbb{R})$ and $(t,x) \in (0,1]\times\mathbb{R}$,

$$D_h\big(\varphi \circ X(t,x)\big) = D_h X(t,x)\varphi' \circ X(t,x).$$

Hence, by (3.1.6),

$$\begin{aligned}\mathbb{E}^{\mathcal{W}}\big[\varphi' \circ X(t,x)\big] &= \mathbb{E}^{\mathcal{W}}\big[\Psi(t,x)D_h\big(\varphi \circ X(t,x)\big)\big] \\ &= \mathbb{E}^{\mathcal{W}}\Big[\big(I(\dot{h})\Psi(t,x) - D_h\Psi(t,x)\big)\varphi \circ X(t,x)\Big],\end{aligned}$$

and so

$$\big|\mathbb{E}^{\mathcal{W}}\big[\varphi' \circ X(t,x)\big]\big| \le \mathbb{E}^{\mathcal{W}}\big[|w(t)\Psi(t,x) - D_h\Psi(t,x)|^p\big]^{\frac{1}{p}}\mathbb{E}^{\mathcal{W}}\big[|\varphi \circ X(t,x)|^{p'}\big]^{\frac{1}{p'}}$$

for $(t,x) \in (0,1] \times \mathbb{R}$. Using the estimates in (5.5.4) and (5.5.5), one concludes that, for each $p \in [2,\infty)$, there is a $C_p < \infty$ such that

$$\big|\mathbb{E}^{\mathcal{W}}\big[\varphi' \circ X(t,x)\big]\big| \le \frac{C_p}{t^{\frac{1}{2}}}\mathbb{E}^{\mathcal{W}}\big[|\varphi \circ X(t,x)|^{p'}\big]^{\frac{1}{p'}}$$

for all $(t,x) \in (0,1] \times \mathbb{R}$. Thus, if $P(t,x,\cdot\,)$ is the distribution of $X(t,x)$, then Lemma 5.5.1 guarantees that $P(t,x,dy) = p(t,x,y)\,dy$, where $p(t,x,\cdot\,)$ is a continuous function that is bounded by a constant times $t^{-\frac{1}{2}}$. Finally, when $t > 1$, one can use the Chapman–Kolmogorov equation to see that $P(t,x,dy) = p(t,x,y)$, where

$$p(t,x,y) = \int p(1,z,y)\,P(t-1,x,dz).$$

Thus, $p(t,x,y)$ is bounded by a constant times $(t \wedge 1)^{-\frac{1}{2}}$.

5.5.2 *Some non-elliptic examples*

Given a differential operator L with smooth coefficients, one says that L is **hypoelliptic** if, for any Schwartz distribution u, u is smooth on any open set where Lu is smooth. In a famous paper, H. Weyl proved in 1940 that, at least in the case when u is a locally square integrable function, u will be smooth on any open set where Δu is smooth, and analysts have subsequently generalized and sharpened Weyl's result by showing that the same conclusion holds when u is a Schwartz distribution on $\mathbb{R}^N$ and L is any strictly elliptic operator of the form in (1.2.5) with smooth coefficients. Further, they have shown that for such operators L, the corresponding parabolic operator $\partial_t + L$ is also hypoelliptic. In particular, this latter result proves that if $t \rightsquigarrow \mu_t$ is a solution to the Kolmogorov's forward equation for a strictly elliptic L with smooth coefficients, then $\mu_t(d\mathbf{y}) = f(t,\mathbf{y})\,d\mathbf{y}$ where $f \in C^\infty\big((0,\infty) \times \mathbb{R}^N;[0,\infty)\big)$.

Powerful as these results are, they do not explain the following simple but striking example discovered by Kolmogorov. Consider the operator

$L = \frac{1}{2}\partial_{x_1}^2 + x_1\partial_{x_2}$. Even though this L is far from being strictly elliptic, Kolmogorov realized that the transition probability function for the associated diffusion admits a smooth density. Indeed, the Itô representation for that diffusion is the solution to

$$X(t,\mathbf{x})_1 = x_1 + w(t) \quad \text{and} \quad X(t,\mathbf{x})_2 = x_2 + \int_0^t X(\tau,\mathbf{x})_1\,d\tau.$$

Hence,

$$X(t,\mathbf{x}) = \begin{pmatrix} x_1 + w(t) \\ x_2 + tx_1 + \int_0^t (t-\tau)\,dw(\tau) \end{pmatrix},$$

and so $X(t,\mathbf{x})$ is an $\mathbb{R}^2$-valued Gaussian random variable with mean value $\mathbf{m}(t,\mathbf{x}) = \left(\begin{smallmatrix} x_1 \\ x_2+tx_1 \end{smallmatrix}\right)$ and covariance

$$C(t) = \begin{pmatrix} t & \frac{t^2}{2} \\ \frac{t^2}{2} & \frac{t^3}{3} \end{pmatrix}.$$

Thus, since $\det\big(C(t)\big) = \frac{t^4}{12} > 0$, the transition probability for this diffusion has density

$$\frac{\sqrt{3}}{\pi t^2} \exp\Big(-\big(\mathbf{y} - \mathbf{m}(t,\mathbf{x}), C(t)^{-1}(\mathbf{y} - \mathbf{m}(t,\mathbf{x}))\big)_{\mathbb{R}^2}\Big).$$

In 1967, Hörmander published a startling article [6] that put Kolmogorov's example in context. To describe his result, let L be given by (4.5.1), where the vector fields V_k are smooth. Next consider the Lie algebra generated by the vector fields $V_0,\dots,V_M$: the smallest linear space $\mathrm{Lie}(V_0,\dots,V_M)$ of vector fields W that contains $\{V_0,\dots,V_M\}$ and has the property that $[W,W']$ is in $\mathrm{Lie}(V_0,\dots,V_M)$ if W and W' are. Then Hörmander's theorem says that L is hypoelliptic if $\mathrm{Lie}(V_0,\dots,V_M)$ has dimension N at all points. As a consequence, $\partial_t + L$ will be hypoelliptic if the dimension of $\mathrm{Lie}\big([V_0,V_1],\dots,[V_0,V_M],V_1\dots,V_M\big)$ is N at all points. In Kolmogorov's example, $N=2$, $V_0 = \left(\begin{smallmatrix} 0 \\ x_1 \end{smallmatrix}\right)$ and $V_1 = \left(\begin{smallmatrix} 1 \\ 0 \end{smallmatrix}\right)$, and so V_1 and $[V_0,V_1]$ span $\mathbb{R}^2$ at every point.

The best understood cases of Hörmander's result are those when the vector field V_0 is not needed; that is, when $\mathrm{Lie}(V_1,\dots,V_M)$ has dimension N. Further, in the case when $V_0 = 0$, L. Rothschild and E. Stein gave a beautiful interpretation of Hörmander's result in terms of sub-Riemannian geometry. On the other hand, when, as in Kolmogorov's example, V_0 plays an essential role, all the analytic proofs of Hörmander's theorem are less than revealing. Thus it is gratifying that, as we are about to see, Malliavin's ideas shed some light in special cases. Namely, let $m \ge 1$, and consider the operator $L = \frac{1}{2}\partial_{x_1}^2 + x_1^m\partial_{x_2}$. Of course, when $m=1$, this is just Kolmogorov's example. However, when $m \geq 2$, the associated diffusion is no longer Gaussian, and so one cannot simply write down the associated transition probability function. Nonetheless, the associated diffusion is given by

$$X(t,\mathbf{x})_1 = x_1 + w(t) \quad \text{and} \quad X(t,\mathbf{x})_2 = x_2 + \int_0^t \bigl(x_1 + w(\tau)\bigr)^m \, d\tau.$$

Notice that if $S_\lambda^{(m)} : \mathbb{R}^2 \longrightarrow \mathbb{R}^2$ is defined by $S_\lambda^{(m)}\mathbf{y} = \bigl(\lambda^{\frac{1}{2}} y_1, \lambda^{\frac{m}{2}+1} y_2\bigr)$, then

$$S_\lambda^{(m)} X\bigl(\lambda^{-1}t, S_{\lambda^{-1}}^{(m)}\mathbf{x}\bigr)(w) = X(t,\mathbf{x})(w_\lambda) \quad \text{where } w_\lambda(t) = \lambda^{\frac{1}{2}} w(\lambda^{-1}t).$$

Hence, since w_λ is again a Brownian motion, $\{S_\lambda^{(m)} X(\lambda^{-1}t, S_{\lambda^{-1}}^{(m)}\mathbf{x}) : t \geq 0\}$ has the same distribution as $\{X(t,\mathbf{x}) : t \geq 0\}$. In particular, if $P(t,\mathbf{x},\,\cdot\,)$ is the distribution of $X(t,\mathbf{x})$, then $P(t,\mathbf{x},\,\cdot\,) = (S_t^{(m)})_* P(1, S_{t^{-1}}^{(m)}\mathbf{x},\,\cdot\,)$, and so, if $P(1,\mathbf{x},d\mathbf{y}) = p(1,\mathbf{x},\mathbf{y})\,d\mathbf{y}$, then $P(t,\mathbf{x},d\mathbf{y}) = p(t,\mathbf{x},\mathbf{y})\,d\mathbf{y}$ where

$$p(t,\mathbf{x},\mathbf{y}) = t^{-\frac{m+3}{2}} p\bigl(1, S_{t^{-1}}^{(m)}\mathbf{x}, S_{t^{-1}}^{(m)}\mathbf{y}\bigr). \tag{5.5.6}$$

For this reason, we need only consider $X(1,\mathbf{x})$.

Clearly

$$D_h X(1,\mathbf{x})_1 = h(1) \quad \text{and} \quad D_h X(1,\mathbf{x})_2 = m \int_0^1 \bigl(x_1 + w(t)\bigr)^{m-1} h(t)\, dt.$$

Now assume that $m = 2n+1$ for some $n \geq 1$, and take $h_1(t) = t \wedge 1$ and $h_2(t) = t(1-t)^+$. Then

$$\begin{aligned} D_{h_1}\bigl(\varphi \circ X(1,\mathbf{x})\bigr) &= (\partial_{y_1}\varphi) \circ X(1,\mathbf{x}) \\ &\qquad + \left((2n+1)\int_0^1 t\bigl(x_1 + w(t)\bigr)^{2n} dt\right)(\partial_{y_2}\varphi) \circ X(1,\mathbf{x}) \end{aligned}$$

and

$$D_{h_2}\bigl(\varphi \circ X(1,\mathbf{x})\bigr) = \left((2n+1)\int_0^1 t(1-t)\bigl(x_1 + w(t)\bigr)^{2n} dt\right)(\partial_{y_2}\varphi) \circ X(1,\mathbf{x}).$$

Therefore, if

$$\Phi_1 = \int_0^1 t\bigl(x_1 + w(t)\bigr)^{2n} dt \quad \text{and} \quad \Phi_2 = \int_0^1 t(1-t)\bigl(x_1 + w(t)\bigr)^{2n} dt,$$

then

$$(\partial_{y_1}\varphi) \circ X(1,\mathbf{x}) = D_{h_1}\bigl(\varphi \circ X(1,\mathbf{x})\bigr) - \frac{\Phi_1}{\Phi_2} D_{h_2}\bigl(\varphi \circ X(1,\mathbf{x})\bigr),$$

$$(\partial_{y_2}\varphi)\circ X(1,\mathbf{x}) = \frac{1}{(2n+1)\Phi_2} D_{h_2}\big(\varphi\circ X(1,\mathbf{x})\big).$$

and so we will be more or less done once we show that Φ_2^{-1} is sufficiently integrable. For this purpose, observe that, by Jensen's inequality,

$$\begin{aligned}\int_0^1 t(1-t)\big(x_1+w(t)\big)^{2n} &\geq 6^{1-n}\left(\int_0^1 t(1-t)\big(x_1+w(t)\big)^2\,dt\right)^n\\ &\geq \frac{3^n\cdot 6^{1-n}}{16^n}\left(\int_{\frac14}^{\frac34}\big(x_1+w(t)\big)^2\,dt\right)^n.\end{aligned}$$

Next, observe that, by the Markov property and (5.2.4),

$$\begin{aligned}&\mathbb{E}^{\mathcal{W}}\left[e^{-\alpha\int_{\frac14}^{\frac34}\left(x_1+w(t)\right)^2\,dt}\right]\\ &\quad=\int_{\mathbb{R}}\mathbb{E}^{\mathcal{W}}\left[e^{-\alpha\int_0^{\frac12}\left(y+w(t)\right)^2\,dt}\right]\gamma_{x_1,\frac14}(dy)\leq\frac{1}{\cosh\sqrt{2^{-1}\alpha}}\leq 2e^{-\sqrt{2^{-1}\alpha}}\end{aligned}$$

for all $\alpha>0$. Hence, if $\beta_n=\frac{2^{5n-1}}{3}$, then

$$\mathcal{W}\big(\Phi_2^{-1}\geq\beta_n R\big)=\mathcal{W}\big(e^{-\alpha\beta_n\Phi_2}\geq e^{-\frac{\alpha}{R}}\big)\leq 2e^{\frac{\alpha}{R}-\sqrt{2^{-1}\alpha}},$$

and so $\mathcal{W}\big(\Phi_2^{-1}\geq\beta_n R\big)\leq 2e^{-\frac{R}{8}}$, which is more than enough to guarantee that $\Phi_2^{-1}\in L^p(\mathcal{W};\mathbb{R})$ for all $p\in[1,\infty)$. Therefore, by Lemma 5.5.2 and (3.1.6), we know that

$$\begin{aligned}&\mathbb{E}^{\mathcal{W}}\big[(\partial_{y_1}\varphi)\circ X(1,\mathbf{x})\big]\\ &\quad=\mathbb{E}^{\mathcal{W}}\left[\left(w(1)-\frac{\Phi_1}{\Phi_2}\int_0^1(1-2t)\,dw(t)+D_{h_2}\frac{\Phi_1}{\Phi_2}\right)\varphi\big(X(1,\mathbf{x})\big)\right]\end{aligned}$$

and

$$\begin{aligned}&\mathbb{E}^{\mathcal{W}}\big[(\partial_{y_2}\varphi)\circ X(1,\mathbf{x})\big]\\ &\quad=(2n+1)^{-1}\mathbb{E}^{\mathcal{W}}\left[\left(\frac{\int_0^1(1-2t)\,dw(t)}{\Phi_2}-D_{h_2}\frac{1}{\Phi_2}\right)\varphi\big(X(1,\mathbf{x})\big)\right].\end{aligned}$$

Finally,

$$D_{h_2}\frac{\Phi_1}{\Phi_2}=\frac{2n\int_0^1 t^2(1-t)\big(x_1+w(t)\big)^{2n-1}\,dt}{\Phi_2}-\Phi_1 D_{h_2}\frac{1}{\Phi_2}$$

and

$$D_{h_2}\frac{1}{\Phi_2}=-\frac{2n\int_0^1 t^2(1-t)^2\big(x_1+w(t)\big)^{2n-1}\,dt}{\Phi_2^2},$$

both of which are in $L^p(\mathcal{W};\mathbb{R})$ for all $p \in [1,\infty)$. Hence, by Lemma 5.5.1, when m is odd, we have shown that $P(1,\mathbf{x},\,\cdot\,) = p(1,\mathbf{x},\,\cdot\,)$, where $p(1,x,\,\cdot\,)$ is bounded and continuous.

When m is even, the preceding line of reasoning breaks down from the start. Indeed, what made it work when m is odd is that we could choose an h for which $D_hX(1,\mathbf{x})_2$ is positive, and obviously no such h will exist when m is even. For this reason, one has to take a more sophisticated approach, one that requires the use of an entire basis of $h \in H^1(\mathbb{R})$. With this in mind, given a differentiable function $w \rightsquigarrow \Phi(w)$, assume that, for each w, there is a $C(w) < \infty$ such that $|D_h\Phi(w)| \le C(w)\|h\|_{H^1(\mathbb{R})}$, in which case there is a $D\Phi(w) \in H^1(\mathbb{R})$ such that $\big(D\Phi(w), h\big)_{H^1(\mathbb{R})} = D_h\Phi(w)$ for all $h \in H^1(\mathbb{R})$. Clearly

$$
\begin{aligned}
\big[DX(1,\mathbf{x})_1(w)\big](t) &= t\wedge 1,\\
\big[DX(1,\mathbf{x})_2(w)\big](t) &= m\int_0^{t\wedge 1}\left(\int_s^1 \big(x_1+w(\tau)\big)^{m-1}\,d\tau\right)ds,
\end{aligned}
$$

and

$$
D\big(\varphi\circ X(t,\mathbf{x})\big) = (\partial_{y_1}\varphi)\circ X(1,\mathbf{x})DX(t,\mathbf{x})_1 + (\partial_{y_2}\varphi)\circ X(1,\mathbf{x})DX(t,\mathbf{x})_2.
$$

Thus, if

$$
\begin{aligned}
\mathcal{A}(x_1) &:= \begin{pmatrix} \big(DX(1,\mathbf{x})_1, DX(1,\mathbf{x})_1\big)_{H^1(\mathbb{R})} & \big(DX(1,\mathbf{x})_1, DX(1,\mathbf{x})_2\big)_{H^1(\mathbb{R})} \\ \big(DX(1,\mathbf{x})_2, DX(1,\mathbf{x})_1\big)_{H^1(\mathbb{R})} & \big(DX(1,\mathbf{x})_2, DX(1,\mathbf{x})_2\big)_{H^1(\mathbb{R})} \end{pmatrix}\\
&= \begin{pmatrix} 1 & m\int_0^1\left(\int_s^1(x_1+w(\tau))^{m-1}\,d\tau\right)ds \\ m\int_0^1\left(\int_s^1(x_1+w(\tau))^{m-1}\,d\tau\right)ds & m^2\int_0^1\left(\int_s^1(x_1+w(\tau))^{m-1}\,d\tau\right)^2 ds \end{pmatrix}
\end{aligned}
$$

then

$$
\begin{pmatrix} \big(D(\varphi\circ X(1,\mathbf{x})), DX(1,\mathbf{x})_1\big)_{H^1(\mathbb{R})} \\ \big(D(\varphi\circ X(1,\mathbf{x})), DX(1,\mathbf{x})_2\big)_{H^1(\mathbb{R})} \end{pmatrix} = \mathcal{A}(x_1)\begin{pmatrix} (\partial_{y_1}\varphi)\circ X(1,x) \\ (\partial_{y_2}\varphi)\circ X(1,x) \end{pmatrix}.
$$

The next step is to invert the matrix $\mathcal{A}(x_1)$. Obviously

$$
\begin{aligned}
\Delta(x_1) := \det(\mathcal{A}(x_1)) = m^2 &\int_0^1\left(\int_s^1\big(x_1+w(\tau)\big)^{m-1}\,d\tau\right)^2 ds\\
&- m^2\left(\int_0^1\left(\int_s^1\big(x_1+w(\tau)\big)^{m-1}\,d\tau\right)ds\right)^2,
\end{aligned}
$$

which is the variance of $s \rightsquigarrow m\int_s^1\big(x_1+w(\tau)\big)^{m-1}\,d\tau$ with respect to $\lambda_{[0,1]}$. Therefore, by writing this variance as half the second moment of the difference between two independent copies, we see that

$$\Delta(x_1) = m^2 \iint\limits_{0\le s<t\le 1} \left(\int_s^t \big(x_1 + w(\tau)\big)^{m-1}\, d\tau \right)^2 ds\, dt,$$

and so $\Delta(x_1) > 0$. Hence we can write

$$\begin{pmatrix} (\partial_{y_1}\varphi)\circ X(1,\mathbf{x}) \\ (\partial_{y_2}\varphi)\circ X(1,\mathbf{x}) \end{pmatrix} = \mathcal{A}(x_1)^{-1} \begin{pmatrix} D\big(\varphi\circ X(1,\mathbf{x})_1\big) \\ D\big(\varphi\circ X(1,\mathbf{x})_2\big) \end{pmatrix}$$

where $\mathcal{A}(x_1)^{-1}$ equals $\frac{1}{\Delta(x_1)}$ times

$$\begin{pmatrix} m^2 \int_0^1 \left(\int_s^1 \big(x_1+w(\tau)\big)^{m-1} d\tau \right)^2 ds & -m\int_0^1 \tau\big(x_1+w(\tau)\big)^{m-1} d\tau \\ -m\int_0^1 \tau\big(x_1+w(\tau)\big)^{m-1} d\tau & 1 \end{pmatrix}.$$

The challenge now is to show that $\frac{1}{\Delta(x_1)}$ is in $L^p(\mathcal{W};\mathbb{R})$ for all $p\in[1,\infty)$, and, to do that, we will need the following facts about Brownian motion.

Lemma 5.5.3. *For $r>0$, set $\zeta_{\pm r} = \inf\{t\ge 0 : \pm w(t)\ge r\}$. Then, for $\alpha>0$,*

$$\mathbb{E}^{\mathcal{W}}\left[e^{-\alpha\zeta_{\pm r}^4}\right] \le 2^{-\frac{1}{2}} \exp\left(-\left(\frac{\alpha r^2}{64}\right)^{\frac{1}{5}} \right).$$

Next, for $r>0$ and $x\in\mathbb{R}$, set $\sigma_r(x) = \inf\{t\ge 0 : |x+w(t)|\ge r\}$. Then, for all $t>0$ and $x\in\mathbb{R}$,

$$\mathcal{W}\big(\sigma_r(x)\ge t\big) \le \sqrt{2}\exp\left(-\frac{\pi^2 t}{32 r^2}\right).$$

Proof. By Brownian scaling and symmetry, $\zeta_{\pm r}$ has the same distribution as $r^2\zeta_1$, and, by the reflection principle (cf. part (vi) of Exercise 2.5), we know that $\mathcal{W}(\zeta_1\le t) = 2\mathcal{W}\big(w(t)\ge 1\big)$. Hence

$$\mathbb{E}^{\mathcal{W}}\left[e^{-\alpha\zeta_{\pm r}^4}\right] = \mathbb{E}^{\mathcal{W}}\left[e^{-\alpha r^8\zeta_1^4}\right],$$

and, because $\frac{1}{4t} + \alpha t^4 \ge \left(\frac{\alpha}{64}\right)^{\frac{1}{5}}$,

$$\begin{aligned} \sqrt{2\pi}\,\mathbb{E}^{\mathcal{W}}\left[e^{-\alpha\zeta_1^4}\right] &= \int_0^\infty t^{-\frac{3}{2}} e^{-\frac{1}{2t}} e^{-\alpha t^4}\, dt \\ &\le e^{-\left(\frac{\alpha}{64}\right)^{\frac{1}{5}}} \int_0^\infty t^{-\frac{3}{2}} e^{-\frac{1}{4t}}\, dt = e^{-\left(\frac{\alpha}{64}\right)^{\frac{1}{5}}} 2^{\frac{3}{2}} \int_0^\infty e^{-\frac{\tau^2}{2}}\, d\tau = 2\sqrt{\pi} e^{-\left(\frac{\alpha}{64}\right)^{\frac{1}{5}}}. \end{aligned}$$

To prove the second estimate, first observe that, by Brownian scaling, $\sigma_r(x)$ has the same distribution as $r^2\sigma_1(\frac{x}{r})$. Thus, what we have to show is that $\mathcal{W}\big(\sigma_1(x)\ge t\big) \le \sqrt{2}e^{-\frac{\pi^2 t}{32}}$ for all $x\in(-1,1)$. To this end, set $u(t,x) = e^{\frac{\pi^2 t}{32}} \sin\frac{\pi(x+2)}{4}$. Then $\big(u(t,x+w(t)),\mathcal{B}_t,\mathcal{W}\big)$ a martingale, and so

$$1 \geq \sin \frac{\pi(x+2)}{4} = \mathbb{E}^{\mathcal{W}}\big[u(t \wedge \sigma_1(x), w(t \wedge \sigma_1(x))\big] \geq 2^{-\frac{1}{2}} e^{\frac{\pi^2 t}{32}} \mathcal{W}\big(\sigma_1(x) \geq t\big).$$

□

By symmetry, one knows that $\Delta(-x_1)$ has the same distribution as $\Delta(x_1)$, and so we will assume that $x_1 \geq 0$. For $k \geq 1$, define

$$\beta_k(x_1) = \inf\left\{t \geq 0 : |x_1 + w(t)| \geq \frac{1}{k}\right\}$$

and

$$\beta_k'(x_1) = \inf\left\{t \geq \beta_k : |x_1 + w(t)| \leq \frac{1}{2k}\right\}.$$

If $\beta_k \leq \frac{1}{2}$, then

$$\begin{aligned}\Delta(x_1) &\geq m^2 \iint\limits_{\beta_k \leq s < t \leq \beta_k' \wedge 1} \left(\int_s^t \big(x_1 + w(\tau)\big)^{m-1} d\tau\right)^2 ds\, dt \\ &\geq \frac{m^2}{12(2k)^{2m-2}} \Big(\big(\beta_k'(x_1) - \beta_k(x_1)\big) \wedge \tfrac{1}{2}\Big)^4,\end{aligned}$$

and therefore

$$\begin{aligned}&\mathbb{E}^{\mathcal{W}}\big[e^{-\alpha\Delta(x_1)}\big] \\ &\leq \mathbb{E}^{\mathcal{W}}\Big[\exp\Big(-\alpha\gamma_m\big((\beta_1'(x_1) - \beta_1(x_1)) \wedge \tfrac{1}{2}\big)^4\Big), \beta_1(x_1) \leq \tfrac{1}{2}\Big] \\ &\quad + \sum_{k=2}^{\infty} \mathbb{E}^{\mathcal{W}}\Big[\exp\Big(-\alpha\gamma_m k^{2-2m}\Big(\big(\beta_k'(x_1) - \beta_k(x_1)\big) \wedge \tfrac{1}{2}\Big)^4\Big), \\ &\qquad\qquad\qquad\qquad\qquad \beta_k(x_1) \leq \tfrac{1}{2} < \beta_{k-1}(x_1)\Big],\end{aligned}$$

where $\gamma_m = 3^{-1}m^2 4^{-m}$. Since $\beta_k(x_1) = \sigma_{\frac{1}{k}}(x_1)$ and, by the Markov property,

$$\mathcal{W}\big(\beta_k'(x_1) - \beta_k(x_1) > t\big) = \mathcal{W}\big(\zeta_{\frac{1}{2k}} > t\big) \quad \text{for all } t > 0,$$

we have that

$$\begin{aligned}\mathbb{E}^{\mathcal{W}}\big[e^{-\alpha\Delta(x_1)}\big] \leq &\,\mathbb{E}^{\mathcal{W}}\Big[e^{-\alpha\gamma_m\left(\zeta_{\frac{1}{2}} \wedge \frac{1}{2}\right)^4}\Big] \\ &+ \sum_{k=2}^{\infty} \mathbb{E}^{\mathcal{W}}\Big[e^{-2\alpha\gamma_m k^{2-2m}\left(\zeta_{\frac{1}{2k}} \wedge \frac{1}{2}\right)^4}\Big]^{\frac{1}{2}} \mathcal{W}\big(\sigma_{\frac{1}{k-1}} > \tfrac{1}{2}\big)^{\frac{1}{2}},\end{aligned}$$

which, after an application of the estimates in Lemma 5.5.3, means that

$$\mathbb{E}^{\mathcal{W}}\big[e^{-\alpha\Delta(x_1)}\big] \leq C_m \sum_{k=1}^{\infty} e^{-\epsilon_m(\alpha k^{2-2m})^{\frac{1}{5}}} e^{-\epsilon_m k^2}$$

for some $C_m < \infty$ and $\epsilon_m > 0$. Since

$$\sum_{k=1}^{\infty} e^{-\epsilon_m(\alpha k^2 - 2m)^{\frac{1}{5}}} e^{-\epsilon_m k^2} \le e^{-\epsilon_m \alpha^{\frac{1}{m+4}}} \sum_{k \le \alpha^{\frac{1}{2m+8}}} e^{-\epsilon_m k^2} + \sum_{k > \alpha^{\frac{1}{2m+8}}} e^{-\epsilon_m k^2},$$

after adjusting, C_m, we obtain

$$\mathbb{E}^{\mathcal{W}}\big[e^{-\alpha\Delta(x_1)}\big] \le C_m e^{-\epsilon_m \alpha^{\frac{1}{m+4}}}$$

and therefore

$$\mathcal{W}\big(\Delta(x_1)^{-1} \ge R\big) = \mathcal{W}\big(e^{-R\Delta(x_1)} \ge e^{-1}\big) \le C_m e^{1-\epsilon_m R^{\frac{1}{m+4}}}.$$

Hence $\sup_{x_1 \in \mathbb{R}} \|\Delta(x_1)^{-1}\|_{L^p(\mathcal{W};\mathbb{R})} < \infty$ for all $p \in [1, \infty)$.

Knowing that $x_1 \rightsquigarrow \frac{1}{\Delta(x_1)}$ is bounded in $L^p(\mathcal{W};\mathbb{R})$ for all $p \in [1,\infty)$, the next step is to show that Lemma 5.5.2 applies and justifies the use of (3.1.6). With that in mind, set $h_0(t) = t \wedge 1$ and

$$h_k(t) = \frac{2^{\frac{1}{2}} \sin \pi k(t \wedge 1)}{\pi k} \quad \text{for } k \ge 1.$$

Although $\{h_k : k \ge 0\}$ is not a basis for $H^1(\mathbb{R})$, it is an orthonormal basis for the subspace $\{h \in H^1(\mathbb{R}) : \dot{h}(t) = 0 \,\text{for}\, t > 1\}$.[4] Thus

$$\Big(D(\varphi \circ X(1,\mathbf{x})), DX(1,\mathbf{x})_j\Big)_{H^1(\mathbb{R})} = \sum_{k=0}^{\infty} D_{h_k}\big(\varphi \circ X(1,\mathbf{x})\big) D_{h_k} X(1,\mathbf{x})_j,$$

and so one would like to apply (3.1.6) to justify writing

$$\mathbb{E}^{\mathcal{W}}\big[(\partial_{y_i}\varphi) \circ X(1,\mathbf{x})\big] = \mathbb{E}^{\mathcal{W}}\big[\Psi_i \varphi \circ X(1,\mathbf{x})\big] \tag{5.5.7}$$

where

$$\Psi_i = \sum_{j=1}^{2} \sum_{k=0}^{\infty} \Big(I(\dot{h}_k)(\mathcal{A}(x_1)^{-1})_{ij} D_{h_k} X(1,\mathbf{x})_j - D_{h_k}\big((\mathcal{A}(x_1)^{-1})_{ij} D_{h_k} X(1,\mathbf{x})_j\big)\Big).$$

However, before we can do so, we must show that the series for Ψ_i converges in $L^p(\mathcal{W};\mathbb{R})$ for all $p \in [1,\infty)$. First observe that, since $D_{h_k} X(1,\mathbf{x})_1 = h_k(1)$,

[4] I use here and below the fact that both $\{\mathbf{1}\} \cup \{2^{\frac{1}{2}} : \cos \pi k t : k \ge 1\}$ and $\{2^{\frac{1}{2}} \sin \pi k t : k \ge 1\}$ are orthonormal bases in $L^2([0,1];\mathbb{R})$.

$$\sum_{k=0}^{\infty}\Big(I(\dot h_k)(\mathcal{A}(x_1)^{-1})_{ij}D_{h_k}X(1,\mathbf{x})_1 - D_{h_k}\big((\mathcal{A}(x_1)^{-1})_{ij}D_{h_k}X(1,\mathbf{x})_1\big)\Big)$$
$$= w(1)(\mathcal{A}(x_1)^{-1})_{i1} - D_{h_0}(\mathcal{A}(x_1)^{-1})_{i1}.$$

Hence, since

$$D_{h_0}(\mathcal{A}(x_1))_{12} = m(m-1)\int_0^1 \tau\big(x_1+w(\tau)\big)^{m-2}\,d\tau,$$
$$D_{h_0}(\mathcal{A}(x_1))_{22} = 2m^2(m-1)\int_0^1\left(\int_s^1 \tau\big(x_1+w(\tau)\big)^{m-2}\,d\tau\right)\times\left(\int_s^1 \big(x_1+w(\tau)\big)^{m-1}\,d\tau\right)ds,$$
$$D_{h_0}\Delta(x_1) = 2m^2(m-1)\iint\limits_{0\le s<t\le 1}\left(\int_s^t \tau\big(x_1+w(\tau)\big)^{m-2}\,d\tau\right)\times\left(\int_s^t \big(x_1+w(\tau)\big)^{m-1}\,d\tau\right)ds\,dt,$$

and

$$D_{h_0}(\mathcal{A}(x_1)^{-1})_{i1} = (-1)^i\left(\frac{D_{h_0}\mathcal{A}_{i2}(x_1)}{\Delta(x_1)} - \frac{\mathcal{A}(x_1)_{i2}D_{h_0}\Delta(x_1)}{\Delta(x_1)^2}\right),$$

the contribution to Ψ_i from the terms with $j=1$ causes no problems.

To handle $\sum_{k=0}^{\infty} I(\dot h_k)\big(\mathcal{A}(x_1)^{-1}\big)_{i2}D_{h_k}X(t,\mathbf{x})_2$, set

$$f(s,x_1) = m\int_s^1 \big(x_1+w(\tau)\big)^{m-1}\,d\tau \text{ and}$$
$$\psi_n(\tau,x_1) = -\sum_{k=0}^{n}\big(\dot f(\,\cdot\,,x_1),h_k\big)_{L^2([0,1];\mathbb{R})}\dot h_k(\tau).$$

Then

$$\sum_{k=0}^{n} I(\dot h_k)D_{h_k}X(1,\mathbf{x})_2 = \int_0^1 \psi_n(\tau,x_1)\,dw(\tau)$$
$$= w(1)\psi_n(1,x_1) - \int_0^1 w(\tau)\dot\psi_n(\tau,x_1)\,d\tau,$$

where the $dw(\tau)$-integral is taken in the sense of Riemann–Stieltjes. Since, by Schwarz's and Bessel's inequalities,

$$\left(\sum_{k=1}^{\infty}\left|\left(\dot f(\,\cdot\,,x_1),h_k\right)_{L^2([0,1];\mathbb{R})}\right|\right)^2$$
$$=\frac{1}{\pi^2}\left(\sum_{k=1}^{n}\frac{2^{\frac12}}{k}\int_0^1 \dot f(\tau,x_1)\sin\pi k\tau\,d\tau\right)^2\le\frac16\int_0^1 \dot f(s,x_1)^2\,ds$$

and

$$-\left(\dot f(\,\cdot\,,x_1),h_k\right)_{L^2([0,1];\mathbb{R})}=\left(f(\,\cdot\,,x_1),\dot h_k\right)_{L^2([0,1];\mathbb{R})},$$

it follows that $\psi_n(\,\cdot\,,x_1)\longrightarrow f(\,\cdot\,,x_1)$ in $L^p\big(\mathcal{W};C([0,1];\mathbb{R})\big)$ for all $p\in[1,\infty)$. At the same time,

$$\dot\psi_n(\tau,x_1)=-2\sum_{k=1}^{n}\left(\int_0^1 \dot f(s,x_1)\sin(\pi ks)\,ds\right)\sin(\pi k\tau),$$

and so

$$\int_0^1 \dot\psi_n(\tau,x_1)^2\,d\tau\le\int_0^1 \dot f(s,x_1)^2\,ds$$

and $\dot\psi_n(\,\cdot\,,x_1)\longrightarrow -\dot f(\,\cdot\,,x_1)$ in $L^p\big(\mathcal{W};L^2([0,1];\mathbb{R})\big)$ for all $p\in[1,\infty)$. Hence

$$\sum_{k=0}^{\infty}I(\dot h_k)\left(\mathcal{A}(x_1)^{-1}\right)_{i2}D_{h_k}X(1,\mathbf{x})$$
$$\longrightarrow\left(\mathcal{A}(x_1)^{-1}\right)_{i2}\left(w(1)f(1,x_1)+\int_0^1 w(\tau)\dot f(\tau,x_1)\,d\tau\right)$$

in $L^p(\mathbb{P};\mathbb{R})$ for all $p\in[1,\infty)$.

Turning to the other term with $j=2$ in the expression for Ψ_i, note that

$$\sum_{k=0}^{n}D_{h_k}\left((\mathcal{A}(x_1)^{-1})_{i2}D_{h_k}X(1,\mathbf{x})_2\right)$$
$$=\sum_{k=0}^{n}\left(D_{h_k}(\mathcal{A}(x_1)^{-1})_{i2}\right)\left(D_{h_k}X(1,\mathbf{x})_2\right)+\sum_{k=0}^{n}(\mathcal{A}(x_1)^{-1})_{i2}D^2_{h_k}X(1,\mathbf{x})_2.$$

Because

$$D^2_{h_k}X(1,\mathbf{x})_2=m(m-1)\int_0^1 h_k(\tau)^2\big(x_1+w(\tau)\big)^{m-2}\,d\tau$$

and $\sum_{k=0}^{n}h_k(\tau)^2\nearrow\tau$, the second sum converges in $L^p(\mathcal{W};\mathbb{R})$ for every $p\in[1,\infty)$ to

$$m(m-1)(\mathcal{A}(x_1)^{-1})_{i2}\int_0^1 \tau\big(x_1+w(\tau)\big)^{m-2}\,d\tau.$$

As for the first sum, we will know that it converges in $L^p(\mathcal{W};\mathbb{R})$ for every $p \in [1,\infty)$ once we show that $\|D(\mathcal{A}(x_1)^{-1})_{i2}\|_{H^1(\mathbb{R})}$ and $\|DX(1,\mathbf{x})_2\|_{H^1(\mathbb{R})}$ are in $L^p(\mathcal{W};\mathbb{R})$ for all $p \in [1,\infty)$. Since

$$DX(1,\mathbf{x})_2 = m\int_0^{t\wedge 1}\left(\int_s^1 \big(x_1+w(\tau)\big)^{m-1}\,d\tau\right)ds,$$

it poses no problem. Next observe that

$$D(\mathcal{A}(x_1)^{-1})_{i2} = (-1)^{i+1}\left(\frac{D\mathcal{A}(x_1)_{i1}}{\Delta(x_1)} - \frac{\mathcal{A}(x_1)_{i1}D\Delta(x_1)}{\Delta(x_1)^2}\right),$$

and so we need only show that $\|D\mathcal{A}(x_1)_{i1}\|_{H^1(\mathbb{R})}$ and $\|D\Delta(x_1)\|_{H^1(\mathbb{R})}$ are in $L^p(\mathcal{W};\mathbb{R})$ for all $p \in [1,\infty)$. But $D\mathcal{A}(x_1)_{11} = 0$ and

$$D\mathcal{A}(x_1)_{21}(t) = m(m-1)\int_0^{t\wedge 1}\left(\int_s^1 \tau\big(x_1+w(\tau)\big)^{m-2}\,d\tau\right)ds,$$

and, using the identity

$$\iint\limits_{0\le s<t\le 1}\left(\int_s^t f(\tau)\,d\tau\right)^2 ds\,dt = \iint\limits_{[0,1]^2} \tau_1\wedge\tau_2(1-\tau_1\vee\tau_2)f(\tau_1)f(\tau_2)\,d\tau_1\,d\tau_2,$$

one sees that $\big[D\Delta(x_1)\big](t)$ equals $2m^2(m-1)$ times

$$\int_0^{t\wedge 1}\left(\int_s^1 \big(x_1+w(\tau_1)\big)^{m-2}\right.$$
$$\left.\times\left(\int_0^1 (\tau_1\wedge\tau_2)(1-\tau_1\vee\tau_2)\big(x_1+w(\tau_2)\big)^{m-1}\,d\tau_2\right)d\tau_1\right)ds,$$

from which the desired estimates are easy consequences.

The preceding calculations justify the use of (3.1.6) to obtain (5.5.7) and show that the Ψ_1 and Ψ_2 there are in $L^p(\mathcal{W};\mathbb{R})$ for all $p \in [1,\infty)$. Thus, by Lemma 5.5.1 and (5.5.6), we have now proved that, for any $m \in \mathbb{Z}^+$, the transition probability for the diffusion corresponding to $\frac{1}{2}\partial_{x_1}^2 + x_1^m\partial_{x_2}$ admits a density $p(t,x,y)$ that satisfies $p(t,x,y) \le Ct^{-\frac{m+3}{2}}$ for some $C < \infty$.

5.6 Exercises

Exercise 5.1. Show that if $\mathbb{P}_\mathbf{x}$ solves the martingale problem for L starting from $\mathbf{x}$ and $a = 0$, then

$$\psi(t) = \mathbf{x} + \int_0^t b\big(\psi(\tau)\big)\,d\tau \quad (\text{a.s.}, \mathbb{P}).$$

Hence, when $a = 0$, solutions to the martingale problem are concentrated on integral curves of b.

Exercise 5.2. Suppose that $\beta \in PM^2(\mathbb{R}^N)$ and that $\big(X(t), \mathcal{F}_t, \mathbb{P}\big)$ is an $\mathbb{R}^N$-valued semi-martingale for which $X(0) = \mathbf{0}$. Set

$$B(t) = X(t) - \int_0^t \beta(\tau)\,d\tau$$

$$\text{and} \quad E(t) = \exp\left(-\int_0^t \big(\beta(\tau), dB(\tau)\big)_{\mathbb{R}^N} - \frac{1}{2}\int_0^t |\beta(\tau)|^2\,d\tau\right),$$

and assume that $\big(B(t), \mathcal{F}_t, \mathbb{P}\big)$ is an $\mathbb{R}^N$-valued Brownian motion and that $\big(E(t), \mathcal{F}_t, \mathbb{P}\big)$ is a martingale. If $\widetilde{\mathbb{P}} \in \mathbf{M}_1\big(\mathcal{P}(\mathbb{R}^N)\big)$ is determined by $d\widetilde{\mathbb{P}} \restriction \mathcal{B}_t = E(t)\,d\mathbb{P} \restriction \mathcal{B}_t$ for all $t \geq 0$, show that $\big(X(t), \mathcal{F}_t, \widetilde{\mathbb{P}}\big)$ is a Brownian motion. This is the statement that is called **Girsanov's theorem.**

Exercise 5.3. Refer to the last part of §5.2. To produce an example in which explosion occurs, consider the integral equation

$$X(t)(w) = 2 + w(t) + \int_0^t X(\tau)(w)^2\,d\tau \quad \text{for } w \in \mathbb{W}(R).$$

Suppose that $w(t) \geq -1$ for $t \in [0,1]$ and that $\mathfrak{e}(w) \geq 1$. If $u(t) = X(t)(w)$ and $U(t) = \int_0^t u(\tau)^2\,d\tau$, observe that $\dot{U}(t) \geq \big(1 + U(t)\big)^2$, and therefore that $u(t) \geq 1 + U(t) \geq \frac{1}{1-t}$ for $t \in [0,1)$. Conclude that $\mathfrak{e}(w) \leq 1$ if $w(t) \geq -1$ for $t \in [0,1]$ and therefore that $\mathcal{W}(\mathfrak{e} \leq 1) > 0$. As a consequence, show that

$$\mathbb{E}^{\mathcal{W}}\left[\exp\left(\int_0^1 w(\tau)^2\,dw(\tau) - \frac{1}{2}\int_0^1 w(\tau)^4\,d\tau\right)\right] < 1$$

and therefore that

$$\left(\exp\left(\int_0^t w(\tau)^2\,dw(\tau) - \frac{1}{2}\int_0^t w(\tau)^4\,d\tau\right), \mathcal{W}_t, \mathcal{W}\right)$$

is not a martingale.

Exercise 5.4. Let $M = \mathbb{S}^{N-1}(r) := \{\mathbf{x} \in \mathbb{R}^N : |\mathbf{x}| = r\}$. Show that

$$\Pi^M(\mathbf{x})\mathbf{y} = \mathbf{y} - \frac{(\mathbf{x}, \mathbf{y})_{\mathbb{R}^N}}{r^2}\mathbf{x} \quad \text{for } \mathbf{x} \in M \text{ and } \mathbf{y} \in \mathbb{R}^N,$$

and conclude that the mean curvature normal in this case equals $-\frac{N-1}{r^2}\mathbf{x}$ at $\mathbf{x} \in M$.

Exercise 5.5. Let H be a separable Hilbert space over $\mathbb{R}$. In this exercise you are to derive some elementary properties of operators on H. See [2] for background and more information.

(**i**) A linear operator K on H is said to be compact if the images of bounded sets under K are relatively compact sets. Using the fact that a bounded subset of H is sequentially relatively compact with respect to the weak topology, show that K is a compact operator if and only if it takes weakly convergent sequences into strongly convergent ones. That is, K is compact if and only if $(h_n, g)_H \longrightarrow (h, g)_H$ for all $g \in H$ implies that $\|Kh_n - Kh\|_H \longrightarrow 0$.

(**ii**) Let $(E, \mathcal{B}, \mu)$ be a measure space, and assume that $\mathcal{B}$ is countably generated and therefore that $L^2(\mu; \mathbb{R})$ is separable. Given a $\mathcal{B} \times \mathcal{B}$-measurable function $k : E \times E \longrightarrow \mathbb{R}$ for which

$$\iint_{E\times E} k(x,y)^2 \, \mu(dx)\mu(dy) < \infty,$$

set $k(x) = \sqrt{\int_E k(x,y)^2 \, \mu(dy)}$, and define

$$K\varphi(x) = \int_E k(x,y)\varphi(y)\,\mu(dy) \quad \text{for } \varphi \in L^2(\mu; \mathbb{R}).$$

Show that $|K\varphi(x)| \le \|\varphi\|_{L^2(\mu;\mathbb{R})} k(x)$. Next, given a sequence $\{\varphi_n : n \ge 1\}$ that converges weakly in $L^2(\mu; \mathbb{R})$ to φ, show that $K\varphi_n \longrightarrow K\varphi$ (a.e., μ), and use the preceding estimate together with Lebesgue's dominated convergence theorem to show that $\|K\varphi_n - K\varphi\|_{L^2(\mu;\mathbb{R})} \longrightarrow 0$. Hence, the operator K is compact.

(**iii**) Let R be a bounded operator on H, and let R^* be its adjoint. Show that $\text{Range}(R) \subseteq \text{Null}(R^*)^\perp$ and $\text{Range}(R)^\perp \subseteq \text{Null}(R^*)$. Conclude from these that $\text{Range}(R) = \text{Null}(R^*)^\perp$ if $\text{Range}(R)$ is closed.

(**iv**) Let K be a compact operator on H, and set $R = I - K$. Given $f \in \overline{\text{Range}(R)}$, show that there exists a sequence $\{h_n : n \ge 1\} \subseteq \text{Null}(R)^\perp$ such that $f = \lim_{n\to\infty} Rh_n$. Assuming that $\{h_n : n \ge 1\}$ is bounded, show that there is a subsequence $\{h_{n_k} : k \ge 1\}$ which converges to an $h \in \text{Null}(R)^\perp$ for which $f = Rh \in \text{Range}(R)$. Next, suppose that $\{h_n : n \ge 1\}$ were unbounded and therefore, after passing to a subsequence, that one could assume that $1 \le \|h_n\|_H \longrightarrow \infty$. Set $g_n = \frac{h_n}{\|h_n\|_H}$, and show that $Rg_n \longrightarrow 0$. Now use the preceding to see that this would lead to the contradiction that there exists a $g \in \text{Null}(R)^\perp \cap \text{Null}(R)$ such that $\|g\|_H = 1$.

In view of the above, we now know that $\text{Range}(R)$ is closed when $R = I - K$ and K is compact. Such operators are said to be *Fredholm operators*, and, in conjunction with (iii), we have shown that when R is a Fredholm operator and $\text{Null}(R^*) = \{0\}$, $\text{Range}(R) = H$.

Exercise 5.6. As (5.4.9) makes clear, the Kalman–Bucy filter predicts the value of $X_1(t)$ at least as well as the prediction $\mathbb{E}^{\mathbb{P}}[X_1(t)]$ that one would make if one didn't have access to the information in $X_2(\,\cdot\,)$. Of course, when $\beta_2 := 0$, $X_2(\,\cdot\,)$ conveys no information about $X_1(\,\cdot\,)$, and so it is not surprising that $\widehat{X}_1(t) = \mathbb{E}^{\mathbb{P}}[X_1(t)]$ and $D(t) = \operatorname{var}\big(X_1(t)\big)$ in this case. In this exercise, you are to examine what happens in various cases when β_2 doesn't vanish.

(i) Assume that α_1 and β_1 are identically 0, and show that

$$D(t) = \frac{V_0}{1 + V_0 \int_0^t \rho_2(\tau)^2\, d\tau}$$

where $V_0 = \operatorname{var}(X_0)$.

(ii) Set $V_0 = \operatorname{var}(X_0)$ and $\gamma = \sqrt{\rho_2^2\alpha_1^2 + \beta_1^2}$, and assume that $\rho_2 = \frac{\alpha_2}{\beta_2}$ and γ are uniformly positive. Show that

$$\dot{D}(t) = -\rho_2^2\left(D(t) + \frac{\gamma - \beta_1}{\rho_2^2}\right)\left(D(t) - \frac{\gamma + \beta_1}{\rho_2^2}\right).$$

Next, set $c = \rho_2^2 D + \gamma - \beta_1$, and note that $c \geq \epsilon$ for some $\epsilon > 0$. Show that

$$D(t) = V_0 e^{-\int_0^t c(\tau)\, d\tau} + e^{-\int_0^t c(\tau)\, d\tau} \int_0^t c(\tau) e^{\int_0^\tau c(\sigma)\, d\sigma} \frac{\gamma(\tau) + \beta_1(\tau)}{\rho_2(\tau)^2}\, d\tau.$$

When the coefficients in (5.4.1) are constant, conclude from this that $D(t) = V_0$ if $\rho_2^2 V_0 = \gamma + \beta_1$,

$$D(t) \searrow \frac{\gamma + \beta_1}{\rho_2^2} \quad \text{if } \rho_2^2 V_0 > \gamma + \beta_1,$$

and

$$D(t) \nearrow \frac{\gamma + \beta_1}{\rho_2^2} \quad \text{if } \rho_2^2 V_0 < \gamma + \beta_1.$$

More generally, even if the coefficients are not constant, show that

$$D(t) \longrightarrow \lim_{t\to\infty} \frac{\gamma(t) + \beta_1(t)}{\rho_2(t)^2} \quad \text{if } \lim_{t\to\infty} \frac{\gamma(t) + \beta_1(t)}{\rho_2(t)^2} \text{ exists in } \mathbb{R}.$$

In particular, $D(t)$ stays bounded even when $\beta_1 \geq 0$, whereas $\operatorname{var}\big(X(t)\big) \longrightarrow \infty$ if either $|\alpha_1|$ is uniformly positive and $\beta_1 \geq 0$ or $V_0 \neq 0$ and β_1 is uniformly positive.

(iii) Continue in the setting of (ii), and assume that the coefficients in (5.4.1) are constant. Begin by showing that

$$\left(\frac{1}{D(t) - \frac{\gamma+\beta_1}{\rho_2^2}} - \frac{1}{D(t) + \frac{\gamma-\beta_1}{\rho_2^2}}\right)\dot{D}(t) = -2\gamma,$$

and conclude that

$$D(t) = \frac{(\gamma+\beta_1)(\rho_2^2 V_0 + \gamma - \beta_1) + (\gamma - \beta_1)(\rho_2^2 V_0 - \gamma - \beta_1)e^{-2\gamma t}}{\rho_2^2\big((\rho_2^2 V_0 + \gamma - \beta_1) - (\rho_2^2 V_0 - \gamma - \beta_1)e^{-2\gamma t}\big)}.$$

Exercise 5.7. In §5.5.2 we encountered the operator D which plays the role for functions on $\mathbb{W}(\mathbb{R}^N)$ that the gradient operator plays for functions on $\mathbb{R}^N$. In this and the next exercise, you are to examine this and related operators more closely.

(i) Given a separable Hilbert space H over $\mathbb{R}$, let $\mathscr{P}(\mathbb{W}(\mathbb{R}^N);H)$ be the vector space spanned by functions $\Phi\colon \mathbb{W}(\mathbb{R}^N) \longrightarrow H$ of the form $\Phi = F\big(I(\dot f_1),\dots,I(\dot f_L)\big)h$, where $L \ge 1$, $F\colon \mathbb{R}^L \longrightarrow \mathbb{R}$ is a polynomial, $f_1,\dots,f_L \in H^1(\mathbb{R}^N)$, $I(\dot f_\ell) = \int_0^\infty \big(\dot f_\ell(\tau), dw(\tau)\big)_{\mathbb{R}^N}$, and $h \in H$ with $\|h\|_H = 1$. By combining Theorem 3.5.3 and Exercise 3.83, show that $\mathscr{P}(\mathbb{W}(\mathbb{R}^N);H)$ is a dense subspace of $L^2(\mathcal{W};H)$. In addition, show that

$$D\Phi = \sum_{\ell=1}^{L} \partial_{x_\ell} F\big(I(\dot f_1),\dots,I(\dot f_L)\big) f_\ell$$

if $\Phi = F\big(I(\dot f_1),\dots,I(\dot f_L)\big)$, and conclude from this that D maps $\mathscr{P}(\mathbb{W}(\mathbb{R}^N);\mathbb{R})$ into $\mathscr{P}\big(\mathbb{W}(\mathbb{R}^N);H^1(\mathbb{R})\big)$.

(ii) Since D is densely defined, there exists an adjoint operator

$$D^\top : \mathscr{P}(\mathbb{W}(\mathbb{R}^N);H^1(\mathbb{R}^N)) \longrightarrow L^2(\mathcal{W};\mathbb{R})$$

whose domain is the set of $\Psi \in L^2\big(\mathcal{W};H^1(\mathbb{R}^N)\big)$ for which there exists a $C < \infty$ such that

$$\big|(\Psi, D\Phi)_{L^2(\mathcal{W};H^1(\mathbb{R}^N))}\big| \le C\|\Phi\|_{L^2(\mathcal{W};\mathbb{R})} \quad \text{for all } \Phi \in \mathscr{P}(\mathbb{W}(\mathbb{R}^N);\mathbb{R}),$$

in which case $D^\top\Psi$ is the unique element of $L^2\big(\mathcal{W};\mathbb{R})\big)$ such that

$$\big(D^\top\Psi, \Phi\big)_{L^2(\mathcal{W};\mathbb{R})} = \big(\Psi, D\Phi\big)_{L^2(\mathcal{W};H^1(\mathbb{R}^N))} \quad \text{for all } \Phi \in \mathscr{P}(\mathbb{W}(\mathbb{R}^N);\mathbb{R}).$$

Using Lemma 5.5.2 and (3.1.6), show that if $\Psi = G\big(I(\dot g_1),\dots,I(\dot g_L)\big)h$ where $h \in H^1(\mathbb{R}^N)$ with $\|h\|_H = 1$, then

$$\big(\Psi, D\Phi\big)_{L^2(\mathcal{W};H^1(\mathbb{R}^N))} = \Big(\big(I(\dot h) - D_h\big)(\Psi, h)_{H^1(\mathbb{R}^N)}, \Phi\Big)_{L^2(\mathcal{W};\mathbb{R})}$$

for all $\Phi \in \mathscr{P}(\mathbb{W}(\mathbb{R}^N);\mathbb{R})$. Conclude that $\mathscr{P}\big(\mathbb{W}(\mathbb{R}^N);H^1(\mathbb{R}^N)\big) \subseteq \mathrm{Dom}(D^\top)$ and that

$$D^\top\Psi = \big(I(\dot h) - D_h\big)(\Psi, h)_H.$$

(iii) Recall that an operator mapping one Hilbert space to another is said to be closed if its graph is closed, and that it is said to be closable if the closure of its graph is the graph of an operator. Further, if an operator is densely defined, then its adjoint is always closed, and it is closable if its adjoint is densely defined. Therefore, we know that $D^\top$ is closed and that D is closable. Continue to use D to denote its own closure, and show that show that

$$\big(\Psi, D\Phi\big)_{L^2(\mathcal{W};H^1(\mathbb{R}^N))} = \big(D^\top\Psi, \Phi\big)_{L^2(\mathcal{W};\mathbb{R})} \tag{5.6.1}$$

for all $\Phi \in \mathrm{Dom}(D)$ and $\Psi \in \mathrm{Dom}(D^\top)$

(iv) Suppose that $\Psi \in \mathscr{P}\big(\mathbb{W}(\mathbb{R}^N); H^1(\mathbb{R}^N)\big)$, and let $\{h_k : k \geq 1\}$ be an orthonormal basis in $H^1(\mathbb{R}^N)$. Show that

$$\sum_{k=1}^\infty I(\dot{h}_k)\big(\Psi, h_k\big)_{H^1(\mathbb{R}^N)} \quad \text{and} \quad \sum_{k=1}^\infty D_{h_k}\big(\Psi, h_k\big)_{H^1(\mathbb{R}^N)}$$

both converge in $L^2(\mathcal{W};\mathbb{R})$ and that

$$D^\top\Psi = \sum_{k=1}^\infty \big(I(\dot{h}_k) - D_{h_k}\big)\big(\Psi, h_k\big)_{H^1(\mathbb{R}^N)}. \tag{5.6.2}$$

Hence, as it should be, $D^\top$ is an analog of minus the divergence operator on $C^1(\mathbb{R}^N;\mathbb{R}^N)$.

Exercise 5.8. This exercise is a continuation of Exercise 5.7. Define the operator $\mathcal{L}$ on $\mathscr{P}(\mathbb{W}(\mathbb{R}^N);\mathbb{R})$ by $\mathcal{L}\Phi = D^\top D\Phi$.

(i) Using (5.6.1), show that

$$\big(\Psi, \mathcal{L}\Phi\big)_{L^2(\mathcal{W};\mathbb{R})} = \big(D\Psi, D\Phi\big)_{L^2(\mathcal{W};\mathbb{R})},$$

and therefore that $\mathcal{L}$ is non-negative definite and symmetric in the sense that

$$\big(\Phi, \mathcal{L}\Phi\big)_{L^2(\mathcal{W};\mathbb{R})} \geq 0 \quad \text{and} \quad \big(\Psi, \mathcal{L}\Phi\big)_{L^2(\mathcal{W};\mathbb{R})} = \big(\Phi, \mathcal{L}\Psi\big)_{L^2(\mathcal{W};\mathbb{R})}$$

for all $\Phi, \Psi \in \mathscr{P}(\mathbb{W}(\mathbb{R}^N);\mathbb{R})$. Thus, because it has a densely defined adjoint, $\mathcal{L}$ is closable, and we will again use $\mathcal{L}$ to denote its closure, which is also a non-negative definite, symmetric operator. Show that $\mathrm{Dom}(\mathcal{L}) \subseteq \mathrm{Dom}(D)$ and that $\big(D\Phi_1, D\Phi_2\big)^2_{L^2(\mathcal{W};H^1(\mathbb{R}^N))} = \big(\Phi_1, \mathcal{L}\Phi_2\big)_{L^2(\mathcal{W};\mathbb{R})}$ for $\Phi_1, \Phi_2 \in \mathrm{Dom}(\mathcal{L})$.

(ii) If $\Phi \in \mathscr{P}(\mathbb{W}(\mathbb{R}^N);\mathbb{R})$ and $\{h_k : k \geq 1\}$ is an orthonormal basis in $H^1(\mathbb{R}^N)$, use (5.6.2) to show that

$$\mathcal{L}\Phi = \sum_{k=1}^\infty \big(I(\dot{h}_k)D_{h_k} - D^2_{h_k}\big)\Phi,$$

where the convergence of the series is in $L^2(\mathcal{W};\mathbb{R})$.

(**iii**) Recall the Hermite polynomial $H_m(x) = (-1)^m e^{\frac{x^2}{2}} \partial_x^m e^{-\frac{x^2}{2}}$. Clearly $H_0 = \mathbf{1}$ and $H_{m+1} = xH_m - H_m'$. Using Taylor's theorem, show that

$$e^{\zeta x - \frac{\zeta^2}{2}} = \sum_{m=0}^{\infty} \frac{\zeta^m}{m!} H_m(x) \quad \text{and} \quad (x-\zeta)e^{\zeta x - \frac{\zeta^2}{2}} = \sum_{m=1}^{\infty} \frac{\zeta^{m-1}}{(m-1)!} H_m(x),$$

where the convergence of both series is absolute and uniform for ζ in compact subsets of $\mathbb{C}$. By combining these with $H_m' = xH_m - H_{m+1}$ for $m \geq 0$, show that

$$\sum_{m=0}^{\infty} \frac{\zeta^m}{m!} H_m' = \zeta \sum_{m=0}^{\infty} \frac{\zeta^m}{m!} H_m = \sum_{m=1}^{\infty} \frac{\zeta^m}{(m-1)!} H_{m-1},$$

and therefore that $H_m' = mH_{m-1}$ for $m \geq 1$. In particular, conclude that (cf. part (iii) in Exercise 3.6)

$$x\partial_x H_m(x,a) - a\partial_x^2 H_m(x,a) = mH_m(x,a) \quad \text{for all } m \geq 0 \text{ and } a \geq 0.$$

(**iv**) Given $f \in H^1(\mathbb{R}^N)$, show that

$$\mathcal{L}\Phi = m\Phi \quad \text{if } \Phi = H_m\big(I(\dot{f}), \|f\|_{H^1(\mathbb{R}^N)}^2\big).$$

Next, recall (cf. (3.5.4)) the space $Z^{(m)}$ of mth order homogeneous chaos, and, using Exercise 3.8 and Theorem 3.5.3, show that the span of

$$\big\{H_m\big(I(\dot{f}), \|f\|_{H^1(\mathbb{R}^N)}^2\big) : f \in H^1(\mathbb{R}^N)\big\}$$

is dense in $Z^{(m)}$. Conclude that $Z^{(m)} \subseteq \operatorname{Dom}(\mathcal{L})$ and $\mathcal{L}\Phi = m\Phi$ for all $\Phi \in Z^{(m)}$. In particular, use this to show that if $\Phi \in \operatorname{Dom}(\mathcal{L})$ and $\mathcal{L}\Phi = m\Phi$, then, for any $\Psi \in Z^{(n)}$, $m(\Phi,\Psi)_{L^2(\mathcal{W};\mathbb{R})} = n(\Phi,\Psi)_{L^2(\mathcal{W};\mathbb{R})}$ and therefore $\Phi \in Z^{(m)}$. Hence,

$$Z^{(m)} = \big\{\Phi \in \operatorname{Dom}(\mathcal{L}) : \mathcal{L}\Phi = m\Phi\big\}.$$

Finally, given $m \geq 1$ and $\Phi \in Z^{(m)}$, show that, for any $h \in H^1(\mathbb{R}^N)$, $D_h\Phi \in Z^{(m-1)}$.

(**v**) Let $\Pi_{Z^{(m)}}$ denote orthogonal projection onto $Z^{(m)}$, and suppose that $\Phi \in L^2(\mathcal{W};\mathbb{R})$ has the property that

$$\sum_{m=0}^{\infty} m^2 \|\Pi_{Z^{(m)}}\Phi\|_{L^2(\mathcal{W};\mathbb{R})}^2 < \infty. \tag{5.6.3}$$

Show that $\Phi \in \operatorname{Dom}(\mathcal{L})$ and that

$$\mathcal{L}\Phi = \sum_{m=0}^{\infty} m\Pi_{Z^{(m)}}\Phi. \tag{5.6.4}$$

Conversely, show that if $\Phi \in \mathrm{Dom}(\mathcal{L})$, then (5.6.3) holds. Hence $\mathrm{Dom}(\mathcal{L})$ is the subspace of $\Phi \in L^2(\mathcal{W};\mathbb{R})$ for which (5.6.3) holds, and $\mathcal{L}\Phi$ is given by (5.6.4) for all $\Phi \in \mathrm{Dom}(\mathcal{L})$. In particular, this shows that $\mathcal{L}$ is a self-adjoint operator, that its spectrum is $\mathbb{N}$, and that $Z^{(m)}$ is the eigenspace for the eigenvalue m.

Part (v) accounts for physicists calling $\mathcal{L}$ the **number operator**, although probabilists are inclined to think of it as the infinite dimensional Ornstein–Uhlenbeck operator.

References

1. Coddington, E.A., Levinson, N.: Theory of Ordinary Differential Equations. McGraw-Hill, New York–Toronto–London (1955)
2. Conway, J.B.: A Course in Functional Analysis, *Graduate Texts in Mathematics*, vol. 96, 2nd edn. Springer, New York (1990)
3. Doob, J.L.: Stochastic Processes. Wiley, New York (1953)
4. Friedman, A.: Partial Differential Equations of Parabolic Type. Prentice-Hall, Englewood Cliffs, NJ (1964)
5. Gilbarg, D., Trudinger, N.S.: Elliptic Partial Differential Equations of Second Order. Classics in Mathematics. Springer, Berlin (2001)
6. Hörmander, L.: Hypoelliptic second order differential equations. Acta Math. **119**, 147–171 (1967)
7. Ito, K.: On stochastic differential equations. Mem. Amer. Math. Soc. **4** (1951)
8. Itô, K.: The Brownian motion and tensor fields on Riemannian manifold. In: Proc. Internat. Congr. Mathematicians (Stockholm, 1962), pp. 536–539. Inst. Mittag-Leffler, Djursholm (1963)
9. Kolmogoroff, A.: über die analytischen Methoden in der Wahrscheinlichkeitsrechnung. Math. Ann. **104**(1), 415–458 (1931)
10. Kunita, H., Watanabe, S.: On square integrable martingales. Nagoya Math. J. **30**, 209–245 (1967)
11. Le Gall, J.-F.: Brownian Motion, Martingales, and Stochastic Calculus, *Graduate Texts in Mathematics*, vol. 274. Springer, Cham (2016)
12. Nadirashvili, N.: Nonuniqueness in the martingale problem and the Dirichlet problem for uniformly elliptic operators. Ann. Scuola Norm. Sup. Pisa Cl. Sci. (4) **24**(3), 537–549 (1997)
13. Nualart, D.: The Malliavin Calculus and Related Topics, 2nd edn. Probability and its Applications. Springer, Berlin (2006)
14. Øksendal, B.: Stochastic Differential Equations. An Introduction with Applications, 5th edn. Universitext. Springer, Berlin (1998)
15. Ruf, J.: A new proof for the conditions of Novikov and Kazamaki. Stochastic Process. Appl. **123**(2), 404–421 (2013)
16. Stratonovich, R.L.: A new form of representing stochastic integrals and equations. Vestnik Moskov. Univ. Ser. I Mat. Meh. **1964**(1), 3–12 (1964)
17. Stroock, D.W.: An Introduction to the Analysis of Paths on a Riemannian Manifold, *Mathematical Surveys and Monographs*, vol. 74. Amer. Math. Soc., Providence, RI (2000)
18. Stroock, D.W.: Markov Processes from K. Itô's Perspective, *Annals of Mathematics Studies*, vol. 155. Princeton Univ. Press, Princeton, NJ (2003)

D. W. Stroock, *Elements of Stochastic Calculus and Analysis*,
CRM Short Courses, https://doi.org/10.1007/978-3-319-77038-3

19. Stroock, D.W.: Essentials of Integration Theory for Analysis, *Graduate Texts in Mathematics*, vol. 262. Springer, New York (2011)
20. Stroock, D.W.: Probability Theory. An Analytic View, 2nd edn. Cambridge Univ. Press, Cambridge (2011)
21. Stroock, D.W., Varadhan, S.R.S.: On the support of diffusion processes with applications to the strong maximum principle. In: Proceedings of the Sixth Berkeley Symposium on Mathematical Statistics and Probability (Univ. California, Berkeley, Calif., 1970/1971), Vol. III: Probability Theory, pp. 333–359. Univ. California Press, Berkeley, CA (1972)
22. Stroock, D.W., Varadhan, S.R.S.: Multidimensional Diffusion Processes. Classics in Mathematics. Springer, Berlin (2006)

Index

D. W. Stroock, *Elements of Stochastic Calculus and Analysis*,
CRM Short Courses, https://doi.org/10.1007/978-3-319-77038-3

Printed in the United States
By Bookmasters